Mark M. Meerschaert, Alla Sikorskii

Stochastic Models for Fractional Calculus

De Gruyter Studies in Mathematics

Volume 43

Mark M. Meerschaert, Alla Sikorskii

Stochastic Models for Fractional Calculus

2nd Edition

DE GRUYTER

Mathematics Subject Classification 2010
60F05, 26A33, 82C31, 60G05, 60H30

Authors
Prof. Dr. Mark M. Meerschaert
Michigan State University
Department of Statistics and Probability
C430 Wells Hall
619 Red Cedar Road
East Lansing, MI 48824-1027
USA
mcubed@stt.msu.edu

Prof. Dr. Alla Sikorskii
Michigan State University
Department of Statistics and Probability
C430 Wells Hall
619 Red Cedar Road
East Lansing, MI 48824-1027
USA
sikorska@stt.msu.edu

ISBN 978-3-11-055907-1
e-ISBN (PDF) 978-3-11-056024-4
e-ISBN (EPUB) 978-3-11-055914-9
ISSN 0179-0986

Library of Congress Control Number: 2019914385

Bibliographic information published by the Deutsche Nationalbibliothek
The Deutsche Nationalbibliothek lists this publication in the Deutsche Nationalbibliografie; detailed bibliographic data are available on the Internet at http://dnb.dnb.de.

Printing and binding: CPI books GmbH, Leck

www.degruyter.com

This book is dedicated to our spouses and families. Your love, support, and patience makes everything possible.

Preface

This book is based on a series of lecture notes for a graduate course in the Department of Statistics and Probability at Michigan State University. The goal is to prepare graduate students for research in the area of fractional calculus, anomalous diffusion, and heavy tails. The book covers basic limit theorems for random variables and random vectors with heavy tails. This includes regular variation, triangular arrays, infinitely divisible laws, random walks, and stochastic process convergence in the Skorokhod topology. The basic ideas of fractional calculus and anomalous diffusion are introduced in the context of probability theory. Each section of the book provides material roughly equivalent to one lecture. Most sections conclude with some additional details, intended for individual reading, and to make the book relatively self-contained.

Heavy tails are applied in finance, insurance, physics, geophysics, cell biology, ecology, medicine, and computer engineering. A random variable has heavy tails if $P(|X| > x) \approx Cx^{-\alpha}$ for some $\alpha > 0$. If $\alpha < 2$, then the second moment of X is undefined, so the usual central limit theorem does not apply. A heavy-tailed version of the central limit theorem leads to a stable distribution. Random walks with heavy tails converge to a stable Lévy motion, similar to Brownian motion. The densities of Brownian motion solve the diffusion equation, which provides a powerful link between differential equations and stochastic processes. Densities of a stable Lévy motion solve a fractional diffusion equation like

$$\frac{\partial}{\partial t}p(x,t) = c\frac{\partial^\alpha}{\partial x^\alpha}p(x,t)$$

using a fractional derivative of order α. Fractional derivatives are limits of fractional difference quotients, using the same fractional difference operator that appears in time series models of long range dependence.

Vector models with heavy tails are useful in many applications, since physical space is not one dimensional. A heavy tailed version of the central limit theorem for random vectors leads to an operator stable limit distribution, which can have a combination of normal and stable components. Vector random walks with heavy tails converge to an operator stable Lévy motion. Probability densities of this random walk limit solve a vector fractional diffusion equation like

$$\frac{\partial}{\partial t}p(x,y,t) = c_1\frac{\partial^\alpha}{\partial x^\alpha}p(x,y,t) + c_2\frac{\partial^\beta}{\partial y^\beta}p(x,y,t)$$

with a different fractional derivative in each coordinate.

Many interesting research problems in this area remain open. This book will guide the motivated reader to gain the essential background, insights, and techniques needed to read and understand current research papers and begin making their own contributions to this rapidly growing field.

In this second edition of the text, we have corrected all known errors and added three sections on applications: The fractional Poisson process; distributed order fractional derivatives; and the correlation structure of fractional processes. We would like to thank all the people who wrote to us about the first edition with helpful ideas and corrections. It is our hope that the second edition continues to be a useful contribution to the literature on this interesting area of theoretical and applied research.

East Lansing, April 2019 Mark M. Meerschaert, Alla Sikorskii

Acknowledgments

The material in this book relies on results from many hundreds of research papers, by a large number of talented authors. These works come from mathematics, physics, statistics, and many related areas of science and engineering. Although we did attempt to highlight connections with the existing literature, it seems inevitable that we should leave out some important contributors, for which we can only beg their forgiveness.

The book in its present form owes a great deal to those students and colleagues who reviewed portions of the book for accuracy and content. A special thanks goes to Farzad Sabzikar, Iowa State University, for reading and re-reading the entire manuscript. We would also like to thank Philippe Barbe from the French Centre National de la Recherche Scientifique (CNRS), Joonyong Choi from the USA University of Washington, Cheng-Gang Li from the Chinese Southwest Jiaotong University, Enrico Scalas from the British University of Sussex, Hans-Peter Scheffler from the German Universität Siegen, and Peter Straka from the UK Facebook London, for many helpful suggestions. Any remaining errors are the sole responsibility of the authors.

Thanks also to Mohsen Zayernouri and the FMATH group at Michigan State University, for many collaborations that helped to provide a real world context for material in this book.

Contents

1 Introduction

Fractional calculus is a rapidly growing field of research, at the interface between probability, differential equations, and mathematical physics. Fractional calculus is used to model anomalous diffusion, in which a cloud of particles spreads in a different manner than traditional diffusion. This book develops the basic theory of fractional calculus and anomalous diffusion, from the point of view of probability.

Traditional diffusion represents the long-time limit of a random walk, where finite variance jumps occur at regularly spaced intervals. Eventually, after each particle makes a series of random steps, a histogram of particle locations follows a bell-shaped normal density. The central limit theorem of probability ensures that this same bell-shaped curve will eventually emerge from any random walk with finite variance jumps, so that this diffusion model can be considered universal. The random walk limit is a Brownian motion, whose probability densities solve the diffusion equation. This link between differential equations and probability is a powerful tool. For example, a method called particle tracking computes approximate solutions of differential equations, by simulating the underlying stochastic process.

However, anomalous diffusion is often observed in real data. The "particles" might be pollutants in ground water, stock prices, sound waves, proteins crossing a cell boundary, or animals invading a new ecosystem. The anomalous diffusion can manifest in asymmetric densities, heavy tails, sharp peaks, and/or different spreading rates. The square root scaling in the central limit theorem implies that the width of a particle histogram should spread like the square root of the elapsed time. Both anomalous super-diffusion (a faster spreading rate) and sub-diffusion have been observed in real applications. In this book, we will develop models for both, based on fractional calculus.

The traditional diffusion equation relates the first time derivative of particle concentration to the second derivative in space. The fractional diffusion equation replaces the space and/or time derivatives with their fractional analogues. We will see that fractional derivatives are related to heavy tailed random walks. Fractional derivatives in space model super-diffusion, related to long power-law particle jumps. Fractional derivatives in time model sub-diffusion, related to long power-law waiting times between particle jumps. Fractional derivatives were invented by Leibnitz soon after their more familiar integer-order cousins, but they have become popular in practical applications only in the past few decades. In this book, we will see how fractional calculus and anomalous diffusion can be understood at a deep and intuitive level, using ideas from probability.

The first chapter of this book presents the basic ideas of fractional calculus and anomalous diffusion in the simplest setting. All of the material introduced here will be developed further in later chapters.

1.1 The traditional diffusion model

The traditional model for diffusion combines elements of probability, differential equations, and physics. A random walk provides the basic physical model of particle motion. The central limit theorem gives convergence to a Brownian motion, whose probability densities solve the diffusion equation. We start with a sequence of independent and identically distributed (iid) random variables $Y, Y_1, Y_2, Y_3, \ldots$ that represent the jumps of a randomly selected particle. The *random walk*

$$S_n = Y_1 + \cdots + Y_n$$

gives the location of that particle after n jumps. Next we recall the well-known *central limit theorem*, which shows that the probability distribution of S_n converges to a normal limit. Here we sketch the argument in the simplest case, using Fourier transforms. Details are provided at the end of this section to make the argument rigorous. A complete proof of the central limit theorem will be given in Theorem 3.36 using different methods. Then in Theorem 4.5, we will use regular variation to show that the same normal limit governs a somewhat broader class of random walks.

Let $F(x) = \mathbb{P}[Y \le x]$ denote the cumulative distribution function (cdf) of the jumps, and assume that the probability density function (pdf) $f(x) = F'(x)$ exists. Then we have

$$P[a \le Y \le b] = \int_a^b f(x)\,dx = F(b) - F(a)$$

for any real numbers $a < b$. The moments of this distribution are given by

$$\mu_p = \mathbb{E}\left[Y^p\right] = \int x^p f(x)\,dx$$

where the integral is taken over the domain of the function f.

The *Fourier transform* (FT) of the pdf is

$$\hat{f}(k) = \mathbb{E}\left[e^{-ikY}\right] = \int e^{-ikx} f(x)\,dx.$$

The FT is closely related to the *characteristic function* $\mathbb{E}\left[e^{ikY}\right] = \hat{f}(-k)$. If the first two moments exist, a Taylor series expansion $e^z = 1 + z + z^2/2! + \cdots$ leads to

$$\hat{f}(k) = \int \left(1 - ikx + \frac{1}{2!}(-ikx)^2 + \cdots\right) f(x)\,dx = 1 - ik\mu_1 - \tfrac{1}{2}k^2\mu_2 + o(k^2) \tag{1.1}$$

since $\int f(x)\,dx = 1$. Here $o(k^2)$ denotes a function that tends to zero faster than k^2 as $k \to 0$. A formal proof of (1.1) is included in the details at the end of this section.

Suppose $\mu_1 = 0$ and $\mu_2 = 2$, i.e., the jumps have mean zero and variance 2. Then we have

$$\hat{f}(k) = 1 - k^2 + o(k^2)$$

as $k \to 0$. The sum $S_n = Y_1 + \cdots + Y_n$ has FT

$$\begin{aligned}\mathbb{E}\left[e^{-ikS_n}\right] &= \mathbb{E}\left[e^{-ik(Y_1+\cdots+Y_n)}\right]\\ &= \mathbb{E}\left[e^{-ikY_1}\right]\cdots\mathbb{E}\left[e^{-ikY_n}\right]\\ &= \mathbb{E}\left[e^{-ikY}\right]^n = \hat{f}(k)^n\end{aligned}$$

and so the normalized sum $n^{-1/2}S_n$ has FT

$$\begin{aligned}\mathbb{E}\left[e^{-ik(n^{-1/2}S_n)}\right] &= \mathbb{E}\left[e^{-i(n^{-1/2}k)S_n}\right] = \hat{f}(n^{-1/2}k)^n\\ &= \left(1 - \frac{k^2}{n} + o(n^{-1})\right)^n \to e^{-k^2}\end{aligned} \tag{1.2}$$

using the general fact that $(1 + (r/n) + o(n^{-1}))^n \to e^r$ as $n \to \infty$ for any $r \in \mathbb{R}$ (see details). The limit

$$e^{-k^2} = \mathbb{E}\left[e^{-ikZ}\right] = \int e^{-ikx}\frac{1}{\sqrt{4\pi}}e^{-x^2/4}\,dx$$

using the standard formula from FT tables [203, p. 524]. Then the continuity theorem for FT (see details) yields the traditional central limit theorem (CLT):

$$n^{-1/2}S_n = \frac{Y_1 + \cdots + Y_n}{\sqrt{n}} \Rightarrow Z \tag{1.3}$$

where $\Rightarrow$ indicates convergence in distribution. The limit Z in (1.3) is normal with mean zero and variance 2.

An easy extension of this argument gives convergence of the rescaled random walk:

$$S_{[ct]} = Y_1 + \cdots + Y_{[ct]}$$

gives the particle location at time $t > 0$ at any time scale $c > 0$. Increasing the time scale c makes time go faster, e.g., multiply c by 60 to change from minutes to hours. The long-time limit of the rescaled random walk is a Brownian motion: As $c \to \infty$ we have

$$\mathbb{E}\left[e^{-ik\,c^{-1/2}S_{[ct]}}\right] = \left(1 - \frac{k^2}{c} + o(c^{-1})\right)^{[ct]} = \left[\left(1 - \frac{k^2}{c} + o(c^{-1})\right)^c\right]^{\frac{[ct]}{c}} \to e^{-tk^2} \tag{1.4}$$

where the limit

$$e^{-tk^2} = \hat{p}(k,t) = \int e^{-ikx}p(x,t)\,dx$$

is the FT of a normal density

$$p(x,t) = \frac{1}{\sqrt{4\pi t}}e^{-x^2/(4t)}$$

with mean zero and variance $2t$. Then the continuity theorem for FT implies that

$$c^{-1/2}S_{[ct]} \Rightarrow Z_t$$

where the Brownian motion Z_t is normal with mean zero and variance $2t$.

Clearly the FT $\hat{p}(k,t) = e^{-tk^2}$ solves a differential equation

$$\frac{d\hat{p}}{dt} = -k^2\hat{p} = (ik)^2\hat{p}. \tag{1.5}$$

If f' exists and if f, f' are integrable, then the FT of $f'(x)$ is $(ik)\hat{f}(k)$ (see details). Using this fact, we can invert the FT on both sides of (1.5) to get (see details)

$$\frac{\partial p}{\partial t} = \frac{\partial^2 p}{\partial x^2}. \tag{1.6}$$

This shows that the pdf of Z_t solves the diffusion equation (1.6). The diffusion equation models the spreading of a cloud of particles. The random walk S_n gives the location of a randomly selected particle, and the long-time limit density $p(x,t)$ gives the relative concentration of particles at location x at time $t > 0$.

More generally, suppose that $\mu_1 = \mathbb{E}[Y_n] = 0$ and $\mu_2 = \mathbb{E}[Y_n^2] = \sigma^2 > 0$. Then

$$\hat{f}(k) = 1 - \tfrac{1}{2}\sigma^2 k^2 + o(k^2)$$

leads to

$$\mathbb{E}[e^{-ik\,n^{-1/2}S_n}] = \left(1 - \frac{\sigma^2 k^2}{2n} + o(n^{-1})\right)^n \to \exp(-\tfrac{1}{2}\sigma^2 k^2)$$

and

$$\mathbb{E}[e^{-ik\,c^{-1/2}S_{[ct]}}] = \left(1 - \frac{\sigma^2 k^2}{2c} + o(c^{-1})\right)^{[ct]} \to \exp(-\tfrac{1}{2}t\sigma^2 k^2) = \hat{p}(k,t). \tag{1.7}$$

This FT inverts to a normal density

$$p(x,t) = \frac{1}{\sqrt{2\pi\sigma^2 t}} e^{-x^2/(2\sigma^2 t)}$$

with mean zero and variance $\sigma^2 t$. The FT solves

$$\frac{d\hat{p}}{dt} = -\frac{\sigma^2}{2}k^2\hat{p} = \frac{\sigma^2}{2}(ik)^2\hat{p}$$

which inverts to

$$\frac{\partial p}{\partial t} = \frac{\sigma^2}{2}\frac{\partial^2 p}{\partial x^2}. \tag{1.8}$$

This form of the diffusion equation shows the relation between the *dispersivity* $D = \sigma^2/2$ and the particle jump variance. Apply the continuity theorem for FT to (1.7) to get random walk convergence:

$$c^{-1/2}S_{[ct]} \Rightarrow Z_t$$

where Z_t is a Brownian motion, normal with mean zero and variance $\sigma^2 t$.

In many applications, it is useful to add a drift: $vt + Z_t$ has FT

$$\mathbb{E}[e^{-ik(vt+Z_t)}] = e^{-ikvt-\frac{1}{2}t\sigma^2k^2} = \hat{p}(k, t),$$

which solves

$$\frac{d\hat{p}}{dt} = \left(-ikv + \frac{\sigma^2}{2}(ik)^2\right)\hat{p}.$$

Invert the FT to obtain the diffusion equation with drift:

$$\frac{\partial p}{\partial t} = -v\frac{\partial p}{\partial x} + \frac{\sigma^2}{2}\frac{\partial^2 p}{\partial x^2}. \tag{1.9}$$

This represents the long-time limit of a random walk whose jumps have a non-zero mean $v = \mu_1$ (see details). Figure 1.1 shows a typical concentration profile, a normal pdf

$$p(x, t) = \frac{1}{\sqrt{2\pi\sigma^2 t}} e^{-(x-vt)^2/(2\sigma^2 t)} \tag{1.10}$$

that solves the diffusion equation with drift (1.9). Figure 1.2 shows how the solution evolves in time. Since $vt+Z_t$ has mean vt, the center of mass is proportional to the time variable. Since $vt + Z_t$ has variance $\sigma^2 t$, the standard deviation is $\sigma\sqrt{t}$, so the particle plume spreads proportional to the square root of time. Setting $x = vt$ in (1.10) shows that the peak concentration falls like the square root of time. The simple R codes used to produce the plots in Figures 1.1 and 1.2 will be presented and discussed in Examples 5.1 and 5.2, respectively.

Details

The FT $\hat{f}(k) = \int e^{-ikx} f(x)\,dx$ is defined for integrable functions f, since $|e^{-ikx}| = 1$. Hence the pdf of any random variable X has a FT. In fact, the FT $\hat{f}(k) = \mathbb{E}[e^{-ikX}]$ exists for all $k \in \mathbb{R}$, for any random variable X, whether or not it has a density. The next two results justify the FT expansion (1.1).

Proposition 1.1. *If* $\mathbb{E}[|X|^p]$ *exists, then*

$$(-i)^p \mu_p = \hat{f}^{(k)}(0) = \frac{d^p}{dk^p}\mathbb{E}\left[e^{-ikX}\right]_{k=0} \tag{1.11}$$

Proof. The first derivative of the FT is

$$\begin{aligned}\hat{f}^{(1)}(k) &= \lim_{h\to 0}\frac{\hat{f}(k+h) - \hat{f}(k)}{h}\\ &= \lim_{h\to 0} h^{-1}\left(\mathbb{E}\left[e^{-i(k+h)X}\right] - \mathbb{E}\left[e^{-ikX}\right]\right) = \lim_{h\to 0}\mathbb{E}[g_h(X)]\end{aligned}$$

where $g_h(x) = h^{-1}(e^{-i(k+h)x} - e^{-ikx}) = h^{-1}(e^{-ihx} - 1)e^{-ikx}$ is the difference quotient for the differentiable function $k \mapsto e^{-ikx}$, so that $g_h(x) \to g(x) = -ixe^{-ikx}$ as $h \to 0$.

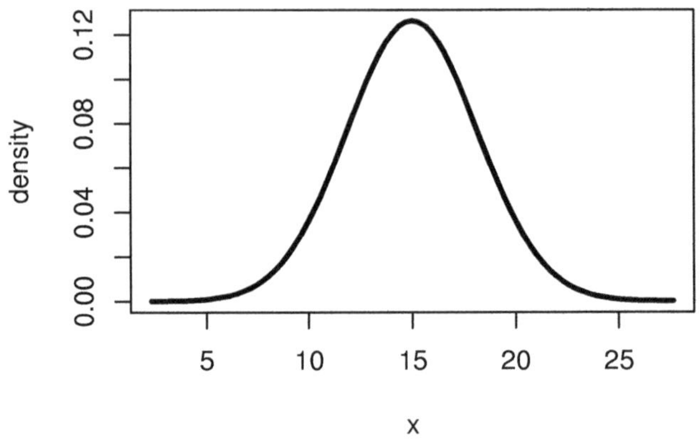

Fig. 1.1: Solution to diffusion equation (1.9) at time $t = 5.0$ with velocity $v = 3.0$ and variance $\sigma^2 = 2.0$.

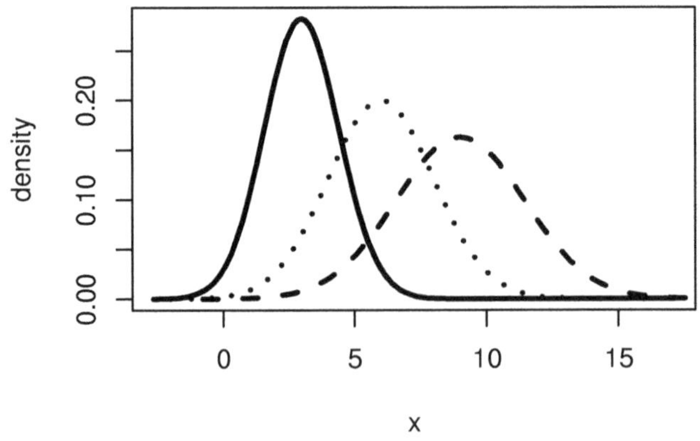

Fig. 1.2: Solution to diffusion equation (1.9) at times $t_1 = 1.0$ (solid line), $t_2 = 2.0$ (dotted line), and $t_3 = 3.0$ (dashed line). The velocity $v = 3.0$ and variance $\sigma^2 = 2.0$.

From the geometric interpretation of e^{iy} as a vector in complex plane, it follows that $|e^{iy} - 1| \leq |y|$ for all $y \in \mathbb{R}$. Then

$$|g_h(x)| = \left|\frac{e^{-ihx} - 1}{h}\right| \cdot |e^{-ikx}| \leq |x|$$

for all $h \in \mathbb{R}$ and all $x \in \mathbb{R}$. The *Dominated Convergence Theorem* states that if $g_h(x) \to g(x)$ for all $x \in \mathbb{R}$ and if $|g_h(x)| \leq r(x)$ for all h and all $x \in \mathbb{R}$, and if $\mathbb{E}[r(X)]$ is finite, then $\mathbb{E}[g_h(X)] \to \mathbb{E}[g(X)]$ and these expectations exist (e.g., see Durrett [62, Theorem 1.6.7, p. 29]). Since $\mathbb{E}[|X|]$ exists, the dominated convergence theorem with $r(x) = |x|$ implies that

$$\hat{f}^{(1)}(k) = \lim_{h\to 0} \mathbb{E}[g_h(X)] = \mathbb{E}[g(X)] = \mathbb{E}\left[(-iX)\, e^{-ikX}\right].$$

Set $k = 0$ to arrive at (1.11) in the case $p = 1$. The case $p > 1$ is similar, using the fact that $g_h(x) = h^{-p}(e^{-ihx} - 1)^p e^{-ikx}$ is the pth order difference quotient for $k \mapsto e^{-ikx}$. Alternatively, the proof for the case $p > 1$ can be completed using an induction argument. □

Proposition 1.2. *If $\mathbb{E}[|X|^p]$ exists, then the FT of X is*

$$\hat{f}(k) = \sum_{j=0}^{p} \frac{(-ik)^j}{j!}\mu_j + o(k^p) \tag{1.12}$$

as $k \to 0$.

Proof. If the FT $\hat{f}(k)$ is p times differentiable, then the Taylor expansion

$$\hat{f}(k) = \sum_{j=0}^{p} \frac{k^j}{j!}\hat{f}^{(j)}(0) + o(k^p)$$

is valid for all $k \in \mathbb{R}$. Apply Proposition 1.1 to arrive at (1.12). □

In equation (1.2) we used the fact that

$$\left(1 + \frac{r}{n} + o(1/n)\right)^n \to e^r \quad \text{as } n \to \infty. \tag{1.13}$$

To verify this, write $o(1/n) = \varepsilon_n/n$ where $\varepsilon_n \to 0$ as $n \to \infty$. Note that $|r + \varepsilon_n| < 1$ for n sufficiently large, and then use the fact that $\ln(1 + z) = z + O(z^2)$ as $z \to 0$. This notation means that for some $\delta > 0$ we have

$$\left|\frac{\ln(1 + z) - z}{z^2}\right| < C$$

for some constant $C > 0$, for all $|z| < \delta$. Then we can write

$$\begin{aligned}\ln\left[\left(1 + \frac{r + \varepsilon_n}{n}\right)^n\right] &= n \ln\left[1 + \frac{r + \varepsilon_n}{n}\right] \\ &= n\left[\frac{r + \varepsilon_n}{n} + O\left(\frac{1}{n^2}\right)\right] = r + \varepsilon_n + O\left(\frac{1}{n}\right) \to r.\end{aligned}$$

Then apply the continuous function $\exp(z)$ to both sides to conclude that (1.13) holds.

In (1.3) we use the idea of weak convergence. Suppose that X_n is a sequence of random variables with cdf $F_n(x) = \mathbb{P}[X_n \le x]$, and X is a random variable with cdf $F(x) = \mathbb{P}[X \le x]$. We write $X_n \Rightarrow X$ if $F_n(x) \to F(x)$ for all $x \in \mathbb{R}$ such that F is continuous at x. This is equivalent to the condition that $\mathbb{E}[h(X_n)] \to \mathbb{E}[h(X)]$ for all bounded, continuous functions $h : \mathbb{R} \to \mathbb{R}$. See for example Billingsley [37].

In (1.3) we use the continuity theorem for the Fourier transform. Let $\hat{f}_n(k) = \mathbb{E}[e^{-ikX_n}]$ and $\hat{f}(k) = \mathbb{E}[e^{-ikX}]$. The Lévy Continuity Theorem [146, Theorem 1.3.6] implies that $X_n \Rightarrow X$ if and only if $\hat{f}_n(k) \to \hat{f}(k)$. More precisely, we have:

Theorem 1.3 (Lévy Continuity Theorem). *If X_n, X are random variables on $\mathbb{R}$, then $X_n \Rightarrow X$ implies that $\hat{f}_n(k) \to \hat{f}(k)$ for each $k \in \mathbb{R}$, uniformly on compact subsets. Conversely, if X_n is a sequence of random variables such that $\hat{f}_n(k) \to \hat{f}(k)$ for each $k \in \mathbb{R}$, and the limit $\hat{f}(k)$ is continuous at $k = 0$, then $\hat{f}(k)$ is the FT of some random variable X, and $X_n \Rightarrow X$.*

In (1.6) we used the fact that the FT of $f'(x)$ is $(ik)\hat{f}(k)$. If $f'(x)$ exists and is integrable, the limits

$$\lim_{x\to\infty} f(x) = f(0) + \lim_{x\to\infty} \int_0^x f'(u)du \quad \text{and} \quad \lim_{x\to-\infty} f(x) = f(0) - \lim_{x\to-\infty} \int_x^0 f'(u)du$$

exist. If f is integrable, then these limits must equal zero. Then we can integrate by parts to get

$$\int_{-\infty}^{\infty} e^{-ikx} f'(x)\, dx = \left[e^{-ikx} f(x)\right]_{x=-\infty}^{\infty} + \int_{-\infty}^{\infty} ike^{-ikx} f(x)\, dx = 0 + (ik)\hat{f}(k). \tag{1.14}$$

Applying this fact to the function f' shows that, if f'' is also integrable, then its FT equals $(ik)^2\hat{f}(k)$, and so

$$(ik)^2\hat{p}(k,t) = \int e^{-ikx} \frac{\partial^2}{\partial x^2} p(x,t)\, dx. \tag{1.15}$$

To arrive at (1.6), we inverted the FT (1.5). This can be justified using the following theorem.

Theorem 1.4 (Fourier inversion theorem). *If $\int |f(x)|\, dx < \infty$, then FT $\hat{f}(k)$ exists. Then if $\int |\hat{f}(k)|\, dk < \infty$, we have*

$$f(x) = \frac{1}{2\pi} \int e^{ikx} \hat{f}(k)\, dk \tag{1.16}$$

for all $x \in \mathbb{R}$.

Proof. See [146, Theorem 1.3.7] or Stein and Weiss [208, Corollary 1.21]. □

Apply Theorem 1.4 to both sides of (1.5) to get

$$\frac{1}{2\pi}\int e^{ikx}\frac{\partial}{\partial t}\hat{p}(k,t)\,dk = \frac{1}{2\pi}\int e^{ikx}(ik)^2\hat{p}(x,t)\,dk. \tag{1.17}$$

By (1.15), the right hand side of (1.17) equals $\partial^2 p(x,t)/\partial x^2$. In order to prove (1.6), it suffices to show that

$$\int e^{ikx}\frac{\partial}{\partial t}\hat{p}(k,t)\,dk = \frac{\partial}{\partial t}\int e^{ikx}\hat{p}(k,t)\,dk$$

for any fixed $t > 0$. Write

$$\frac{\partial}{\partial t}\int e^{ikx}\hat{p}(k,t)\,dk = \lim_{h\to 0}\int e^{ikx}\frac{\hat{p}(k,t+h)-\hat{p}(k,t)}{h}\,dk,$$

where $\hat{p}(k,t) = e^{-tk^2}$. Since $h \to 0$, consider h small such that $|h| < t/2$ ($t > 0$ is fixed in this argument). Then the mean value theorem yields $|1 - e^{-hk^2}| \le |h|k^2 e^{tk^2/2}$, and therefore

$$\left|\frac{\hat{p}(k,t+h)-\hat{p}(k,t)}{h}\right| = e^{-tk^2}\left|\frac{1-e^{-hk^2}}{h}\right| \le k^2 e^{-tk^2/2}.$$

Another version of the Dominated Convergence Theorem (e.g., see Rudin [181, Theorem 11.32]) states that if $f_n(y) \to f(y)$ as $n \to \infty$ and if $|f_n(y)| \le g(y)$ for all n and all y, where $\int g(y)\,dy$ exists, then $\int f_n(y)\,dy \to \int f(y)\,dy$ and these integrals exist. Since for any $t > 0$, the function $k^2 e^{-tk^2/2}$ is integrable with respect to k, the dominated convergence theorem implies

$$\frac{\partial}{\partial t}\int e^{ikx}\hat{p}(k,t)\,dk = \int e^{ikx}\lim_{h\to 0}\frac{\hat{p}(k,t+h)-\hat{p}(k,t)}{h}\,dk = \int e^{ikx}\frac{\partial}{\partial t}\hat{p}(k,t)\,dk.$$

Similar arguments justify (1.8) and (1.9).

To show that (1.9) governs the limit of a random walk with drift, suppose that $Y, Y_1, Y_2, Y_3, \ldots$ are iid with mean $\mu_1 = v = \mathbb{E}[Y]$ and finite variance $\sigma^2 = \mu_2 - \mu_1^2 = \mathbb{E}[(Y-\mu_1)^2]$. Write

$$S_{[ct]} = \sum_{j=1}^{[ct]} Y_j = \sum_{j=1}^{[ct]}(Y_j - v) + \sum_{j=1}^{[ct]} v$$

and note that the first sum grows like $c^{1/2}$ while the second grows like c as $c \to \infty$. Hence, in order to get convergence, we must normalize at two scales. Since $Y - v$ has FT

$$\hat{f}(k) = 1 - \tfrac{1}{2}\sigma^2 k^2 + o(k^2)$$

as $k \to 0$, the sum of the mean-centered jumps $(Y_1 - v) + \cdots + (Y_n - v)$ has FT $\hat{f}(k)^n$, and then the centered and normalized sum

$$S^{(c)}(t) = c^{-1/2}\sum_{j=1}^{[ct]}(Y_j - v) + c^{-1}\sum_{j=1}^{[ct]} v$$

has FT

$$\left(1 - \tfrac{1}{2}\sigma^2 \frac{k^2}{c} + o(c^{-1})\right)^{[ct]} \cdot e^{-ikc^{-1}v[ct]} \to \exp(-ikvt - \tfrac{1}{2}t\sigma^2 k^2)$$

The limit inverts to a normal density with mean vt and variance $\sigma^2 t$. Physically, we follow a cloud of particles (iid copies of the random walk $S_{[ct]}$) in a moving coordinate system with origin at $x = vt$. In this coordinate system, the cloud spreads according to the diffusion equation. Translating back to the original coordinates, we see a diffusion with drift.

1.2 Fractional diffusion

The diffusion model presented in Section 1.1 describes random walk limits with finite variance jumps. In many real world applications, particles follow a heavy-tailed jump distribution, and a different model emerges. Here we outline the argument for the simplest case, a Pareto distribution. Additional details are provided at the end of this section. The formal proof for Pareto jumps will be given in Theorem 3.37. Then in Theorem 4.5, we will use regular variation to show that the same limit governs a broad class of random walks whose probability tails fall off like a power law.

As in Section 1.1, the random walk

$$S_n = Y_1 + \cdots + Y_n$$

gives the location of a particle after n independent and identically distributed (iid) jumps. Suppose that the jump variables Y_n follow a Pareto distribution, centered to mean zero: Suppose $\mathbb{P}[X > x] = Cx^{-\alpha}$ where $C > 0$ and $1 < \alpha < 2$. Then the first moment $\mu = \mathbb{E}[X]$ exists, but the second moment $\mathbb{E}[X^2] = \infty$. Now take Y_n iid with $X - \mu$, so that $\mu_1 = \mathbb{E}[Y_n] = 0$. Since the variance of Y_n is infinite, the central limit theorem (1.3) does not apply. Instead, we will see that a different limit occurs, with a different scaling. For suitably chosen C, the FT of Y_n is (see details)

$$\hat{f}(k) = 1 + (ik)^\alpha + O(k^2) \tag{1.18}$$

as $k \to 0$. The sum $S_n = Y_1 + \cdots + Y_n$ has FT $\hat{f}(k)^n$ and the normalized sum $n^{-1/\alpha}S_n$ has FT

$$\hat{f}(n^{-1/\alpha}k)^n = \left(1 + \frac{(ik)^\alpha}{n} + O(n^{-2/\alpha})\right)^n \to e^{(ik)^\alpha} \tag{1.19}$$

since $2/\alpha > 1$, where the limit

$$e^{(ik)^\alpha} = \mathbb{E}\left[e^{-ikZ}\right]$$

is the FT of a *stable density* (see details). The continuity theorem for FT yields the *extended central limit theorem*:

$$n^{-1/\alpha}S_n = \frac{Y_1 + \cdots + Y_n}{n^{1/\alpha}} \Rightarrow Z. \tag{1.20}$$

The family of stable distributions includes the normal as a special case, when $\alpha = 2$. They represent all possible limits in the extended central limit theorem, see Theorem 4.5 for details.

Now we show convergence of the random walk. As $c \to \infty$ we have

$$\mathbb{E}[e^{-ik\,c^{-1/\alpha}S_{[ct]}}] = \left(1 + \frac{(ik)^\alpha}{c} + O(c^{-2/\alpha})\right)^{[ct]} \to e^{t(ik)^\alpha}$$

where the limit

$$e^{t(ik)^\alpha} = \mathbb{E}[e^{-ikZ_t}] = \hat{p}(k,t) = \int e^{-ikx}p(x,t)\,dx$$

is the FT of a stable density. Then the continuity theorem for FT implies

$$c^{-1/\alpha}S_{[ct]} \Rightarrow Z_t.$$

Unlike the normal case $\alpha = 2$, the stable FT $\hat{p}(k,t) = e^{t(ik)^\alpha}$ cannot be inverted in closed form when $1 < \alpha < 2$.

Clearly the FT $\hat{p}(k,t) = e^{t(ik)^\alpha}$ solves

$$\frac{d\hat{p}}{dt} = (ik)^\alpha \hat{p}. \tag{1.21}$$

Recalling that $(ik)^n\hat{f}(k)$ is the FT of the nth derivative, we define the *fractional derivative* $d^\alpha f(x)/dx^\alpha$ to be the function whose FT is $(ik)^\alpha\hat{f}(k)$ (see details). Then we can invert the FT in (1.21) to see that the stable densities solve a *fractional diffusion equation*

$$\frac{\partial p}{\partial t} = \frac{\partial^\alpha p}{\partial x^\alpha}. \tag{1.22}$$

The fractional diffusion equation models the spreading of a cloud of particles with a power-law jump distribution.

The stable pdf $p(x,t)$ is positively skewed, with a heavy power-law tail. In fact, we have $p(x,t) = Ax^{-\alpha-1} + o(x^{-\alpha-1})$ as $x \to \infty$ for some $A > 0$ depending on C, t, and α, so that the limit retains the power-law jump distribution (e.g., see Zolotarev [228], p. 143). This is in stark contrast to the traditional CLT, in which the tail behavior of the individual jumps disappears in the limit.

The fractional diffusion equation (1.22) models *super-diffusion*. In fact, we have $Z_{ct} \simeq c^{1/\alpha}Z_t$ (same distribution) since

$$\begin{aligned}\mathbb{E}[e^{-ikZ_{ct}}] &= \hat{p}(k,ct) = e^{ct(ik)^\alpha}\\ &= e^{t(ikc^{1/\alpha})^\alpha} = \hat{p}(c^{1/\alpha}k,t) = \mathbb{E}[e^{-ik\,c^{1/\alpha}Z_t}]\end{aligned}$$

This property is called *self-similarity*. The index $H = 1/\alpha$ of self-similarity is often called the *Hurst exponent* (e.g., see Embrechts and Maejima [64]). This also implies that solutions $p(x,t)$ to the fractional diffusion equation (1.22) satisfy a scaling relation

$$p(x,ct) = c^{-1/\alpha}p(c^{-1/\alpha}x,t) \quad \text{for all } x \in \mathbb{R} \text{ and all } t > 0.$$

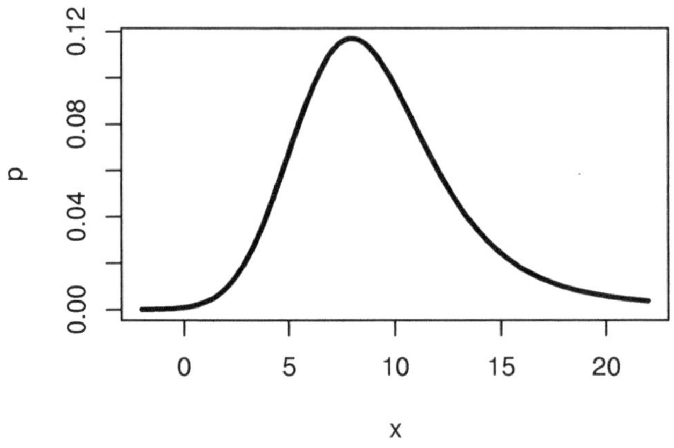

Fig. 1.3: Solution to the fractional diffusion equation (1.23) at time $t = 5.0$ with velocity $v = 2.0$ and dispersion $D = 1.0$, for $\alpha = 1.5$.

In particular, the spreading rate is $t^{1/\alpha}$, and the peak falls at the same rate, which is faster than the $t^{1/2}$ rate in the traditional diffusion equation (1.6).

Next we add scale and drift. The FT of $vt + D^{1/\alpha}Z_t$ is

$$\hat{p}(k,t) = \mathbb{E}\left[e^{-ik(vt+D^{1/\alpha}Z_t)}\right] = e^{-ikvt+Dt(ik)^\alpha}$$

which solves

$$\frac{d\hat{p}}{dt} = (-ikv + D(ik)^\alpha)\,\hat{p}.$$

Invert the FT to obtain the fractional diffusion equation with drift:

$$\frac{\partial p}{\partial t} = -v\frac{\partial p}{\partial x} + D\frac{\partial^\alpha p}{\partial x^\alpha}. \tag{1.23}$$

In applications to ground water hydrology, equation (1.23) is also called the fractional advection dispersion equation (FADE), see Benson et al. [29]. Advection is the displacement of suspended particles in moving water, and dispersion is the particle spreading caused by particles following different flow paths through a porous medium. The particle density $p(x,t)$ that solves (1.23) has center of mass $x = vt$, and it spreads out from the center of mass at the super-diffusive rate $t^{1/\alpha}$ due to self-similarity.

Figure 1.3 shows a stable pdf that solves the FADE (1.23). Note the skewness and the heavy right tail. Figure 1.4 shows how the solution evolves in time. Since the limit

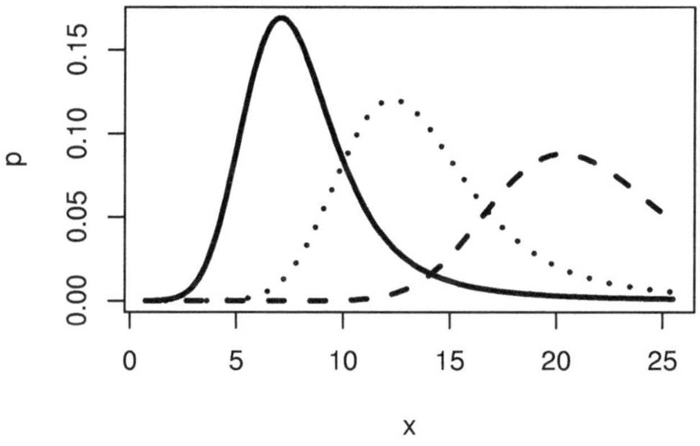

Fig. 1.4: Solution to fractional diffusion equation (1.23) at times $t_1 = 3.0$ (solid line), $t_2 = 5.0$ (dotted line), and $t_3 = 8.0$ (dashed line) with velocity $v = 3.0$ and dispersion $D = 1.0$, for $\alpha = 1.5$.

process is self-similar with index $1/\alpha > 1/2$, the plume spreads faster than a traditional Brownian motion. The R codes used to produce the plots in Figures 1.3 and 1.4 will be presented and discussed in Examples 5.9 and 5.11, respectively.

In ground water hydrology, the FADE (1.23) models concentration of a contaminant that is transported along with moving water under the ground. Particles must find their way through a porous medium consisting of sand, gravel, clay, etc. Some particles will find a relatively direct path, while others will take a more tortuous route. This causes dispersion, traditionally modeled by the second derivative term. A fractional derivative indicates a power-law distribution of particle velocities, thought to be related to a fractal model of the porous medium, see Wheatcraft and Tyler [218]. The plume center of mass moves at a constant rate, modeled by the first derivative term. Concentration measurements are taken at different points x at the same time $t > 0$ to form a histogram, which is then fit to the stable density $p(x, t)$ that solves the FADE (1.23).

To fit the parameter α, the fact that $p(x, t) \approx Ax^{-\alpha-1}$ is used. Since $\log p \approx \log A - (\alpha + 1) \log x$, a log-log plot of the concentration profile should resemble a straight line with slope $-(\alpha + 1)$ for x large, and this can provide a rough estimate of α. Figure 1.5 shows concentration measurements taken at a distance x meters downstream from the initial injection point, from an experiment documented in Benson et al. [29]. A tracer is injected at location $x = 0$ at time $t = 0$ and transported downstream by the natural flow of the ground water. Concentration measurements taken at $t = 224$ and $t = 328$

days after injection were fit to the FADE (1.23) with constant coefficients (black line), and to the traditional advection dispersion equation (ADE) in (1.9) where D is allowed to vary with time (grey line). It is commonly noted in hydrological studies that the best fitting D grows with time like a power law (e.g., see Wheatcraft and Tyler [218]). The popularity of the fractional ADE is partly due to the fact that it can fit the same plume at different times using constant coefficients. The fitted stable density has $\alpha = 1.1$ with $v = 0.12$ meters per day and $D = 0.14$ meters$^\alpha$ per day. It is thought that α reflects the heterogeneity of the porous medium, see Clarke et al. [50]. The power law tail of the stable density is confirmed by the straight line asymptotics on the right hand side of each plot. The best fitting normal density underestimates concentrations by six orders of magnitude at the leading (right) edge of the plume. If the plume represents a pollutant heading towards a municipal water supply well, the ADE would seriously underestimate the risk of downstream contamination. The stable density that solves the FADE, on the other hand, captures the super-diffusive spreading and power-law leading tail observed in the data.

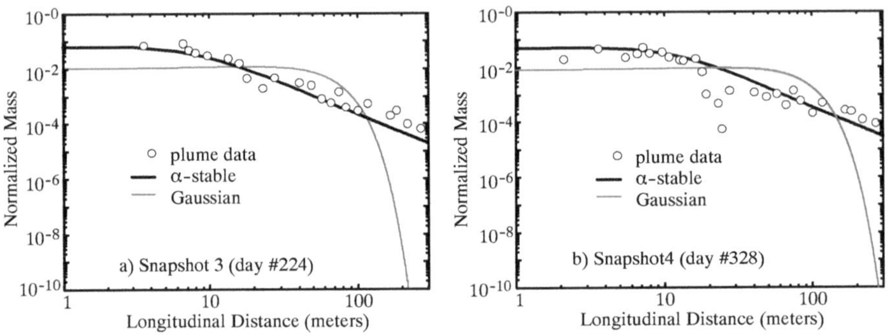

Fig. 1.5: Concentration measurements from Benson et al. [29], and fitted stable density with $\alpha = 1.1$.

A more general fractional diffusion equation pertains when random walk jumps follow a two-sided Pareto distribution. Suppose $\mathbb{P}[X > x] = pCx^{-\alpha}$ and $\mathbb{P}[X < -x] = qCx^{-\alpha}$ for some $1 < \alpha < 2$ and $0 \le p, q \le 1$ with $p + q = 1$. Then $\mu_1 = \mathbb{E}[X]$ exists, and we take (Y_n) iid with $X - \mu_1$. Now for some constant $D > 0$ depending on α and C we have (see details)

$$\hat{f}(k) = 1 + pD(ik)^\alpha + qD(-ik)^\alpha + O(k^2) \tag{1.24}$$

and then we get

$$\begin{aligned}\mathbb{E}[e^{-ik\,c^{-1/\alpha}S_{[ct]}}] &= \left(1 + \frac{pD(ik)^\alpha + qD(-ik)^\alpha}{c} + O(c^{-2/\alpha})\right)^{[ct]} \\ &\to e^{t[pD(ik)^\alpha + qD(-ik)^\alpha]} = \hat{p}(k, t).\end{aligned}$$

This FT solves

$$\frac{d\hat{p}}{dt} = [pD(ik)^\alpha + qD(-ik)^\alpha]\,\hat{p}. \tag{1.25}$$

Now we define the *negative fractional derivative* $d^\alpha f(x)/d(-x)^\alpha$ to be the function whose FT is $(-ik)^\alpha \hat{f}(k)$. Invert the FT in (1.25) to see that the two-sided stable densities solve a two-sided fractional diffusion equation

$$\frac{\partial p(x,t)}{\partial t} = pD\frac{\partial^\alpha p(x,t)}{\partial x^\alpha} + qD\frac{\partial^\alpha p(x,t)}{\partial(-x)^\alpha}. \tag{1.26}$$

The random walk limit $c^{-1/\alpha}S_{[ct]} \Rightarrow Z_t$ is a two-sided stable process. Its densities solve the fractional diffusion equation (1.26), which therefore models the spreading of a cloud of particles with power-law jumps in both directions. The weights p and q represent the relative likelihood of positive or negative jumps. The family of two-sided stable densities $p(x,t)$ for the limit process Z_t spreads at the super-diffusive rate $t^{1/\alpha}$, and has power-law tails in both directions (see details). The skewness $\beta = p - q$ indicates whether the pdf is positively skewed ($\beta > 0$) due to the preponderance of positive jumps, negatively skewed ($\beta < 0$), or symmetric ($\beta = 0$). The two-sided FADE

$$\frac{\partial p(x,t)}{\partial t} = -v\frac{\partial p(x,t)}{\partial x} + pD\frac{\partial^\alpha p(x,t)}{\partial x^\alpha} + qD\frac{\partial^\alpha p(x,t)}{\partial(-x)^\alpha} \tag{1.27}$$

governs the process $vt + Z_t$, the scaling limit of a random walk with mean jump $v = \mathbb{E}[Y_n]$ (see details).

In applications to ground water hydrology, concentration profiles show a power-law leading edge, and we typically find $\beta = 1$, since fast-moving particles jump downstream, as noted by Benson et al. [29]. In a classical study of turbulence by Solomon, Weeks and Swinney [206], velocity measurements follow a symmetric power-law distribution with $\beta = 0$. In a fractional model for invasive species developed by Baeumer, Kovács and Meerschaert [16], animals and plants take power law jumps with $\beta > 0$, indicating a preference for motion in the direction of new territories . In finance, price jumps follow a power law with $\beta \approx 0$, while trading volume follows a power law with $\beta = 1$, see for example Mandelbrot [132]. In medical ultrasound, power law dispersal is observed with $\beta = 1$, see Kelly, McGough and Meerschaert [103]. In river flows, retention of contaminant particles in river beds and eddy pools causes a power-law trailing edge in the concentration profile, modeled by a FADE with $\beta = -1$, see for example Deng, Singh and Bengtsson [60]. This fit is controversial, since the random walk model with $\beta = -1$ suggests that particles are taking long jumps upstream, see discussion in Chakraborty, Meerschaert and Lim [47]. The paper [47] also discusses more advanced statistical methods for fitting a stable pdf to data.

Remark 1.5. The random variable Z with FT

$$\mathbb{E}[e^{-ikZ}] = e^{pD(ik)^\alpha + qD(-ik)^\alpha}$$

is called stable because, if (Y_n) are iid with Z, then the FT $e^{n[pD(ik)^\alpha + qD(-ik)^\alpha]}$ of $n^{1/\alpha}Z$ is also the FT of $Y_1 + \cdots + Y_n$. It follows that

$$\frac{Y_1 + \cdots + Y_n}{n^{1/\alpha}} \simeq Z$$

for all $n \geq 1$, i.e., (1.20) holds with Y_n replaced by Z_n, and convergence in distribution strengthened to equality in distribution.

Remark 1.6. The most cited paper of Einstein [63] concerns the connection between random walks, Brownian motion, and the diffusion equation. Sokolov and Klafter [205] review the history, and the development of fractional diffusion, based on random walks with power law jumps, to address empirical observations of anomalous diffusion.

Details

A Pareto random variable X satisfies $\mathbb{P}[X > x] = Cx^{-\alpha}$ for $x > C^{1/\alpha}$, where $C > 0$ and $1 < \alpha < 2$. It has cdf

$$F(x) = \mathbb{P}[X \leq x] = \begin{cases} 1 - Cx^{-\alpha} & \text{if } x \geq C^{1/\alpha} \\ 0 & \text{if } x < C^{1/\alpha} \end{cases} \tag{1.28}$$

and pdf

$$f(x) = \begin{cases} C\alpha x^{-\alpha-1} & \text{if } x \geq C^{1/\alpha} \\ 0 & \text{if } x < C^{1/\alpha} \end{cases} \tag{1.29}$$

The pth moment

$$\begin{aligned} \mu_p = \mathbb{E}[X^p] &= \int x^p f(x)\, dx \\ &= C\alpha \int_{C^{1/\alpha}}^{\infty} x^{p-\alpha-1}\, dx \\ &= C\alpha \left[\frac{x^{p-\alpha}}{p-\alpha} \right]_{C^{1/\alpha}}^{\infty} = \frac{\alpha}{\alpha - p} C^{p/\alpha} \end{aligned} \tag{1.30}$$

when $0 < p < \alpha$. For $p \geq \alpha$, the pth moment μ_p does not exist, since the integral in (1.30) diverges at infinity. Hence for $1 < \alpha < 2$, the first moment μ_1 exists, but the second moment μ_2 is undefined.

In this book, we define the fractional power z^α of any complex number $z = re^{i\theta}$ as $z^\alpha = r^\alpha e^{i\theta\alpha}$, where $\alpha \geq 0$, $r \geq 0$, $-\pi < \theta \leq \pi$, and $e^{i\theta} = \cos\theta + i \sin\theta$.

Proposition 1.7. *A Pareto random variable X with pdf (1.29) for some $1 < \alpha < 2$ has FT*

$$\mathbb{E}[e^{-ikX}] = 1 - ik\mu_1 + D(ik)^\alpha + O(k^2) \quad \text{as } k \to 0. \tag{1.31}$$

where μ_1 is given by (1.30) with $p = 1$, and $D = C\Gamma(2-\alpha)/(\alpha - 1)$.

Proof. Write

$$
\begin{aligned}
\mathbb{E}[e^{-ikX}] &= \int_{C^{1/\alpha}}^{\infty} e^{-ikx} C\alpha x^{-\alpha-1}\,dx \\
&= \int_{C^{1/\alpha}}^{\infty} [1 - ikx + (e^{-ikx} - 1 + ikx)]C\alpha x^{-\alpha-1}\,dy \\
&= 1 - ik\mu_1 + \int_0^{\infty} (e^{-ikx} - 1 + ikx)C\alpha x^{-\alpha-1}\,dy \\
&\quad - \int_0^{C^{1/\alpha}} (e^{-ikx} - 1 + ikx)C\alpha x^{-\alpha-1}\,dy
\end{aligned}
$$

where $\mu_1 = C^{1/\alpha}\alpha/(\alpha - 1)$ by (1.30) with $p = 1$. It follows by an elementary but lengthy computation (see the proof of Proposition 3.12) that

$$
\int_0^{\infty} (e^{-ikx} - 1 + ikx)C\alpha x^{-\alpha-1}\,dx = C\frac{\Gamma(2-\alpha)}{\alpha - 1}(ik)^{\alpha}
$$

when $1 < \alpha < 2$. For the remaining integral, a Taylor series expansion shows that

$$
|e^{-ikx} - 1 + ikx| \le \frac{(kx)^2}{2!} \quad \text{for all } x \in \mathbb{R} \text{ and } k \in \mathbb{R}.
$$

Then

$$
\begin{aligned}
\left| \int_0^{C^{1/\alpha}} (e^{-ikx} - 1 + ikx)C\alpha x^{-\alpha-1}\,dy \right| &\le \frac{k^2}{2} \int_0^{C^{1/\alpha}} C\alpha x^{1-\alpha}\,dy \\
&= \frac{k^2}{2} C\alpha \left[\frac{x^{2-\alpha}}{2-\alpha} \right]_0^{C^{1/\alpha}} = \frac{k^2}{2}\frac{\alpha}{2-\alpha}C^{2/\alpha}
\end{aligned}
$$

since $1 < \alpha < 2$. Then it follows that X has FT (1.31). □

Setting $C = (\alpha - 1)/\Gamma(2 - \alpha)$ and using the Taylor series for e^z, it follows from (1.31) that $Y = X - \mu_1$ has FT

$$
\begin{aligned}
\mathbb{E}[e^{-ik(X-\mu_1)}] &= \left[1 - ik\mu_1 + (ik)^{\alpha} + O(k^2)\right] \cdot \left[1 + ik\mu_1 + \tfrac{1}{2!}(ik\mu_1)^2 + O(k^3)\right] \\
&= 1 + (ik)^{\alpha} + O(k^2)
\end{aligned}
$$

which justifies the FT expansion in (1.18).

Inverting the FT (1.21) to arrive at (1.22) also requires

$$
\int e^{ikx} \frac{\partial}{\partial t} \hat{p}(k,t)\,dk = \frac{\partial}{\partial t} \int e^{ikx} \hat{p}(k,t)\,dk
$$

for the FT $\hat{p}(k,t) = e^{t(ik)^\alpha}$ of a stable density. Write

$$\frac{\partial}{\partial t}\int e^{ikx}\hat{p}(k,t)\,dk = \lim_{h\to 0}\int e^{ikx}\frac{\hat{p}(k,t+h)-\hat{p}(k,t)}{h}\,dk,$$

where

$$\left|\frac{\hat{p}(k,t+h)-\hat{p}(k,t)}{h}\right| = \left|e^{t(ik)^\alpha}\right|\left|\frac{1-e^{h(ik)^\alpha}}{h(ik)^\alpha}\right|\left|(ik)^\alpha\right|. \tag{1.32}$$

Note that $|e^z| = e^{\mathrm{Re}(z)}$, $i = \cos(\pi/2) + i\sin(\pi/2)$, and $(ik)^\alpha = (i\,\mathrm{sgn}(k)|k|)^\alpha = |k|^\alpha \exp(i\,\mathrm{sgn}(k)\pi\alpha/2) = |k|^\alpha[\cos(\pi\alpha/2) + i\,\mathrm{sgn}(k)\sin(\pi\alpha/2)]$, where $\mathrm{sgn}(k)$ is sign of k. Then the first term in (1.32) reduces to

$$\left|e^{t(ik)^\alpha}\right| = e^{t|k|^\alpha\cos(\pi\alpha/2)}$$

where $\cos(\pi\alpha/2) < 0$, since $1 < \alpha < 2$. Also, the third term is

$$\left|(ik)^\alpha\right| = |k|^\alpha$$

since $|e^{i\theta}| = 1$ for all real θ. To bound the second term, use the Taylor series for e^z to write

$$\left|\frac{1-e^z}{z}\right| \le 1 + \frac{|z|}{2} + \frac{|z|^2}{3!} + \cdots = \frac{e^{|z|}-1}{|z|}.$$

Fix $t > 0$ and consider $z = h(ik)^\alpha$ for $|h| < -(t/2)\cos(\pi\alpha/2)$. Then $|z| = |h||k|^\alpha$, and the mean value theorem implies that

$$e^{|h||k|^\alpha} - 1 \le |h||k|^\alpha e^{-(t/2)\cos(\pi\alpha/2)|k|^\alpha}$$

for all $|h| < -(t/2)\cos(\pi\alpha/2)$. Then the second term in (1.32) is bounded by

$$\left|\frac{1-e^{h(ik)^\alpha}}{h(ik)^\alpha}\right| \le \frac{e^{|h||k|^\alpha}-1}{|h||k|^\alpha} \le e^{-(t/2)\cos(\pi\alpha/2)|k|^\alpha}$$

Putting all three terms together, it follows that

$$\left|\frac{\hat{p}(k,t+h)-\hat{p}(k,t)}{h}\right| \le |k|^\alpha e^{(t/2)|k|^\alpha\cos(\pi\alpha/2)}$$

for all $|h| < -(t/2)\cos(\pi\alpha/2)$. Since the function $|k|^\alpha e^{(t/2)|k|^\alpha\cos(\pi\alpha/2)}$ is integrable with respect to k for any $t > 0$, the dominated convergence theorem implies that

$$\frac{\partial}{\partial t}\int e^{ikx}\hat{p}(k,t)\,dk = \int e^{ikx}\lim_{h\to 0}\frac{\hat{p}(k,t+h)-\hat{p}(k,t)}{h}\,dk = \int e^{ikx}\frac{\partial}{\partial t}\hat{p}(k,t)\,dk.$$

Similar arguments justify (1.23) and (1.26).

A two-sided Pareto random variable X with index $1 < \alpha < 2$ satisfies $\mathbb{P}[X > x] = pCx^{-\alpha}$ and $\mathbb{P}[X < -x] = qCx^{-\alpha}$ for all $x > C^{1/\alpha}$, where $C > 0$, and $0 \le p, q \le 1$ with $p + q = 1$. Then X has pdf

$$f(x) = \begin{cases} pC\alpha x^{-\alpha-1} & \text{if } x > C^{1/\alpha};\\ 0 & \text{if } -C^{1/\alpha} < x < C^{1/\alpha};\\ qC\alpha|x|^{-\alpha-1} & \text{if } x < -C^{1/\alpha}.\end{cases}$$

Noting that $|x| = -x$ for $x < 0$, a substitution $y = -x$ along with (1.30) shows that the nth moment of X is

$$\begin{aligned}\mathbb{E}[X^n] &= \int x^n f(x)\,dx \\ &= pC\alpha \int_{C^{1/\alpha}}^{\infty} x^{n-\alpha-1}\,dx + qC\alpha \int_{-\infty}^{-C^{1/\alpha}} x^n(-x)^{-\alpha-1}\,dx \\ &= pC^{n/\alpha}\frac{\alpha}{\alpha-n} + qC\alpha\int_{C^{1/\alpha}}^{\infty}(-1)^n y^{n-\alpha-1}\,dy \\ &= (p+(-1)^n q)C^{n/\alpha}\frac{\alpha}{\alpha-n}\end{aligned} \tag{1.33}$$

when $0 < n < \alpha$. For $n \ge \alpha$, the nth moment does not exist.

The FT of X follows easily from Proposition 1.7. A change of variables $y = -x$ together with (1.31) leads to

$$\begin{aligned}\mathbb{E}[e^{-ikX}] &= p\int_{C^{1/\alpha}}^{\infty} e^{-ikx}C\alpha x^{-\alpha-1}dx + q\int_{-\infty}^{-C^{1/\alpha}} e^{-ikx}C\alpha(-x)^{-\alpha-1}dx \\ &= p\left[1 - ik\mu + D(ik)^\alpha + O(k^2)\right] + q\int_{C^{1/\alpha}}^{\infty} e^{iky}C\alpha y^{-\alpha-1}dy \\ &= p\left[1 - ik\mu + D(ik)^\alpha + O(k^2)\right] + q\left[1 + ik\mu + D(-ik)^\alpha + O(k^2)\right] \\ &= 1 - (p-q)ik\mu + pD(ik)^\alpha + qD(-ik)^\alpha + O(k^2)\end{aligned} \tag{1.34}$$

where $\mu = C^{1/\alpha}\alpha/(\alpha-1)$ and $D = C\Gamma(2-\alpha)/(\alpha-1)$. Since $\mu_1 = (p-q)\mu = \mathbb{E}[X]$ by (1.33), it follows from (1.34) that $Y = X - \mu_1$ has FT

$$\begin{aligned}\mathbb{E}[e^{-ikY}] &= \left[1 - ik\mu_1 + pD(ik)^\alpha + qD(-ik)^\alpha + O(k^2)\right]\cdot\left[1 + ik\mu_1 + O(k^2)\right] \\ &= 1 + pD(ik)^\alpha + qD(-ik)^\alpha + O(k^2)\end{aligned}$$

which justifies the FT expansion in (1.24).

The Lévy Continuity Theorem 1.3 shows that the limit $e^{(ik)^\alpha}$ in (1.19) is the FT of some probability measure, since it is continuous at $k = 0$. In Section 4.5 we will prove that this probability distribution has a density, using the FT inversion formula (Theorem 1.4).

In Proposition 2.5 we will use the FT inversion formula prove that if f and its derivatives up to some integer order $n > 1 + \alpha$ exist and are absolutely integrable, then the fractional derivative $d^\alpha f(x)/dx^\alpha$ exists, and its FT equals $(ik)^\alpha \hat{f}(k)$.

To show that (1.23) governs the limit of a random walk with drift, take X_j iid with X, where $\mathbb{P}[X > x] = Cx^{-\alpha}$ for some $1 < \alpha < 2$ and some $C > 0$. Proposition 1.7 shows that X_j has FT (1.31), and it follows that the FT of $X_j - \mu_1$ is $1 + D(ik)^\alpha + O(k^2)$. Take

$S_n = X_1 + \cdots + X_n$ and consider the normalized random walk

$$c^{-1/\alpha}(S_{[ct]} - v[ct]) + c^{-1}v[ct] = c^{-1/\alpha}\sum_{j=1}^{[ct]}(X_j - v) + c^{-1}\sum_{j=1}^{[ct]} v$$

where $v = \mu_1 = \mathbb{E}[X]$. Take FT to get

$$\left(1 + D\frac{(ik)^\alpha}{c} + O(c^{-2/\alpha})\right)^{[ct]} \cdot e^{-ikc^{-1}v[ct]} \to \exp(-ikvt + Dt(ik)^\alpha).$$

Remark 1.8. The FT expansion (1.18) can also be proven using Tauberian theorems from Pitman [169]. These Tauberian theorems relate the asymptotic behavior of the probability tail $G(x) = \mathbb{P}[Y > x]$ at infinity to that of the FT at zero. We will write $f(x) \sim g(x)$ to mean that the ratio $f(x)/g(x) \to 1$. Suppose $Y > 0$ with $G(x) \sim Cx^{-\alpha}$ as $x \to \infty$ for some $1 < \alpha < 2$, and let $\hat{f}(k) = \mathbb{E}[e^{-ikY}]$. Then the real and imaginary parts of the FT satisfy

$$\begin{aligned}
\mathrm{Re}\hat{f}(-k) &= 1 - \frac{\pi}{2\Gamma(\alpha)\sin(\pi\alpha/2)}G(1/k) + o(k^\alpha)\\
\mathrm{Im}\hat{f}(-k) &= \frac{\pi}{2\Gamma(\alpha)\cos(\pi\alpha/2)}G(1/k) + o(k^\alpha)
\end{aligned}$$

as $k \to 0$, by [169, Theorem 1] and [169, Theorem 8], respectively. Putting the real and imaginary parts together, and using the formula

$$(-i)^\alpha = (e^{-i\pi/2})^\alpha = e^{-i\pi\alpha/2} = \cos(\pi\alpha/2) - i\sin(\pi\alpha/2)$$

we have

$$\begin{aligned}
\hat{f}(-k) &= 1 - Ck^\alpha\frac{\pi}{2\Gamma(\alpha)}\left[\frac{1}{\sin(\pi\alpha/2)} - \frac{i}{\cos(\pi\alpha/2)}\right] + o(k^\alpha)\\
&= 1 - C\pi k^\alpha\frac{\cos(\pi\alpha/2) - i\sin(\pi\alpha/2)}{2\sin(\pi\alpha/2)\cos(\pi\alpha/2)\Gamma(\alpha)} + o(k^\alpha)\\
&= 1 - \frac{C\pi}{\Gamma(\alpha)\sin(\pi\alpha)}(-ik)^\alpha + o(k^\alpha)
\end{aligned}$$

as $k \to 0$. This shows that

$$\hat{f}(k) = 1 + D(ik)^\alpha + o(k^\alpha)$$

as $k \to 0$, where

$$D = -\frac{C\pi}{\Gamma(\alpha)\sin(\pi\alpha)} > 0$$

since $1 < \alpha < 2$. Using Euler's formula

$$\Gamma(\alpha)\Gamma(1-\alpha) = \frac{\pi}{\sin(\pi\alpha)}$$

we have

$$D = -C\Gamma(1-\alpha) = \frac{C\Gamma(2-\alpha)}{\alpha - 1}$$

which agrees with Proposition 1.7.

2 Fractional Derivatives

Fractional derivatives were invented by Leibnitz soon after their integer-order cousins. In this chapter, we develop the main ideas and mathematical techniques for dealing with fractional derivatives.

2.1 The Grünwald formula

In the first chapter of this book, we defined the fractional derivative $d^\alpha f(x)/dx^\alpha$ as the function with FT $(ik)^\alpha \hat{f}(k)$. Our present goal is to develop a more familiar and intuitive definition in terms of difference quotients. Given a function $f(x)$, we can define the first derivative

$$\frac{df(x)}{dx} = \lim_{h\to 0} \frac{f(x) - f(x-h)}{h}$$

when the limit exists. Higher order derivatives are defined by

$$\frac{d^n f(x)}{dx^n} = \lim_{h\to 0} \frac{\Delta^n f(x)}{h^n}$$

where

$$\begin{aligned}
\Delta f(x) &= f(x) - f(x-h)\\
\Delta^2 f(x) &= \Delta\Delta f(x) = \Delta[f(x) - f(x-h)]\\
&= f(x) - 2f(x-h) + f(x-2h)\\
\Delta^3 f(x) &= f(x) - 3f(x-h) + 3f(x-2h) - f(x-3h)\\
&\vdots\\
\Delta^n f(x) &= \sum_{j=0}^{n} \binom{n}{j} (-1)^j f(x-jh)
\end{aligned}$$

using the binomial formula: Using the backward shift operator $Bf(x) = f(x-h)$ we can write

$$\Delta f(x) = (I - B)f(x)$$

where $If(x) = f(x)$ is the identity operator; then we have

$$\Delta^n f(x) = (I - B)^n f(x) = \sum_{j=0}^{n} \binom{n}{j} (-B)^j I^{n-j} f(x).$$

The fractional difference operator

$$\Delta^\alpha f(x) = (I - B)^\alpha f(x) = \sum_{j=0}^{\infty} \binom{\alpha}{j} (-B)^j f(x) = \sum_{j=0}^{\infty} \binom{\alpha}{j} (-1)^j f(x-jh)$$

is also used in time series analysis to model long range correlation. Here

$$\binom{\alpha}{j} = \frac{\Gamma(\alpha+1)}{j!\Gamma(\alpha-j+1)}$$

extends the usual definition, since $\Gamma(n+1) = n!$ for positive integers n. Now we write the *Grünwald-Letnikov finite difference* form

$$\frac{d^\alpha f(x)}{dx^\alpha} = \lim_{h\to 0} \frac{\Delta^\alpha f(x)}{h^\alpha} \tag{2.1}$$

for the fractional derivative. Our next result shows that this definition agrees with our original definition of the fractional derivative in terms of Fourier transforms.

Proposition 2.1. *For a bounded function f, such that f and its derivatives up to some order $n > 1 + \alpha$ exist and are absolutely integrable, the Grünwald fractional derivative* (2.1) *exists, and its FT is $(ik)^\alpha \hat{f}(k)$.*

Proof. The binomial series

$$(1+z)^\alpha = \sum_{j=0}^{\infty} \binom{\alpha}{j} z^j \tag{2.2}$$

converges for any complex $|z| \le 1$ and any $\alpha > 0$ (e.g., see Hille [89, p. 147]). Equation (2.12) in the details at the end of this section shows that

$$\sum_{j=0}^{\infty} \left| \binom{\alpha}{j} (-1)^j \right| < \infty.$$

Hence, if f is bounded, the series

$$\Delta^\alpha f(x) = \sum_{j=0}^{\infty} \binom{\alpha}{j} (-1)^j f(x - jh)$$

converges, uniformly on $-\infty < x < \infty$.

Proposition 2.5 in the details at the end of this section shows that $d^\alpha f(x)/dx^\alpha$ exists as the function with FT $(ik)^\alpha \hat{f}(k)$. A substitution $y = x - a$ shows that $f(x - a)$ has FT

$$\begin{aligned}\int e^{-ikx} f(x-a)\,dx &= \int e^{-ik(y+a)} f(y)\,dy \\ &= e^{-ika} \int e^{-iky} f(y)\,dy = e^{-ika}\hat{f}(k).\end{aligned}$$

Then $\Delta^\alpha f(x)$ has FT

$$\begin{aligned}\int e^{-ikx} \sum_{j=0}^{\infty} \binom{\alpha}{j} (-1)^j f(x-jh)\,dx &= \sum_{j=0}^{\infty} \binom{\alpha}{j} (-1)^j \int e^{-ikx} f(x-jh)\,dx \\ &= \sum_{j=0}^{\infty} \binom{\alpha}{j} (-1)^j e^{-ikjh} \hat{f}(k) \\ &= (1 - e^{-ikh})^\alpha \hat{f}(k).\end{aligned}$$

The first equality above can be justified using the uniform convergence of the series $\Delta^\alpha f(x)$, and the integrability of each term (e.g., see Rudin [181, Theorem 7.16, p. 151]).

If $k \neq 0$, then the FT of $\Delta^\alpha f(x)/h^\alpha$ is

$$
\begin{aligned}
h^{-\alpha}(ikh)^\alpha \left(\frac{1-e^{-ikh}}{ikh}\right)^\alpha \hat{f}(k) &= (ik)^\alpha \left(\frac{1-[1-ikh+\frac{1}{2!}(-ikh)^2+\cdots]}{ikh}\right)^\alpha \hat{f}(k)\\
&= (ik)^\alpha \left(\frac{ikh-\frac{1}{2!}(-ikh)^2+\cdots}{ikh}\right)^\alpha \hat{f}(k)\\
&= (ik)^\alpha \left(1-\tfrac{1}{2!}(ikh)+\cdots\right)^\alpha \hat{f}(k)\\
&\to (ik)^\alpha \hat{f}(k)
\end{aligned}
$$

as $h \to 0$. If $k = 0$, then obviously $(1-e^{-ikh})^\alpha \hat{f}(k) = (ik)^\alpha \hat{f}(k)$. Hence the FT of $\Delta^\alpha f(x)/h^\alpha$ converges pointwise to that of $d^\alpha f(x)/dx^\alpha$. Then Proposition 2.6 in the details at the end of this section shows that (2.1) holds. □

Remark 2.2. A similar argument shows that for any fixed integer $p > 0$ we have

$$
\frac{d^\alpha f(x)}{dx^\alpha} = \lim_{h\to 0} h^{-\alpha} \sum_{j=0}^{\infty} \binom{\alpha}{j} (-1)^j f(x-(j-p)h)
$$

which is useful in numerical methods, see for example Meerschaert and Tadjeran [155].

The negative fractional derivative can be defined by

$$
\frac{d^\alpha f(x)}{d(-x)^\alpha} = \lim_{h\to 0} h^{-\alpha} \sum_{j=0}^{\infty} \binom{\alpha}{j} (-1)^j f(x+jh). \tag{2.3}
$$

An argument similar to Proposition 2.1 shows that this expression has FT $(-ik)^\alpha \hat{f}(k)$.

The fractional difference is a discrete convolution with the *Grünwald weights*

$$
\begin{aligned}
w_j = (-1)^j \binom{\alpha}{j} &= \frac{(-1)^j \Gamma(\alpha+1)}{\Gamma(j+1)\Gamma(\alpha-j+1)}\\
&= \frac{(-1)^j \alpha(\alpha-1)\cdots(\alpha-j+1)}{\Gamma(j+1)}\\
&= \frac{-\alpha(1-\alpha)\cdots(j-1-\alpha)}{\Gamma(j+1)}\\
&= \frac{-\alpha\Gamma(j-\alpha)}{\Gamma(j+1)\Gamma(1-\alpha)}
\end{aligned} \tag{2.4}
$$

using the property $\Gamma(x+1) = x\Gamma(x)$. Write $f(x) \sim g(x)$ to mean that $f(x)/g(x) \to 1$. Apply Stirling's approximation $\Gamma(x+1) \sim \sqrt{2\pi x}\, x^x e^{-x}$ as $x \to \infty$ to get

$$
\begin{aligned}
w_j &\sim \frac{-\alpha}{\Gamma(1-\alpha)} \frac{\sqrt{2\pi(j-\alpha-1)}\,(j-\alpha-1)^{j-\alpha-1} e^{-(j-\alpha-1)}}{\sqrt{2\pi j}\, j^j e^{-j}}\\
&= \frac{-\alpha}{\Gamma(1-\alpha)} \sqrt{\frac{j-\alpha-1}{j}} \left(\frac{j-\alpha-1}{j}\right)^{j-\alpha-1} j^{-\alpha-1} e^{\alpha+1}
\end{aligned}
$$

and note that

$$\sqrt{\frac{j-\alpha-1}{j}} \to 1$$

and

$$\begin{aligned}\left(\frac{j-\alpha-1}{j}\right)^{j-\alpha-1} &= \left(1-\frac{\alpha+1}{j}\right)^{j}\cdot\left(\frac{j-\alpha-1}{j}\right)^{-\alpha-1}\\ &\to e^{-(\alpha+1)}\cdot 1\end{aligned}$$

as $j \to \infty$. It follows that the Grünwald weights follow a power law asymptotically:

$$w_j \sim \frac{-\alpha}{\Gamma(1-\alpha)} j^{-\alpha-1} \quad \text{as } j \to \infty. \tag{2.5}$$

The Grünwald formula (2.1) gives a concrete interpretation to the fractional derivative. Suppose that $p(x, t)$ represents the relative concentration of particles in the fractional diffusion equation $\partial p/\partial t = \partial^\alpha p/\partial x^\alpha$. Suppose that $1 < \alpha < 2$, so that $w_j > 0$ for all $j \geq 2$. Since

$$\frac{\Delta p(x,t)}{\Delta t} \approx (\Delta x)^{-\alpha} \sum_{j=0}^{\infty} w_j p(x - j\Delta x, t)$$

we see that the change in concentration at location x at time t is increased by an amount $w_j p(x - j\Delta x, t)$ that is transported to location x from location $x - j\Delta x$. Since w_j falls off like a power law $j^{-\alpha-1}$, the fraction of particles at any location that moves j steps to the right follows a power law distribution. This deterministic model is completely consistent with the random power law model of particle jumps assumed in the last chapter, leading to the extended central limit theorem, and a stable density that solves this fractional diffusion equation. This connection between the deterministic (Euler) picture and the random (Lagrange) picture of diffusion is fundamental.

Remark 2.3. Here we explain the Eulerian picture. We give a physical derivation of the deterministic model for diffusion, and show how it extends to the fractional case. Let $p(x, t)$ represent the mass concentration at location x at time t. The conservation of mass law is

$$\frac{\partial p}{\partial t} = -\frac{\partial q}{\partial x} \tag{2.6}$$

where $q(x, t)$ is the flux. Consider a small cube of side Δx in three dimensions, with flow from left to right in the x direction. The flux

$$\text{flux} = \frac{\text{mass}}{\text{area} \cdot \text{time}} \tag{2.7}$$

at location x is the rate at which mass passes through the face of the cube at location x. Since the face of the cube has area $A = (\Delta x)^2$ the change in mass in the cube over time Δt can be approximated by

$$\Delta M = q(x,t)\, A\, \Delta t - q(x + \Delta x, t) A\, \Delta t.$$

Then the approximate change in concentration is

$$\Delta p = \frac{\Delta M}{A\Delta x} = \frac{-[q(x+\Delta x, t) - q(x, t)]A\Delta t}{A\Delta x} = -\frac{-\Delta q(x, t)\Delta t}{\Delta x}$$

and so

$$\frac{\Delta p}{\Delta t} = -\frac{\Delta q}{\Delta x}$$

which leads to (2.6) in the limit as $\Delta x \to 0$. See Figure 2.1 for an illustration.

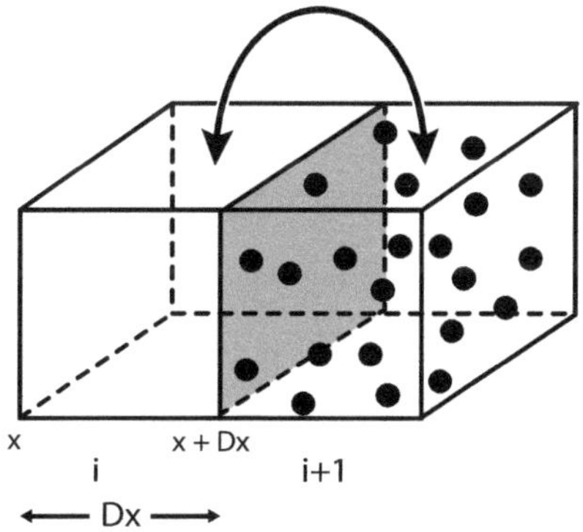

Fig. 2.1: Eulerian picture for diffusion, from Schumer et al. [195].

The diffusion equation comes from combining the conservation of mass equation (2.6) with Fick's Law for the flux

$$q = -D\frac{\partial p}{\partial x} \tag{2.8}$$

which states that particle flux is proportional to the concentration gradient. Fick's law is based on empirical observation. If fluid to the left of the point x contains a higher concentration of dissolved mass than fluid to the right of the point x, then random motion will send more particles to the right than to the left. In this case, we have $\partial p/\partial x < 0$ and $q > 0$, i.e., the sign of the flux is the opposite of the sign of the concentration gradient. Experiments indicate that flux is generally a linear function of the gradient. The dispersivity constant D in (2.8) depends on physical parameters such as temperature (a higher temperature increases D).

The diffusion equation comes from combining (2.6) with (2.8):

$$\frac{\partial p}{\partial t} = -\frac{\partial}{\partial x}\left[-D\frac{\partial p}{\partial x}\right] = D\frac{\partial^2 p}{\partial x^2}$$

assuming D is a constant independent of x. The fractional diffusion equation with $1 < \alpha < 2$ can be derived from a fractional Fick's Law

$$q = -D\frac{\partial^{\alpha-1} p}{\partial x^{\alpha-1}} \tag{2.9}$$

combined with the classical conservation of mass equation

$$\frac{\partial p}{\partial t} = -\frac{\partial}{\partial x}\left[-D\frac{\partial^{\alpha-1} p}{\partial x^{\alpha-1}}\right] = D\frac{\partial^{\alpha} p}{\partial x^{\alpha}}$$

where (2.9) can be understood in terms of the Grünwald formula. In the traditional derivation of Fick's Law, we consider particle movements between adjacent cubes of side Δx, as illustrated in Figure 2.1. The fractional Fick's Law for the flux allows particles to jump into the box at location x from a box at location $x - j\Delta x$. The proportion of particles that make a jump this long drops off as a power of the separation distance. See Schumer et al. [195] for more details. An alternative derivation uses the traditional Fick's Law (2.8) along with a fractional conservation of mass equation

$$\frac{\partial p}{\partial t} = -\frac{\partial^{\alpha-1} q}{\partial x^{\alpha-1}} \tag{2.10}$$

which leads to the same fractional diffusion equation

$$\frac{\partial p}{\partial t} = -\frac{\partial^{\alpha-1}}{\partial x^{\alpha-1}}\left[-D\frac{\partial p}{\partial x}\right] = D\frac{\partial^{\alpha} p}{\partial x^{\alpha}},$$

see Meerschaert, Mortensen and Wheatcraft [139] for additional details. The physical interpretation of (2.10) is similar to (2.9), using the Grünwald interpretation of the fractional derivative. Both lead to the same fractional diffusion equation when the dispersivity D is a constant. For a combination of positive and negative fractional derivatives, particles can also jump into the box at location x from a box at location $x + j\Delta x$. See Figure 2.2 for an illustration. For the case where D varies with x, see for example Zhang, Benson, Meerschaert and LaBolle [224]. A more general model of flux as a convolution was developed by Cushman and Ginn [54]. It was shown in Cushman and Ginn [55] that this more general model reduces to the fractional diffusion equation when the convolution is a power law. Note that the physical derivation also explains why we focus on the case $1 < \alpha \leq 2$.

Details

Here we collect some mathematical details needed to check the arguments in this section. The gamma function is defined for $\alpha > 0$ by

$$\Gamma(\alpha) = \int_0^\infty e^{-x} x^{\alpha-1}\, dx.$$

Note that $e^{-x}x^{\alpha-1} \sim x^{\alpha-1}$ as $x \to 0+$, so that the integral exists. A simple *integration by parts*

$$\int_a^b u(x)v'(x)dx = u(x)v(x)\Big|_a^b - \int_a^b v(x)u'(x)\, dx$$

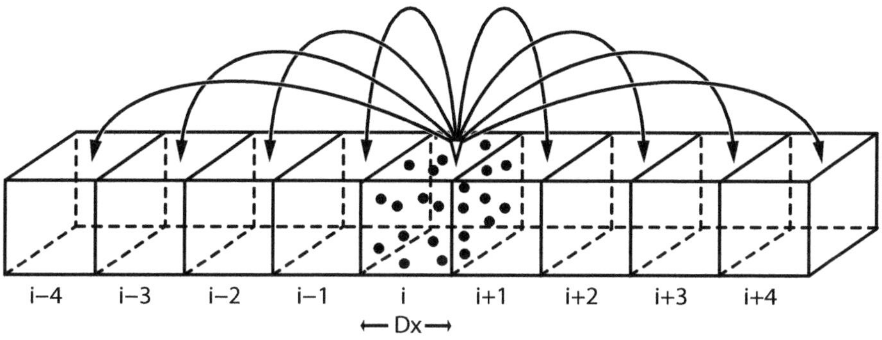

Fig. 2.2: Eulerian picture for fractional diffusion, from Schumer et al. [195].

with $u = x^\alpha$ shows that

$$\Gamma(\alpha+1) = \left[-x^\alpha e^{-x}\right]_0^\infty + \alpha \int_0^\infty e^{-x} x^{\alpha-1}\,dx = \alpha\Gamma(\alpha)$$

for $\alpha > 0$. Now use the formula $\Gamma(\alpha+1) = \alpha\Gamma(\alpha)$ to extend the definition of the gamma function to non-integer values of $\alpha < 0$. For example, $\Gamma(-0.7) = \Gamma(0.3)/(-0.7)$, and $\Gamma(-1.7) = \Gamma(-0.7)/(-1.7)$. Since

$$\Gamma(1) = \int_0^\infty e^{-x}\,dx = 1$$

it follows that $\Gamma(n+1) = n!$ Apply the formula $\Gamma(\alpha+1) = \alpha\Gamma(\alpha)$ j times to see that

$$\binom{\alpha}{j} = \frac{\Gamma(\alpha+1)}{\Gamma(j+1)\Gamma(\alpha-j+1)} = \frac{\alpha\Gamma(\alpha)}{j!\Gamma(\alpha-j+1)} = \cdots = \frac{\alpha(\alpha-1)\cdots(\alpha-j+1)}{\Gamma(j+1)}.$$

Eventually $(j-1-\alpha) > 0$ for all j large, and then

$$w_j = (-1)^j \binom{\alpha}{j} = \frac{-\alpha}{\Gamma(j+1)}(1-\alpha)\cdots(j-1-\alpha)$$

has the same sign for all j large. Since

$$\sum_{j=0}^{\infty} w_j = \sum_{j=0}^{\infty} \binom{\alpha}{j}(-1)^j = (1+(-1))^\alpha = 0 \tag{2.11}$$

by the binomial formula (2.2), it follows that

$$\sum_{j=0}^{\infty} |w_j| < \infty. \tag{2.12}$$

Since $e^{-z} = 1 - z + O(z^2)$, we have for any fixed $k \neq 0$ that

$$\left(\frac{1 - e^{-ikh}}{ikh}\right)^{\alpha} = \left(\frac{1 - [1 - ikh + O(h^2)]}{ikh}\right)^{\alpha} = \left(1 + \frac{1}{ik}O(h)\right)^{\alpha} \to 1$$

as $h \to 0$.

Our next goal is to prove that, under certain technical conditions, the fractional derivative $d^{\alpha}f(x)/dx^{\alpha}$ exists as the function with FT $(ik)^{\alpha}\hat{f}(k)$. This requires the following useful lemma.

Lemma 2.4. *If $f(x)$ and all of its derivatives up to order n exist and are absolutely integrable, then*

$$|\hat{f}(k)| \leq \frac{C}{1 + |k|^n} \tag{2.13}$$

for all $k \in \mathbb{R}$.

Proof. When $k = 0$, $|\hat{f}(0)| \leq \int_{-\infty}^{\infty} |f(x)|dx := C_0$, and similarly for $|k| < 1$ we have

$$|\hat{f}(k)| \leq \frac{2C_0}{1 + |k|^n},$$

since $1 + |k|^n < 2$ in that case. A straightforward extension of the argument for (1.15) shows that, if $f(x)$ and all of its derivatives up to order n exist and are absolutely integrable, then the FT of the nth derivative $f^{(n)}(x)$ equals $(ik)^n\hat{f}(k)$. Then we have

$$\hat{f}(k) = (ik)^{-n} \int_{-\infty}^{\infty} e^{-ikx} f^{(n)}(x)\, dx,$$

and it follows that $|\hat{f}(k)| \leq C_1/|k|^n$ where $C_1 = \int_{-\infty}^{\infty} |f^{(n)}(x)|\, dx$. For $|k| \geq 1$ we have

$$|\hat{f}(k)| \leq \frac{2C_1}{1 + |k|^n},$$

since $2|k|^n \geq 1 + |k|^n$ in that case. Then by choosing C to be the larger of $2C_1$ or $2C_0$, we have that (2.13) holds for all k. □

The classical theory of FT is most clearly stated using the function spaces $L^1(\mathbb{R})$ and $L^2(\mathbb{R})$. We give a brief summary here, see Stein and Weiss [208, Sections 1 and 2] for complete details. A function is in $L^1(\mathbb{R})$ if it is absolutely integrable, meaning that $\int |f(x)|\, dx < \infty$. A function is in $L^2(\mathbb{R})$ if it is square integrable, meaning that $\int |f(x)|^2 dx < \infty$. Theorem 1.4 shows that every $f \in L^1(\mathbb{R})$ has a FT. Then the FT can be extended to $L^2(\mathbb{R})$ by taking limits. The function space $L^2(\mathbb{R})$ is a *Hilbert space* with inner product $\langle f, g\rangle_2 = \int f(x)\overline{g(x)}\, dx$, a special kind of Banach space with the norm $\|f\|_2 = \sqrt{\langle f, f\rangle_2}$. Here $\overline{a + ib} = a - ib$ denotes the complex conjugate. We will give more details about Banach spaces in Section 3.3. If $f \in L^2(\mathbb{R})$ then the function $f_n(x) = f(x)I(|x| \leq n)$ is in $L^1(\mathbb{R}) \cap L^2(\mathbb{R})$ for any positive integer n. Then the FT $\hat{f}(k)$

is defined as the limit of $\hat{f}_n(k)$ in the L^2 sense, so that $\|f - f_n\|_2 \to 0$ as $n \to \infty$. The FT maps $L^2(\mathbb{R})$ onto itself, so that every element of $L^2(\mathbb{R})$ is also the FT of another $L^2(\mathbb{R})$ function. Furthermore, the Plancherel Theorem [208, Theorem 2.3] states that $\langle \hat{f}, \hat{g} \rangle_2 = 2\pi \langle f, g \rangle_2$, and hence $\|\hat{f}\|_2 = \sqrt{2\pi}\|f\|_2$. Then it follows that $\|f_n - f\|_2 \to 0$ if and only if $\|\hat{f}_n - \hat{f}\|_2 \to 0$ as $n \to \infty$.

Proposition 2.5. *If f and its derivatives up to some integer order $n > 1 + \alpha$ exist and are absolutely integrable, then the fractional derivative*

$$\frac{d^\alpha f(x)}{dx^\alpha} = \frac{1}{2\pi} \int_{-\infty}^{\infty} e^{ikx} (ik)^\alpha \hat{f}(k)\, dk$$

exists, and its FT equals $(ik)^\alpha \hat{f}(k)$ in $L^2(\mathbb{R})$.

Proof. Lemma 2.4 implies that (2.13) holds, and then

$$|(ik)^\alpha \hat{f}(k)| \le \frac{C|k|^\alpha}{1 + |k|^n}$$

for all k. Since $n > 1 + \alpha$, the function $(ik)^\alpha \hat{f}(k) \in L^1(\mathbb{R}) \cap L^2(\mathbb{R})$. Define

$$g(x) = \frac{1}{2\pi} \int_{-\infty}^{\infty} e^{ikx} (ik)^\alpha \hat{f}(k)\, dk, \qquad x \in \mathbb{R}.$$

By the Plancherel theorem, we deduce that $g \in L^2(\mathbb{R})$ and its FT (in $L^2(\mathbb{R})$) equals $(ik)^\alpha \hat{f}(k)$. Hence $g(x) = d^\alpha f(x)/dx^\alpha$ in $L^2(\mathbb{R})$. □

Proposition 2.6. *For a bounded function f, such that f and its derivatives up to some order $n > 1 + \alpha$ exist and are absolutely integrable, the Grünwald fractional difference quotient $g_h(x) := \Delta^\alpha f(x)/h^\alpha$ converges to the fractional derivative $d^\alpha f(x)/dx^\alpha$, defined as the function with FT $(ik)^\alpha \hat{f}(k)$, for all $x \in \mathbb{R}$.*

Proof. It was shown in the proof of Proposition 2.1 that the function $g_h(x)$ has FT

$$\widehat{g_h}(k) = \left(\frac{1 - e^{-ikh}}{h} \right)^\alpha \hat{f}(k) \to (ik)^\alpha \hat{f}(k) \quad \text{as } h \to 0 \tag{2.14}$$

where the convergence is pointwise. Moreover, by the mean value theorem and the dominated convergence theorem, (2.14) also holds in $L^1(\mathbb{R})$ and $L^2(\mathbb{R})$. Indeed,

$$\begin{aligned} \left| \left[\left(\frac{1 - e^{-ikh}}{h} \right)^\alpha - (ik)^\alpha \right] \hat{f}(k) \right| &= \left| (e^{-ikc\alpha} - 1)(ik)^\alpha \hat{f}(k) \right| \\ &\le 2 \left| (ik)^\alpha \hat{f}(k) \right| \\ &\in L^1 \cap L^2. \end{aligned}$$

In the first equality, c is a number between 0 and h.

By the L^2-convergence in (2.14), Parseval's identity implies that $g_h \to g$ in L^2 for some function $g \in L^2$. By the inversion formula, we may take g as

$$g(x) := \frac{1}{2\pi} \int_{-\infty}^{\infty} e^{ikx} (ik)^\alpha \hat{f}(k)\, dk$$

since $(ik)^\alpha \hat{f}(k) \in L^1$. By Proposition 2.5, this g is equal to the fractional derivative $d^\alpha f(x)/dx^\alpha$. Now by the pointwise and L^1-convergence,

$$\begin{aligned} |g_h(x) - g(x)| &= \frac{1}{2\pi} \left| \int_{-\infty}^{\infty} e^{ikx} (\widehat{g_h}(k) - \hat{g}(k))\, dk \right| \\ &\le \frac{1}{2\pi} \| \widehat{g_h} - \hat{g} \|_{L^1} \\ &\to 0 \end{aligned} \tag{2.15}$$

for any $x \in \mathbb{R}$ as $h \to 0$. Note that since g_h is continuous by the uniform convergence of the series $\Delta^\alpha f(x)$ and g is continuous by definition, the equality in (2.15) holds everywhere. This proves the pointwise convergence. In fact, the convergence is uniform in x. □

Since we define $\partial^\alpha p/\partial x^\alpha$ as the function with FT $(ik)^\alpha \hat{p}$, it is clear that

$$\frac{\partial^{\alpha-1}}{\partial x^{\alpha-1}} \frac{\partial p}{\partial x} = \frac{\partial}{\partial x} \frac{\partial^{\alpha-1} p}{\partial x^{\alpha-1}} = \frac{\partial^\alpha p}{\partial x^\alpha}$$

for any $\alpha > 0$. Since $\Delta^\alpha f(x)$ has FT $(1 - e^{-ikh})^\alpha \hat{f}(k)$, it is also true that

$$\Delta \Delta^{\alpha-1} f(x) = \Delta^{\alpha-1} \Delta f(x) = \Delta^\alpha f(x).$$

2.2 More fractional derivatives

In this section, we develop some alternative integral forms for the fractional derivative. From equation (2.1) we have

$$\frac{\partial^\alpha f(x)}{\partial x^\alpha} = \lim_{\Delta x \to 0} \frac{\Delta^\alpha f(x)}{\Delta x^\alpha} \tag{2.16}$$

where

$$\Delta^\alpha f(x) = \sum_{j=0}^{\infty} w_j f(x - j\Delta x)$$

is a discrete convolution with the Grünwald weights w_j. Recall from (2.5) that

$$w_j \sim \frac{-\alpha}{\Gamma(1-\alpha)} j^{-\alpha-1} \quad \text{as } j \to \infty.$$

Since $w_0 = 1$ we can write

$$\frac{\Delta^\alpha f(x)}{\Delta x^\alpha} = (\Delta x)^{-\alpha}\left[f(x) + \sum_{j=1}^{\infty} w_j f(x - j\Delta x)\right].$$

From the binomial formula (2.2) it follows that $\sum_{j=0}^{\infty} w_j = 0$, see (2.11). Consider the simplest case $0 < \alpha < 1$. Then it follows from (2.4) that $w_j < 0$ for all $j \geq 1$, and so $\sum_{j=1}^{\infty} w_j = -1$. Define $b_j = -w_j$ for $j \geq 1$, so that

$$b_j \sim \frac{\alpha}{\Gamma(1-\alpha)} j^{-\alpha-1} \quad \text{as } j \to \infty, \quad \text{and} \quad \sum_{j=1}^{\infty} b_j = 1.$$

Then

$$\begin{aligned}\frac{\Delta^\alpha f(x)}{\Delta x^\alpha} &= (\Delta x)^{-\alpha} \sum_{j=1}^{\infty} [f(x) - f(x - j\Delta x)]\, b_j \\ &\approx \sum_{j=1}^{\infty} [f(x) - f(x - j\Delta x)] \frac{\alpha}{\Gamma(1-\alpha)} (j\Delta x)^{-\alpha-1} \Delta x \\ &\approx \int_0^{\infty} [f(x) - f(x-y)] \frac{\alpha}{\Gamma(1-\alpha)} y^{-\alpha-1} dy\end{aligned}$$

which motivates the *generator form* of the fractional derivative:

$$\frac{d^\alpha f(x)}{dx^\alpha} = \int_0^{\infty} [f(x) - f(x-y)] \frac{\alpha}{\Gamma(1-\alpha)} y^{-\alpha-1} dy. \tag{2.17}$$

Integrate by parts with $u = f(x) - f(x-y)$ to get the *Caputo form*

$$\frac{d^\alpha f(x)}{dx^\alpha} = \frac{1}{\Gamma(1-\alpha)} \int_0^{\infty} f'(x-y) y^{-\alpha} dy = \frac{1}{\Gamma(1-\alpha)} \int_0^{\infty} \frac{d}{dx} f(x-y)\, y^{-\alpha} dy. \tag{2.18}$$

Take the derivative outside the integral to get the *Riemann-Liouville form*

$$\frac{d^\alpha f(x)}{dx^\alpha} = \frac{1}{\Gamma(1-\alpha)} \frac{d}{dx} \int_0^{\infty} f(x-y) y^{-\alpha} dy. \tag{2.19}$$

[Are these forms equivalent?] These forms are valid for $0 < \alpha < 1$. For $1 < \alpha < 2$ we can write the generator form

$$\frac{d^\alpha f(x)}{dx^\alpha} = \frac{\alpha(\alpha-1)}{\Gamma(2-\alpha)} \int_0^{\infty} \left[f(x-y) - f(x) + y f'(x)\right] y^{-1-\alpha} dy. \tag{2.20}$$

Integrate by parts twice to get the Caputo form for $1 < \alpha < 2$:

$$\frac{d^\alpha f(x)}{dx^\alpha} = \frac{1}{\Gamma(2-\alpha)} \int_0^{\infty} \frac{d^2}{dx^2} f(x-y) y^{1-\alpha} dy. \tag{2.21}$$

Move the derivative outside to get the Riemann-Liouville form for $1 < \alpha < 2$:

$$\frac{d^\alpha f(x)}{dx^\alpha} = \frac{1}{\Gamma(2-\alpha)} \frac{d^2}{dx^2} \int_0^\infty f(x-y) y^{1-\alpha}\, dy. \tag{2.22}$$

In Chapter 3 we will provide a rigorous proof that the generator form satisfies the FT definition of the fractional derivative. The equivalence of the generator form and the Caputo form will be discussed in the details at the end of this section. The general relation between the Caputo and Riemann-Liouville forms will be discussed further in Section 2.3.

Example 2.7. Let $f(x) = e^{\lambda x}$ for some $\lambda > 0$, so that $f'(x) = \lambda e^{\lambda x}$. Using the Caputo form for $0 < \alpha < 1$, a substitution $u = \lambda y$, and the definition of the gamma function, we get

$$\begin{aligned}
\frac{d^\alpha}{dx^\alpha}\left[e^{\lambda x}\right] &= \frac{1}{\Gamma(1-\alpha)} \int_0^\infty \lambda e^{\lambda(x-y)} y^{-\alpha}\, dy \\
&= \frac{\lambda e^{\lambda x}}{\Gamma(1-\alpha)} \int_0^\infty e^{-\lambda y} y^{-\alpha}\, dy \\
&= \frac{\lambda e^{\lambda x}}{\Gamma(1-\alpha)} \int_0^\infty e^{-u} \left(\frac{u}{\lambda}\right)^{-\alpha} \frac{du}{\lambda} \\
&= \frac{\lambda e^{\lambda x}}{\Gamma(1-\alpha)} \lambda^{\alpha-1} \int_0^\infty e^{-u} u^{(1-\alpha)-1}\, du \\
&= \frac{\lambda e^{\lambda x}}{\Gamma(1-\alpha)} \lambda^{\alpha-1} \Gamma(1-\alpha) = \lambda^\alpha e^{\lambda x}
\end{aligned}$$

which agrees with the integer order case. For example, we have

$$\frac{d^2}{dx^2}\left[e^{\lambda x}\right] = \lambda^2 e^{\lambda x}$$

and so forth. Using the Riemann-Liouville form we get

$$\begin{aligned}
\frac{d^\alpha}{dx^\alpha}\left[e^{\lambda x}\right] &= \frac{d}{dx}\left[\frac{1}{\Gamma(1-\alpha)} \int_0^\infty e^{\lambda(x-y)} y^{-\alpha}\, dy\right] \\
&= \frac{d}{dx}\left[\frac{e^{\lambda x}}{\Gamma(1-\alpha)} \int_0^\infty e^{-\lambda y} y^{-\alpha}\, dy\right] \\
&= \frac{d}{dx}\left[\frac{e^{\lambda x}}{\Gamma(1-\alpha)} \lambda^{\alpha-1} \Gamma(1-\alpha)\right] \\
&= \frac{d}{dx}\left[\lambda^{\alpha-1} e^{\lambda x}\right] = \lambda^\alpha e^{\lambda x}
\end{aligned}$$

which agrees with the Caputo. In this case, both forms lead to the same result.

Before we consider our next example, we first develop some equivalent forms. Make a substitution $u = x - y$ in the Caputo form to get

$$\frac{d^\alpha f(x)}{dx^\alpha} = \frac{1}{\Gamma(1-\alpha)} \int_{-\infty}^{x} f'(u)(x-u)^{-\alpha} du. \tag{2.23}$$

The same substitution gives an alternative Riemann-Liouville derivative for $0 < \alpha < 1$:

$$\frac{d^\alpha f(x)}{dx^\alpha} = \frac{1}{\Gamma(1-\alpha)} \frac{d}{dx} \int_{-\infty}^{x} f(u)(x-u)^{-\alpha} du. \tag{2.24}$$

Example 2.8. For $p > 0$, define $f(x) = x^p$ for $x \geq 0$, and $f(x) = 0$ for $x < 0$. Then $f'(x) = px^{p-1}$ for $x > 0$ and $f'(x) = 0$ for $x < 0$. Recall the formula for the beta density

$$\int_0^x y^{a-1}(x-y)^{b-1}\, dy = \frac{\Gamma(a)\Gamma(b)}{\Gamma(a+b)} x^{a+b-1}$$

for $a > 0$ and $b > 0$. Then the Caputo form is

$$\begin{aligned}
\frac{d^\alpha}{dx^\alpha}[x^p] &= \frac{1}{\Gamma(1-\alpha)} \int_0^x p y^{p-1}(x-y)^{-\alpha} dy \\
&= \frac{p}{\Gamma(1-\alpha)} \int_0^x y^{p-1}(x-y)^{(1-\alpha)-1} dy \\
&= \frac{p}{\Gamma(1-\alpha)} \frac{\Gamma(p)\Gamma(1-\alpha)}{\Gamma(p+1-\alpha)} x^{p+(1-\alpha)-1} \\
&= \frac{p\Gamma(p)}{\Gamma(p+1-\alpha)} x^{p-\alpha} = \frac{\Gamma(p+1)}{\Gamma(p-\alpha+1)} x^{p-\alpha}
\end{aligned}$$

which agrees with the integer order case. For example, we have

$$\frac{d^2\,[x^p]}{dx^2} = p(p-1)x^{p-2} = \frac{\Gamma(p+1)}{\Gamma(p-1)} x^{p-2}$$

since $\Gamma(p+1) = p(p-1)\Gamma(p-1)$. Using the Riemann-Liouville form we get

$$\begin{aligned}
\frac{d^\alpha}{dx^\alpha}[x^p] &= \frac{1}{\Gamma(1-\alpha)} \frac{d}{dx} \left[\int_0^x y^p (x-y)^{-\alpha} dy \right] \\
&= \frac{1}{\Gamma(1-\alpha)} \frac{d}{dx} \left[\int_0^x y^{(p+1)-1} (x-y)^{(1-\alpha)-1} dy \right] \\
&= \frac{1}{\Gamma(1-\alpha)} \frac{d}{dx} \left[\frac{\Gamma(p+1)\Gamma(1-\alpha)}{\Gamma(p+2-\alpha)} x^{p+1-\alpha} \right] \\
&= \frac{\Gamma(p+1)}{\Gamma(p+2-\alpha)} (p+1-\alpha) x^{p-\alpha} = \frac{\Gamma(p+1)}{\Gamma(p-\alpha+1)} x^{p-\alpha}
\end{aligned}$$

which again agrees with the Caputo form.

Our next example shows the Caputo and Riemann-Liouville forms need not agree.

Example 2.9. Let $f(x) = 1$ for $x \geq 0$ and $f(x) = 0$ for $x < 0$. Then $f'(x) = 0$ for $x \neq 0$, so the Caputo fractional derivative is zero. In fact, the Caputo fractional derivative of a constant function is always zero, just like the integer order derivative. But the Riemann-Liouville derivative is not. For $x > 0$ and $0 < \alpha < 1$, use (2.24) to get

$$\begin{aligned}\frac{d^\alpha}{dx^\alpha} f(x) &= \frac{1}{\Gamma(1-\alpha)} \frac{d}{dx}\left[\int_0^x 1\,(x-y)^{-\alpha}\,dy\right]\\ &= \frac{1}{\Gamma(1-\alpha)} \frac{d}{dx}\left[\int_0^x u^{-\alpha}\,du\right]\\ &= \frac{1}{\Gamma(1-\alpha)} \frac{d}{dx}\left[\frac{x^{1-\alpha}}{1-\alpha}\right]\\ &= \frac{x^{-\alpha}}{\Gamma(1-\alpha)} \neq 0.\end{aligned}$$

Since $f(x) = f(x-y)$ unless $y > x > 0$, the generator form is

$$\begin{aligned}\frac{d^\alpha}{dx^\alpha} f(x) &= \int_0^\infty [f(x) - f(x-y)]\,\frac{\alpha}{\Gamma(1-\alpha)} y^{-\alpha-1}\,dy\\ &= \int_x^\infty [1-0]\,\frac{\alpha}{\Gamma(1-\alpha)} y^{-\alpha-1}\,dy\\ &= \frac{\alpha}{\Gamma(1-\alpha)}\left[\frac{x^{-\alpha}}{\alpha}\right] = \frac{x^{-\alpha}}{\Gamma(1-\alpha)}\end{aligned}$$

the same as the Riemann-Liouville form. Recall that we obtained the Caputo form from the generator form via integration by parts. In this case, integration by parts with $u = f(x) - f(x-y)$ in the generator form (2.17) gives

$$\begin{aligned}&\int_0^\infty [f(x) - f(x-y)]\,\frac{\alpha}{\Gamma(1-\alpha)} y^{-\alpha-1}\,dy\\ &= \left[\frac{-y^{-\alpha}}{\Gamma(1-\alpha)}(f(x) - f(x-y))\right]_x^\infty + \int_x^\infty f'(x-y)\frac{1}{\Gamma(1-\alpha)} y^{-\alpha}\,dy\\ &= \frac{x^{-\alpha}}{\Gamma(1-\alpha)} + \int_0^\infty f'(x-y)\frac{1}{\Gamma(1-\alpha)} y^{-\alpha}\,dy\end{aligned}$$

so the difference between these forms comes from the boundary terms.

Details

The generator form (2.17) of the fractional derivative of order $0 < \alpha < 1$ is an improper integral. If $f(x)$ is twice continuously differentiable, then $f(x-y) = f(x) - yf'(x) + O(y^2)$ as $y \to 0$, and hence $[f(x) - f(x-y)]\, y^{-\alpha-1} = O(y^{-\alpha})$ is integrable at $y = 0$. If f is bounded, then $[f(x) - f(x-y)]\, y^{-\alpha-1} = O(y^{-1-\alpha})$ as $y \to \infty$ is also integrable at infinity, so that the generator form of the fractional derivative exists. A similar argument pertains to the generator form (2.20) when $1 < \alpha < 2$.

To derive the Caputo form (2.18) from the generator form (2.17), integrate by parts in (2.17) with

$$u = f(x) - f(x-y) \quad \text{and} \quad dv = \frac{\alpha}{\Gamma(1-\alpha)} y^{-\alpha-1} dy$$

which leads to

$$\left[[f(x) - f(x-y)] \frac{-y^{-\alpha}}{\Gamma(1-\alpha)} \right]_{y=0}^{\infty} + \int_0^{\infty} f'(x-y) \frac{1}{\Gamma(1-\alpha)} y^{-\alpha} dy.$$

If $f(x)$ is continuously differentiable and bounded, then $[f(x) - f(x-y)]y^{-\alpha} = O(y^{1-\alpha})$ as $y \to 0$ and $[f(x) - f(x-y)]y^{-\alpha} = O(y^{-\alpha})$ as $y \to \infty$, so that the Caputo and generator forms are equivalent. Many probability density functions $f(x)$ satisfy these conditions.

2.3 The Caputo derivative

The transform method for solving partial differential equations uses the FT for the space variable x along with the formula

$$\int_{-\infty}^{\infty} e^{-ikx} f'(x)\, dx = (ik)\hat{f}(k).$$

The Laplace transform (LT)

$$\tilde{f}(s) = \int_0^{\infty} e^{-st} f(t)\, dt \tag{2.25}$$

is usually used for the time variable t, along with the formula

$$\int_0^{\infty} e^{-st} f'(t)\, dt = s\tilde{f}(s) - f(0). \tag{2.26}$$

The Laplace transform (2.25) may also be considered as the integral over the entire real line, where the function $f(t) = 0$ for $t < 0$, and then we replace $f(0)$ by $f(0+)$ in (2.26). See Remark 2.13 for more details. The formula (2.26) differs from the FT analogue because of the boundary term from integration by parts: Check (2.26) using $u = e^{-st}$ and

$dv = f'(t)dt$, which leads to

$$\int_0^\infty e^{-st} f'(t)\,dt = e^{-st} f(t)\big|_{t=0}^{\infty} + \int_0^\infty s e^{-st} f(t)\,dt$$
$$= -e^{-0} f(0) + s\tilde{f}(s),$$

assuming $e^{-st}f(t) \to 0$ as $t \to \infty$. Since the FT integrates over the entire real line $-\infty < x < \infty$, the boundary term in that integration by parts vanishes, assuming that $f(x) \to 0$ as $|x| \to \infty$. See Remark 2.13 for additional discussion.

For $0 < \alpha < 1$, the Riemann-Liouville fractional derivative $\mathbb{D}_t^\alpha f(t)$ has LT $s^\alpha \tilde{f}(s)$, while the Caputo fractional derivative $\partial_t^\alpha f(t)$ has LT $s^\alpha \tilde{f}(s) - s^{\alpha-1} f(0)$ (see details at the end of this section). Check using integration by parts that the LT of $f''(t)$ is $s^2\tilde{f}(s) - sf(0) - f'(0)$. For $1 < \alpha < 2$, $\mathbb{D}_t^\alpha f(t)$ has LT $s^\alpha \tilde{f}(s)$, while $\partial_t^\alpha f(t)$ has LT $s^\alpha \tilde{f}(s) - s^{\alpha-1} f(0) - s^{\alpha-2} f'(0)$, and so forth (see details). Since the Caputo derivative incorporates the initial condition in the usual way, it is the preferred form of the fractional time derivative in practical applications.

Example 2.10. Let $p > -1$ and define $f(t) = t^p$ for $t \geq 0$. Substitute $u = st$ and use the definition of the gamma function to see that

$$\tilde{f}(s) = \int_0^\infty e^{-st} t^p \,dt$$
$$= \int_0^\infty e^{-u} \left(\frac{u}{s}\right)^p \frac{du}{s}$$
$$= s^{-p-1} \int_0^\infty e^{-u} u^{(p+1)-1}\,du = s^{-p-1}\Gamma(p+1). \tag{2.27}$$

Then the Riemann-Liouville fractional derivative $\mathbb{D}_t^\alpha f(t)$ has LT

$$\int_0^\infty e^{-st}\left[\frac{d^\alpha}{dt^\alpha} t^p\right] dt = s^{\alpha-p-1}\Gamma(p+1) = \left[s^{-(p-\alpha)-1}\Gamma(p-\alpha+1)\right]\frac{\Gamma(p+1)}{\Gamma(p-\alpha+1)}$$

and inverting the LT shows that

$$\frac{d^\alpha}{dt^\alpha}[t^p] = t^{p-\alpha}\frac{\Gamma(p+1)}{\Gamma(p-\alpha+1)} \tag{2.28}$$

for $p - \alpha > -1$, which agrees with Example 2.8. Since $f(0) = 0$ for $p > 0$, the Caputo and Riemann-Liouville derivatives are equal in this case.

Example 2.11. Suppose $f(t) = 1$ for all $t \geq 0$. It is easy to check that $\tilde{f}(s) = 1/s$. Then the Caputo fractional derivative of order $0 < \alpha < 1$ has LT $s^\alpha(1/s) - s^{\alpha-1}1 = 0$ so that $\partial_t^\alpha f(t) = 0$. The Riemann-Liouville fractional derivative has LT $s^\alpha(1/s) = s^{\alpha-1}$ so that $\mathbb{D}_t^\alpha f(t) = t^{-\alpha}/\Gamma(1-\alpha)$ using (2.27), which agrees with Example 2.9.

Derivatives are linear operators on some space of functions. We say that $f \neq 0$ is an *eigenfunction* of the linear operator $\frac{d}{dt}$ provided that $\frac{d}{dt}f(t) = \lambda f(t)$ for some real (or complex) number λ, called the eigenvalue. The function $f(t) = e^{\lambda t}$ is an eigenfunction since $\frac{d}{dt}[e^{\lambda t}] = \lambda e^{\lambda t}$, which is also reflected in the LT:

$$\tilde{f}(s) = \int_0^\infty e^{-st}e^{\lambda t}\,dt = \int_0^\infty e^{(\lambda-s)t}\,dt = \frac{1}{s-\lambda}$$

for $s > \lambda$. Then $f'(t)$ has LT

$$s\tilde{f}(s) - f(0) = s\left(\frac{1}{s-\lambda}\right) - 1 = \frac{\lambda}{s-\lambda} = \lambda\tilde{f}(s).$$

We have used this fact to solve the diffusion equation: From $\partial p/\partial t = \partial^2 p/\partial x^2$ the FT yields $\frac{d}{dt}\hat{p} = -k^2\hat{p}$, so that the FT solution is an eigenfunction of $\frac{d}{dt}$ with eigenvalue $-k^2$, and hence we can take $\hat{p} = e^{-k^2 t}$, which inverts to a normal density.

The Mittag-Leffler function is defined by a power series

$$E_\beta(z) = \sum_{j=0}^\infty \frac{z^j}{\Gamma(1+\beta j)} \tag{2.29}$$

that converges absolutely for every complex z. Note that $E_\beta(0) = 1$. The Mittag-Leffler function reduces to the exponential function when $\beta = 1$. The eigenfunctions of the Caputo fractional derivative are $f(t) = E_\beta(\lambda t^\beta)$: Differentiate term-by-term using (2.28) to see that

$$\begin{aligned} \partial_t^\beta f(t) &= \partial_t^\beta\left[\sum_{j=0}^\infty \frac{\lambda^j t^{\beta j}}{\Gamma(1+\beta j)}\right] \\ &= \sum_{j=1}^\infty \frac{\lambda^j}{\Gamma(1+\beta j)}\frac{\Gamma(\beta j+1)}{\Gamma(\beta j+1-\beta)}t^{\beta j-\beta} \\ &= \lambda\sum_{j=1}^\infty \frac{\lambda^{j-1}}{\Gamma(1+\beta(j-1))}t^{\beta(j-1)} = \lambda f(t). \end{aligned} \tag{2.30}$$

For a complete and detailed proof, see Mainardi and Gorenflo [129].

Remark 2.12. Another proof uses LT: Use (2.27) to see that $f(t) = E_\beta(\lambda t^\beta)$ has LT

$$\begin{aligned} \sum_{j=0}^\infty \frac{\lambda^j}{\Gamma(1+\beta j)}s^{-\beta j-1}\Gamma(\beta j+1) &= s^{-1}\sum_{j=0}^\infty (\lambda s^{-\beta})^j \\ &= s^{-1}\frac{1}{1-\lambda s^{-\beta}} = \frac{s^{\beta-1}}{s^\beta-\lambda} \end{aligned} \tag{2.31}$$

when $s^\beta > |\lambda|$. Then $\partial_t^\beta f(t)$ has LT

$$s^\beta\left(\frac{s^{\beta-1}}{s^\beta-\lambda}\right) - s^{\beta-1}1 = \frac{s^{2\beta-1}}{s^\beta-\lambda} - \frac{s^{\beta-1}(s^\beta-\lambda)}{s^\beta-\lambda} = \lambda\left(\frac{s^{\beta-1}}{s^\beta-\lambda}\right).$$

Invert the LT to see that $\partial_t^\beta f(t) = \lambda f(t)$.

Eigenfunctions of Caputo fractional derivatives are useful for solving time-fractional diffusion equations. Starting from

$$\partial_t^\beta p(x,t) = D\frac{\partial^2}{\partial x^2}p(x,t)$$

take FT to get

$$\partial_t^\beta \hat{p}(k,t) = -Dk^2\hat{p}(k,t)$$

which shows that $\hat{p}(k,t)$ is an eigenfunction of ∂_t^β with eigenvalue $-Dk^2$. Then

$$\hat{p}(k,t) = E_\beta([-Dk^2]t^\beta)$$

and in order to solve this time-fractional diffusion equation, we need to invert this FT. In the next section, we will solve this problem, and we will also develop a stochastic interpretation for time-fractional diffusion. [Recall that space-fractional diffusion reflects power law jumps in space. What random process do you think is reflected in a time-fractional diffusion?]

Details

In this section, we have used the uniqueness theorem for LT: If $f(t)$ and $g(t)$ are continuous, and if $\tilde{f}(s) = \tilde{g}(s)$ for all $s > s_0$, then $f(t) = g(t)$ for all $t > 0$, see for example Feller [68, p. 433].

In Remark 2.12 we took the LT of the infinite series (2.31) term-by term. This can be justified as follows. Theorem 8.1 in Rudin [181] states that if the power series $\sum_{j=0}^\infty c_j u^j$ converges for $|u| < R$, then $\sum_{j=0}^\infty c_j u^j$ converges uniformly on $|u| < R - \varepsilon$ for any $0 < \epsilon < R$. Then, since the power series (2.29) converges for all z, for any fixed $s > 0$, $\lambda \in \mathbb{R}$, and $\beta > 0$, a substitution $u = t^\beta$ shows that the series

$$\sum_{j=0}^\infty \frac{(\lambda t^\beta)^j}{\Gamma(1+\beta j)}e^{-st} = e^{-st}\sum_{j=0}^\infty \frac{(\lambda t^\beta)^j}{\Gamma(1+\beta j)}$$

converges uniformly in $t \in [0, x]$ for any real number $x > 0$. Next we apply [181, Theorem 7.16]: If a sequence of functions $f_n(t)$ is integrable on $[a, b]$ and converges to $f(t)$ uniformly on $t \in [a, b]$, then $f(t)$ is integrable and

$$\int_a^b f(t)\,dt = \lim_{n\to\infty}\int_a^b f_n(t)\,dt.$$

Define

$$f_n(t) = \sum_{j=0}^n \frac{(\lambda t^\beta)^j}{\Gamma(1+\beta j)}e^{-st}$$

and

$$f(t)=\sum_{j=0}^{\infty}\frac{(\lambda t^{\beta})^j}{\Gamma(1+\beta j)}e^{-st}$$

and apply these two theorems to get

$$\begin{aligned}\int_0^x\sum_{j=0}^{\infty}\frac{(\lambda t^{\beta})^j}{\Gamma(1+\beta j)}e^{-st}dt&=\int_0^x f(t)\,dt\\&=\lim_{n\to\infty}\int_0^x f_n(t)\,dt\\&=\lim_{n\to\infty}\sum_{j=0}^{n}\int_0^x\frac{(\lambda t^{\beta})^j}{\Gamma(1+\beta j)}e^{-st}dt\\&=\sum_{j=0}^{\infty}\int_0^x\frac{(\lambda t^{\beta})^j}{\Gamma(1+\beta j)}e^{-st}dt.\end{aligned}$$

Now let $x\to\infty$ to get

$$\int_0^{\infty}\sum_{j=0}^{\infty}\frac{(\lambda t^{\beta})^j}{\Gamma(1+\beta j)}e^{-st}dt=\lim_{x\to\infty}\sum_{j=0}^{\infty}\int_0^x\frac{(\lambda t^{\beta})^j}{\Gamma(1+\beta j)}e^{-st}dt.\tag{2.32}$$

It remains to show that the limit on the right hand side of (2.32) can be taken inside the sum. Theorem 7.10 in [181] states that, if $|g_j(x)|\le C_j$ for all x and all j, and if $\sum_j C_j<\infty$, then $\sum_j g_j(x)$ converges uniformly in x. Fix $s>0$ such that $s^{\beta}>|\lambda|$ and define

$$g_j(x)=\int_0^x\frac{(\lambda t^{\beta})^j}{\Gamma(1+\beta j)}e^{-st}dt$$

and

$$C_j=\int_0^{\infty}\frac{(|\lambda|t^{\beta})^j}{\Gamma(1+\beta j)}e^{-st}dt.$$

Since $|g_j(x)|\le C_j$ and

$$\sum_{j=0}^{\infty}C_j=\frac{s^{\beta-1}}{s^{\beta}-|\lambda|}<\infty,$$

it follows that $\sum_{j=0}^{\infty}g_j(x)$ converges uniformly in x. Lastly, Theorem 7.11 in [181] implies that if $h_n(x)\to h(x)$ uniformly in x, and if $h_n(x)\to D_n$ as $x\to\infty$ for all n, then

$$\lim_{x\to\infty}\lim_{n\to\infty}h_n(x)=\lim_{n\to\infty}\lim_{x\to\infty}h_n(x).$$

Then with

$$h_n(x)=\sum_{j=1}^{n}g_j(x)=\sum_{j=1}^{n}\int_0^x\frac{(\lambda t^{\beta})^j}{\Gamma(1+\beta j)}e^{-st}dt$$

and

$$h(x) = \sum_{j=1}^{\infty} g_j(x) = \sum_{j=1}^{\infty} \int_0^x \frac{(\lambda t^\beta)^j}{\Gamma(1+\beta j)} e^{-st} dt$$

it follows that

$$\begin{aligned}\lim_{x\to\infty} \lim_{n\to\infty} h_n(x) &= \lim_{x\to\infty} \sum_{j=1}^{\infty} \int_0^x \frac{(\lambda t^\beta)^j}{\Gamma(1+\beta j)} e^{-st} dt \\ &= \lim_{n\to\infty} \lim_{x\to\infty} h_n(x) \\ &= \sum_{j=0}^{\infty} \int_0^{\infty} \frac{(\lambda t^\beta)^j}{\Gamma(1+\beta j)} e^{-st} dt.\end{aligned}$$

This completes the proof of term-by-term integration of the series in Remark 2.12.

We now derive the expression for the LT of the Caputo fractional derivative of order $0 < \alpha < 1$. For a function $f(x)$ defined on $x \geq 0$, the *Caputo fractional derivative* is defined by

$$\partial_x^\alpha f(x) = \frac{1}{\Gamma(1-\alpha)} \int_0^x f'(x-y) y^{-\alpha} dy, \tag{2.33}$$

which is equivalent to (2.18) with $f(x) = 0$ for $x < 0$. Assuming that $e^{-sx} f'(x-y) y^{-\alpha}$ is integrable as a function of two variables, x and y, substitute $x - y = z$, change the order of integration, and apply (2.27) to get

$$\begin{aligned}\frac{1}{\Gamma(1-\alpha)} \int_0^{\infty} e^{-sx} \int_0^x f'(x-y) y^{-\alpha} dy\, dx &= \frac{1}{\Gamma(1-\alpha)} \int_0^{\infty} y^{-\alpha} \int_y^{\infty} e^{-sx} f'(x-y) dx\, dy \\ &= \frac{1}{\Gamma(1-\alpha)} \int_0^{\infty} e^{-sy} y^{-\alpha} dy \int_0^{\infty} e^{-sz} f'(z) dz \\ &= s^{\alpha-1} \left(s\tilde{f}(s) - f(0) \right) \\ &= s^\alpha \tilde{f}(s) - s^{\alpha-1} f(0).\end{aligned}$$

To derive the expression for the LT of the Riemann-Liouville fractional derivative, note that for a function $f(x)$ defined on $x \geq 0$ the *Riemann-Liouville fractional derivative* (2.19) reduces to

$$\mathbb{D}_x^\alpha f(x) = \frac{1}{\Gamma(1-\alpha)} \frac{d}{dx} \int_0^x f(x-y) y^{-\alpha} dy. \tag{2.34}$$

To compute its LT, integrate by parts to get

$$\frac{1}{\Gamma(1-\alpha)} \int_0^{\infty} e^{-sx} \left(\frac{d}{dx} \int_0^x f(x-y) y^{-\alpha} dy \right) dx = I_1 + I_2$$

where

$$I_1 = \frac{1}{\Gamma(1-\alpha)} \left[e^{-sx} \int_0^x f(x-y)y^{-\alpha}\,dy \right]_{x=0}^{\infty} = 0$$

assuming $f(x)$ is bounded, and

$$\begin{aligned}
I_2 &= \frac{s}{\Gamma(1-\alpha)} \int_0^\infty e^{-sx} \int_0^x f(x-y)y^{-\alpha}\,dy\,dx \\
&= \frac{s}{\Gamma(1-\alpha)} \int_0^\infty y^{-\alpha} \int_y^\infty e^{-sx} f(x-y)\,dx\,dy \\
&= \frac{s}{\Gamma(1-\alpha)} \int_0^\infty e^{-sy} y^{-\alpha}\,dy \times \int_0^\infty e^{-sz} f(z)\,dz \\
&= s\, s^{\alpha-1} \tilde{f}(s) = s^\alpha \tilde{f}(s)
\end{aligned}$$

assuming $e^{-sx}f(x-y)y^{-\alpha}$ is integrable. It follows that the Caputo and Riemann-Liouville fractional derivatives of order $0 < \alpha < 1$ are related by

$$\begin{aligned}
\partial_x^\alpha f(x) &= \mathbb{D}_x^\alpha f(x) - f(0)\frac{x^{-\alpha}}{\Gamma(1-\alpha)} \\
&= \frac{1}{\Gamma(1-\alpha)} \frac{d}{dx} \left[\int_0^x f(x-y)y^{-\alpha}\,dy \right] - f(0)\frac{x^{-\alpha}}{\Gamma(1-\alpha)}. \qquad (2.35)
\end{aligned}$$

Some authors use the last line of (2.35) as the definition of the Caputo fractional derivative, since it exists for a broader class of functions (e.g., see Kochubei [105]).

Remark 2.13. This remark explains the connection between Fourier and Laplace transforms in more detail, and introduces the *Fourier-Stieltjes transform* and the *weak derivative*. Suppose that $f(t)$ is a real-valued function defined for $t \geq 0$, and extend to the entire real line by setting $f(t) = 0$ when $t < 0$. Then the two-sided Laplace transform

$$\tilde{f}(s) = \int e^{-st} f(t)\,dt = \int_{-\infty}^{\infty} e^{-st} f(t)\,dt$$

agrees with the definition (2.25). If $f'(t)$ exists at every $t > 0$, then we can write

$$\begin{aligned}
\int e^{-st} f'(t)\,dt &= \lim_{a\downarrow 0} \int_a^\infty e^{-st} f'(t)\,dt \\
&= \lim_{a\downarrow 0} \left[e^{-st} f(t) \Big|_a^\infty - \int_a^\infty (-s) e^{-st} f(t)\,dt \right] \\
&= s\tilde{f}(s) - f(0+) \qquad (2.36)
\end{aligned}$$

using integration by parts with $u = e^{-st}$ and $dv = f'(t)\,dt$. This formula reduces to (2.26) when $f(t)$ is continuous from the right at $t = 0$.

We have noted previously that $f'(t)$ has FT $(ik)\hat{f}(k)$. This was proven in the details at the end of Section 1.1, assuming that $f'(t)$ exists for all $t \in \mathbb{R}$ and f, f' are integrable. These conditions do not hold in the present case, since $f'(t)$ may be undefined at $t = 0$. In fact, let us suppose that $f(0) \neq 0$, so that $f(t)$ is not even continuous at $t = 0$. The usual interpretation of the FT in this case is the Fourier-Stieltjes transform, using the idea of a weak derivative: Suppose that $f(t)$ is a right-continuous function of *bounded variation*, so that $f(t)$ can be written as the difference of two monotone nondecreasing functions, $f(t) = f_1(t) - f_2(t)$ where $f_i(t) \leq f_i(t')$ whenever $t \leq t'$, for $i = 1, 2$. Then we can define a Borel measure μ on $\mathbb{R}$ such that $\mu(a, b] = f(b) - f(a)$, and write the Lebesgue-Stieltjes integral $\int g(t)f(dt) = \int g(t)\mu(dt)$ for any suitable Borel measurable function $g(t)$. The Lebesgue integral is a standard construction in analysis and probability (e.g., see [35, 62, 180]). A brief review of Lebesgue integrals, Lebesgue-Stieltjes integrals, and their connection to Riemann integrals will be included in the details at the end of Section 7.9. Now we can interpret the FT of the derivative as a Fourier-Stieltjes transform

$$\int e^{-ikt} \partial_t f(t)\,dt = \int e^{-ikt} f(dt). \tag{2.37}$$

If the traditional first derivative $f'(t)$ exists for all $t \in \mathbb{R}$, then we have $\partial_t f(t)\,dt = f(dt)$ as an equivalence of measures, but the Fourier-Stieltjes transform also exists for functions with jumps. The canonical example is the *Heaviside function* $f(t) = H(t) := I(t \geq 0)$, so that $H(t) = 0$ when $t < 0$, and $H(t) = 1$ when $t \geq 0$. Here μ is a point mass at $t = 0$ and

$$\int e^{-ikt} \partial_t f(t)\,dt = \int e^{-ikt} f(dt) = e^{-ik0}\mu\{0\} = 1$$

for all $k \in \mathbb{R}$. In functional analysis, it is common to write $\partial_t f(t) = \delta(t)$ in this case, where $\delta(t)$ is the *Dirac delta function*. The Dirac delta function is a distribution, or generalized function, defined as a linear operator on a suitable space of test functions $g(t)$ by the formula

$$\int g(t)\delta(t)\,dt = g(0),$$

another notation for the Lebesgue-Stieltjes integral $\int g(t)\partial_t f(t)\,dt = \int g(t)f(dt)$ when $f(t) = H(t)$ and $\partial_t f(t) = \delta(t)$. The *generalized function* $\partial_t f(t) = \delta(t)$ is also called the weak (or distributional) derivative of the Heaviside function $f(t) = H(t)$. Now apply the integration by parts formula for functions F, G of bounded variation with no common points of discontinuity (e.g., see [85, Theorem 19.3.13]):

$$\int_a^b F(t)G(dt) = F(b)G(b) - F(a)G(a) - \int_a^b G(t)F(dt).$$

Define $F(t) = e^{-ikt}$ and $G(t) = f(t)$, and note that both are functions of bounded variation on any finite interval $[a, b]$, with no common points of discontinuity, since $F(t)$

is continuous. Then

$$\int_a^b e^{-ikt}\partial_t f(t)\,dt = \int_a^b e^{-ikt}f(dt)$$
$$= e^{-ikb}f(b) - e^{-ika}f(a) - \int_a^b (-ik)e^{-ikt}f(t)\,dt.$$

Suppose that $f(t) \to 0$ as $t \to \infty$. Since $f(t) = 0$ for $t < 0$, we can take limits as $a \to -\infty$ and $b \to \infty$ to conclude that

$$\int_{-\infty}^{\infty} e^{-ikt}\partial_t f(t)\,dt = (ik)\hat{f}(k). \tag{2.38}$$

This extends the usual FT formula, using the weak derivative. This notation is commonly used in the physics literature. Recall that we are assuming $f(t) = 0$ for $t < 0$, $f(0) \neq 0$, f is continuous from the right, of bounded variation, and $f'(t)$ exists in the traditional sense for all $t > 0$. Then in the sense of distributions, we can use the physics notation to write

$$\partial_t f(t) = \begin{cases} 0 & t < 0 \\ f(0)\,\delta(t) & t = 0 \\ f'(t) & t > 0 \end{cases}$$

and so we have

$$\int_{-\infty}^{\infty} e^{-ikt}\partial_t f(t)\,dt = 0 + \int_{-\infty}^{\infty} e^{-ikt}f(0)\,\delta(t)\,dt + \int_0^{\infty} e^{-ikt}f'(t)\,dt$$
$$= 0 + e^{-ik0}f(0) + \left[ik\hat{f}(k) - f(0)\right]$$
$$= ik\hat{f}(k).$$

In the third term, we have used integration by parts in exactly the same manner as (2.36). Indeed, this integral may be viewed as the LT of $f'(t)$ evaluated at $s = ik$. In some applications, it is quite natural to consider Laplace transforms where s is a complex number (e.g., see Arendt, Batty, Hieber and Neubrander [8]). In summary, the difference between the formulas for the FT and the LT of the first derivative reflects the fact that these two transforms interpret the first derivative in a different manner at the boundary point $t = 0$.

2.4 Time-fractional diffusion

The simplest time-fractional diffusion equation

$$\partial_t^\beta p(x,t) = D\frac{\partial^2}{\partial x^2}p(x,t) \tag{2.39}$$

employs a Caputo fractional derivative (2.33) of order $0 < \beta < 1$. We will solve this fractional partial differential equation using the Fourier-Laplace transform (FLT):

$$\bar{p}(k,s) = \int_0^\infty \int_{-\infty}^\infty e^{-st} e^{-ikx} p(x,t)\,dx\,dt = \int_0^\infty e^{-st} \hat{p}(k,t)\,dt.$$

To illustrate the method, first consider the traditional Brownian motion solution

$$p(x,t) = \frac{1}{\sqrt{4\pi t}} e^{-x^2/(4t)} \tag{2.40}$$

to the diffusion equation $\partial p/\partial t = \partial^2 p/\partial x^2$. Take FT in (2.40) to get $\hat{p}(k,t) = e^{-k^2 t}$ and then take LT to get

$$\bar{p}(k,s) = \int_0^\infty e^{-st} e^{-k^2 t}\,dt = \int_0^\infty e^{-(s+k^2)t}\,dt = \frac{1}{s+k^2}$$

for all $s > 0$. Note that $\hat{p}(k,0) = 1$ for all k, reflecting the fact that the Brownian motion $B(t) = 0$ with probability one when $t = 0$. Rewrite in the form

$$s\bar{p}(k,s) - 1 = -k^2 \bar{p}(k,s)$$

and invert the LT to get

$$\frac{d}{dt}\hat{p}(k,t) = -k^2 \hat{p}(k,t).$$

Then invert the FT to recover the diffusion equation

$$\frac{\partial}{\partial t} p(x,t) = \frac{\partial^2}{\partial x^2} p(x,t). \tag{2.41}$$

Now we apply the FLT method to the time-fractional diffusion equation (2.39). Take FT to get

$$\partial_t^\beta \hat{p}(k,t) = -Dk^2 \hat{p}(k,t)$$

and assume the point source initial condition $\hat{p}(k,0) \equiv 1$. Take LT to get

$$s^\beta \bar{p}(k,s) - s^{\beta-1} = -Dk^2 \bar{p}(k,s) \tag{2.42}$$

and rearrange to get

$$\bar{p}(k,s) = \frac{s^{\beta-1}}{s^\beta + Dk^2}, \tag{2.43}$$

then invert using (2.31) to get

$$\hat{p}(k,t) = E_\beta(-Dk^2 t^\beta).$$

In order to invert this FT, we will need a stochastic model for time-fractional diffusion.

Remark 2.14. The time-fractional diffusion equation (2.39) can also be written in terms of the Riemann-Liouville fractional derivative $\mathbb{D}_t^\beta$. Recall that $\mathbb{D}_t^\beta f(t)$ has LT $s^\beta \tilde{f}(s)$. Recall (2.27), and substitute $p = -\beta$ to see that $s^{\beta-1}$ is the LT of $t^{-\beta}/\Gamma(1-\beta)$. Invert the LT in (2.42) to get

$$\mathbb{D}_t^\beta \hat{p}(k,t) - \frac{t^{-\beta}}{\Gamma(1-\beta)} = -Dk^2 \hat{p}(k,t)$$

and then invert the FT to arrive at

$$\mathbb{D}_t^\beta p(x,t) = D\frac{\partial^2}{\partial x^2} p(x,t) + \frac{t^{-\beta}}{\Gamma(1-\beta)}\delta(x). \tag{2.44}$$

Here $\delta(x)$ is the Dirac delta function, whose Fourier transform $\hat{\delta}(k) \equiv 1$ (see the details at the end of this section). Equation (2.44) is the fractional kinetic equation for Hamiltonian chaos introduced by Zaslavsky [222] in the physics literature. The mathematical study of (2.44) was initiated by Kochubei [105, 106] and Schneider and Wyss [192].

Now we will outline the stochastic model for time-fractional diffusion. Additional details and precise mathematical proofs will be provided later in Chapter 4 of this book. The random walk $S(n) = Y_1 + \cdots + Y_n$ gives the location of a particle after n iid jumps. Now suppose that the nth jump occurs at time $T_n = J_1 + \cdots + J_n$ where the iid waiting times $J_n > 0$ between jumps have a power law probability tail $\mathbb{P}[J_n > t] = Bt^{-\beta}$ for t large, with $0 < \beta < 1$ and $B > 0$. For suitable choice of B, an argument similar to Section 1.2 shows that

$$c^{-1/\beta} T_{[ct]} \Rightarrow D_t$$

where the limit process D_t is stable with index β, and LT

$$\mathbb{E}[e^{-sD_t}] = e^{-ts^\beta} = \tilde{q}(s,t),$$

where $q(u,t)$ is the density of D_t. Since

$$\frac{d}{dt}\tilde{q}(s,t) = -s^\beta \tilde{q}(s,t)$$

this density solves

$$\frac{\partial}{\partial t} q(u,t) = -\frac{\partial^\beta}{\partial u^\beta} q(u,t)$$

using the Riemann-Liouville fractional derivative. Let

$$N_t = \max\{n \geq 0 : T_n \leq t\}$$

denote the number of jumps by time $t \geq 0$. The continuous time random walk (CTRW) $S(N_t)$ gives the particle location at time t. These are inverse processes: $\{N_t \geq n\} =$

$\{T_n \le t\}$, and in fact, $\{N_t \ge u\} = \{T_{\lceil u \rceil} \le t\}$ where $\lceil u \rceil$ is the smallest integer $n \ge u$. The inverse process has an inverse weak limit:

$$c^{-\beta} N_{ct} \Rightarrow E_t$$

where $\{E_t \le u\} = \{D_u \ge t\}$. We can define $E_t = \inf\{u > 0 : D_u > t\}$, the first passage time of D_u above the level $u > 0$. The scaling $c^{1/\beta} D_t = D_{ct}$ in distribution implies the inverse scaling $c^\beta E_t = E_{ct}$ in distribution. The CTRW scaling limit as the time scale $c \to \infty$ is

$$c^{-\beta/2} S(N_{ct}) = (c^\beta)^{-1/2} S(c^\beta\, c^{-\beta} N_{ct}) \approx (c^\beta)^{-1/2} S(c^\beta\, E_t) \approx B(E_t)$$

a time-changed Brownian motion. Since

$$\mathbb{P}[E_t \le u] = \mathbb{P}[D_u \ge t] = \int_t^\infty q(w, u)\, dw$$

the inner process E_t has density

$$h(u, t) = \frac{d}{du} \mathbb{P}[E_t \le u] = \frac{d}{du}\left[1 - \int_0^t q(w, u)\, dw\right]$$

with LT

$$\begin{aligned} \tilde{h}(u, s) &= -\frac{d}{du}\left[s^{-1} \tilde{q}(s, u)\right] \\ &= -\frac{d}{du}\left[s^{-1} e^{-us^\beta}\right] = s^{\beta-1} e^{-us^\beta} \end{aligned}$$

using the fact that integration corresponds to multiplication by s^{-1} in LT space. Since $B(E_t) = B(u)$ where $u = E_t$ is independent of $x = B(u)$, a simple conditioning argument shows that the process $B(E_t)$ has density

$$m(x, t) = \int_0^\infty p(x, u) h(u, t)\, du \approx \sum_u \mathbb{P}(B(u) = x | E_t = u) \mathbb{P}(E_t = u).$$

Take FLT ($x \mapsto k$ and $t \mapsto s$) to get

$$\bar{m}(k, s) = \int_0^\infty e^{-uDk^2} s^{\beta-1} e^{-us^\beta}\, du = s^{\beta-1} \int_0^\infty e^{-u(s^\beta + Dk^2)}\, du = \frac{s^{\beta-1}}{s^\beta + Dk^2}$$

which agrees with (2.43). This shows that the limit density $m(x, t)$ solves the time-fractional diffusion equation (2.39). Also note that $\hat{m}(k, t) = E_\beta([-Dk^2]t^\beta)$.

The CTRW model provides a physical explanation for fractional diffusion. A power law jump distribution with $\mathbb{P}[Y_n > x] = Cx^{-\alpha}$ leads to a fractional derivative in space

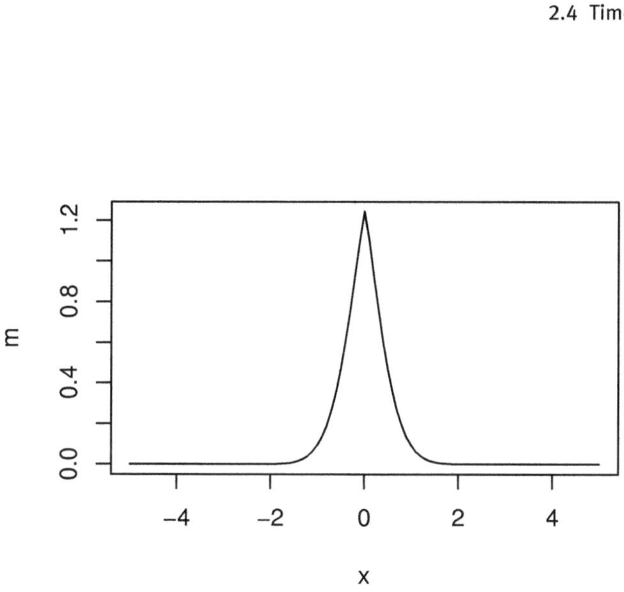

Fig. 2.3: Solution to time-fractional diffusion equation (2.39) at time $t = 0.1$ with $\beta = 0.75$ and dispersion $D = 1.0$

$\partial^\alpha/\partial x^\alpha$ of the same order. A power law waiting time distribution $\mathbb{P}[J_n > t] = Bx^{-\beta}$ leads to a fractional time derivative ∂_t^β of the same order. Long power-law jumps reflect a heavy tailed velocity distribution, which allows particles to make occasional long jumps, leading to anomalous super-diffusion. Long waiting times model particle sticking and trapping, leading to anomalous sub-diffusion:

$$B(E_{ct}) \simeq B(c^\beta E_t) \simeq c^{\beta/2} B(E_t).$$

Since $\beta < 1$, the density of this process spreads slower than a Brownian motion. Figure 2.3 plots a typical density $m(x, t)$ for the process $B(E_t)$. As compared to a normal density, this curve has a sharper peak, and heavier tails. The R code used to produce Figure 2.3 will be discussed in Example 5.13.

Remark 2.15. Continuous time random walks were proposed by Montroll and Weiss [160], and developed further by Scher and Lax [190], Klafter and Silbey [104], and Hilfer and Anton [86]. An interesting CTRW model for the migration of cancer cells was presented in Fedotov and Iomin [67]. See Berkowitz, Cortis, Dentz and Scher [31] for a review of continuous time random walks in hydrology. Scalas [188] reviews applications of the CTRW model in finance. Schumer and Jerolmack [196] develop an interesting CTRW model for sediment deposition in the geological record.

Details

To prove the inverse scaling, recall that $D_{cu} \simeq c^{1/\beta} D_u$ and write

$$\begin{aligned}\mathbb{P}[E_{ct} \le u] &= \mathbb{P}[D_u \ge ct] = \mathbb{P}[c^{-1} D_u \ge t] \\ &= \mathbb{P}[(c^{-\beta})^{1/\beta} D_u \ge t] = \mathbb{P}[D_{c^{-\beta} u} \ge t] \\ &= \mathbb{P}[E_t \le c^{-\beta} u] = \mathbb{P}[c^{\beta} E_t \le u]\end{aligned}$$

so that $E_{ct} \simeq c^{\beta} E_t$.

To prove the inverse limit, recall that $c^{-1/\beta} T_{[ct]} \Rightarrow D_t$ and $\{N_t \ge u\} = \{T_{\lceil u \rceil} \le t\}$ and write

$$\begin{aligned}\mathbb{P}[c^{-\beta} N_{ct} \le u] &= \mathbb{P}[N_{ct} \le c^{\beta} u] = \mathbb{P}[T_{\lceil c^{\beta} u \rceil} \ge ct] \\ &= \mathbb{P}[c^{-1} T_{\lceil c^{\beta} u \rceil} \ge t] = \mathbb{P}[(c^{\beta})^{-1/\beta} T_{\lceil c^{\beta} u \rceil} \ge t] \to \mathbb{P}[D_u \ge t] = \mathbb{P}[E_t \le u]\end{aligned}$$

so that $c^{-\beta} N_{ct} \Rightarrow E_t$.

The Dirac delta function $\delta(x)$ was introduced in Remark 2.13. It is a generalized function, or distribution, defined for suitable test functions $g(t)$ (e.g., bounded continuous functions) by $\int g(x)\delta(x)\,dx = g(0)$. One way to understand equation (2.44) is that $p(x, t)$ is a *weak solution*, sometimes called a distributional solution, to the differential equation, meaning that

$$\int D_t^{\beta} p(x, t) g(x)\,dx = \int D \frac{\partial^2}{\partial x^2} p(x, t) g(x)\,dx + \int \frac{t^{-\beta}}{\Gamma(1-\beta)} \delta(x) g(x)\,dx$$

for suitable test functions $g(x)$. This functional analysis construction is equivalent to using cumulative distribution functions and Fourier-Stieltjes transforms. Let

$$P(x, t) = \int_{-\infty}^{x} p(y, t)\,dy$$

be the cumulative distribution function of a Brownian motion $B(t)$ with pdf $p(x, t)$ given by (2.40). Then $P(x, t)$ is the unique solution to the diffusion equation

$$\frac{\partial}{\partial t} P(x, t) = \frac{\partial^2}{\partial x^2} P(x, t) \tag{2.45}$$

with initial condition $P(x, 0) = I(x \ge 0)$, the Heaviside function. To see this, apply the Fourier-Stieltjes transform

$$\hat{P}(k, t) = \int e^{-ikx} P(dx, t)$$

on both sides of equation (2.45) to get

$$\frac{d}{dt} \hat{P}(k, t) = (ik)^2 \hat{P}(k, t) = -k^2 \hat{P}(k, t)$$

with initial condition $\hat{P}(k,0) = \int e^{-ikx} P(dx,0) = 1$ for all $k \in \mathbb{R}$, since $P(dx,0)$ is a point mass at $x = 0$, i.e., the probability distribution of $B(0)$. Taking derivatives with respect to x on both sides of (2.45) recovers the diffusion equation (2.41) with the Dirac delta function initial condition $p(x,0) = \delta(x)$. Since

$$\int e^{-ikx} P(dx,t) = \int e^{-ikx} p(x,t)\,dx$$

for all $t > 0$, these Fourier transform calculations are completely equivalent. Hence, equation (2.44) is equivalent to

$$D_t^{\beta} P(x,t) = D\frac{\partial^2}{\partial x^2} P(x,t) + \frac{t^{-\beta}}{\Gamma(1-\beta)} H(x)$$

where $p(x,t) = \partial_x P(x,t)$, and $H(x) = I(x \geq 0)$ is the Heaviside function.

3 Stable Limit Distributions

In this chapter, we develop the fundamental mathematical tools for fractional diffusion. The Fourier transform of a stable law is computed from the Lévy representation for infinitely divisible laws. The extended central limit theorem for a random walk with power law jumps follows from the convergence criteria for triangular arrays. The theory of semigroups leads naturally to the generator form of the fractional derivative.

3.1 Infinitely divisible laws

Infinitely divisible laws are a class of probability distributions that includes the normal and stable laws. The Lévy representation for infinitely divisible laws is the basis for both the stable FT, and the generator form of the fractional derivative. Recall that the generator form of the fractional derivative is

$$\frac{d^\alpha f(x)}{dx^\alpha} = \int_0^\infty [f(x) - f(x-y)] \frac{\alpha}{\Gamma(1-\alpha)} y^{-\alpha-1}\,dy \tag{3.1}$$

for $0 < \alpha < 1$, or

$$\frac{d^\alpha f(x)}{dx^\alpha} = \int_0^\infty \left[f(x-y) - f(x) + yf'(x)\right] \frac{\alpha(\alpha-1)}{\Gamma(2-\alpha)} y^{-1-\alpha}\,dy \tag{3.2}$$

for $1 < \alpha < 2$. The stable FT $\hat{p}(k,t) = e^{tD(ik)^\alpha}$ leads to the space-fractional diffusion equation $\partial p/\partial t = D\partial^\alpha p/\partial x^\alpha$. How do these forms connect? The answer lies in some deep mathematical theory, which we now begin to develop.

We start by establishing some notation. Given a random variable Y, we define the cumulative distribution function (cdf) $F(x) = \mathbb{P}[Y \le x]$, the probability density function (pdf) $f(y) = F'(y)$, and the probability measure $\mu(a,b] = F(b) - F(a) = \mathbb{P}[a < Y \le b]$. We write $Y \simeq \mu$ or $Y \simeq F$, and we will also write $X \simeq Y$ if two random variables X, Y have the same distribution. The characteristic function

$$\hat{\mu}(k) = \mathbb{E}[e^{ikY}] = \int e^{ikx}\mu(dx) = \int e^{ikx}F(dx) = \int e^{ikx}f(x)\,dx = \hat{f}(-k)$$

is related to the Fourier transform (FT) by an obvious change of sign. Characteristic functions with e^{ikx} are used in probability, because they simplify the formula (1.11) for moments. Fourier transforms with e^{-ikx} are used in differential equations, because they simplify the formula (1.14) for derivatives. See the details and the end of this section for more information.

We say that (the distribution of) Y is infinitely divisible if $Y \simeq X_1 + \cdots + X_n$ for every positive integer n, where (X_n) are independent and identically distributed (iid) random variables. If $X_n \simeq \mu_n$, then we also have

$$\hat{\mu}(k) = \mathbb{E}[e^{ikY}] = \mathbb{E}[e^{ik(X_1+\cdots+X_n)}] = \mathbb{E}[e^{ikX_1}]\cdots\mathbb{E}[e^{ikX_n}] = \hat{\mu}_n(k)^n$$

since $X_1, \ldots, X_n$ are independent.

Example 3.1. If $Y \simeq \mathcal{N}(a, \sigma^2)$ (normal with mean a and variance σ^2), then $\hat{\mu}(k) = \exp(ika + \frac{1}{2}\sigma^2 k^2)$. If we take $\hat{\mu}_n(k) = \exp(ik(a/n) + \frac{1}{2}(\sigma^2/n)k^2)$ then clearly $\hat{\mu}(k) = \hat{\mu}_n(k)^n$ so Y is infinitely divisible. In fact $Y \simeq X_1 + \cdots + X_n$ where $X_j \simeq \mathcal{N}(a/n, \sigma^2/n)$ are iid. The sum of independent normal random variables is also normal, the means add, and the variances add.

Example 3.2. If Y is Poisson with mean λ, then $\mathbb{P}[Y = j] = \mu\{j\} = e^{-\lambda}\lambda^j/j!$ for $j = 0, 1, 2, \ldots$ and

$$\begin{aligned}\hat{\mu}(k) = \int e^{ikx}\mu(dx) &= \sum_{j=0}^{\infty} e^{ikj}\mathbb{P}[Y = j] \\ &= \sum_{j=0}^{\infty} e^{ikj} e^{-\lambda}\frac{\lambda^j}{j!} \\ &= e^{-\lambda}\sum_{j=0}^{\infty}\frac{(\lambda e^{ik})^j}{j!} \\ &= \exp(-\lambda)\exp(\lambda e^{ik}) = \exp(\lambda[e^{ik} - 1])\end{aligned}$$

so $\hat{\mu}(k) = \hat{\mu}_n(k)^n$ where $\hat{\mu}_n(k) = \exp((\lambda/n)[e^{ik} - 1])$. The sum of independent Poisson random variables is also Poisson, and the means add.

Example 3.3. A compound Poisson random variable $Y = W_1 + \cdots + W_N = S_N$ is a random sum, where $S_n = W_1 + \cdots + W_n$, (W_j) are iid with probability measure $\omega(dy)$, and N has a Poisson distribution with mean λ, independent of (W_j). Then

$$\begin{aligned}F(y) = \mathbb{P}[Y \le y] &= \mathbb{P}[S_N \le y] \\ &= \sum_{j=0}^{\infty}\mathbb{P}[S_N \le y | N = j]\mathbb{P}[N = j] \\ &= \sum_{j=0}^{\infty}\mathbb{P}[S_j \le y]e^{-\lambda}\frac{\lambda^j}{j!}.\end{aligned}$$

Then Y has characteristic function

$$\begin{aligned}\hat{\mu}(k) &= \sum_{j=0}^{\infty}\hat{\omega}(k)^j e^{-\lambda}\frac{\lambda^j}{j!} \\ &= e^{-\lambda}\sum_{j=0}^{\infty}\frac{[\lambda\hat{\omega}(k)]^j}{j!} \\ &= e^{-\lambda}e^{\lambda\hat{\omega}(k)} = e^{\lambda[\hat{\omega}(k)-1]}.\end{aligned}$$

Take $\hat{\mu}_n(k) = e^{(\lambda/n)[\hat{\omega}(k)-1]}$ to see that Y is infinitely divisible. The sum of independent compound Poisson random variables with the same jump distribution are also compound Poisson.

To motivate what comes next, write the compound Poisson characteristic function

$$\begin{aligned}\hat{\mu}(k) &= e^{\lambda[\hat{\omega}(k)-1]}\\ &= \exp\left(\lambda\left[\int e^{ikx}\omega(dx) - 1\right]\right)\\ &= \exp\left(\lambda\left[\int \left(e^{ikx} - 1\right)\omega(dx)\right]\right)\\ &= \exp\left(\left[\int \left(e^{ikx} - 1\right)\lambda\,\omega(dx)\right]\right)\\ &= \exp\left(\int \left(e^{ikx} - 1\right)\phi(dx)\right)\end{aligned}$$

where the *Lévy measure* $\phi(dx) = \lambda\,\omega(dx)$. This is also called the *jump intensity*. The random variable $Y = W_1 + \cdots + W_N$ is the accumulation of a random number of jumps. The number of these jumps that lie in any Borel set B is Poisson with mean $\phi(B) = \lambda\,\omega(B)$. To see this, note that $\omega(B) = \mathbb{P}[W_n \in B]$ and split the Poisson process of jumps into two parts, depending on whether or not the jump lies in B. A general theorem on Poisson processes (e.g., see Ross [179, Proposition 5.2]) shows that an independent splitting produces two independent Poisson processes, and then the number of jumps that lie in B follows a Poisson with mean $\phi(B) = \lambda\,\omega(B)$.

The *Lévy representation* gives the general form of the characteristic function for an infinitely divisible law. This form reflects the normal and compound Poisson cases. We say that a σ-finite Borel measure $\phi(dy)$ on $\{y : y \neq 0\}$ is a *Lévy measure* if $\phi\{y : |y| > R\} < \infty$ and

$$\int\limits_{0<|y|\leq R} y^2\phi(dy) < \infty \tag{3.3}$$

for all $R > 0$. See the details at the end of this section for more information.

Theorem 3.4 (Lévy representation). *A random variable $Y \simeq \mu$ is infinitely divisible if and only if its characteristic function $\hat{\mu}(k) = \mathbb{E}[e^{ikY}] = e^{\psi(k)}$ where*

$$\psi(k) = ika - \tfrac{1}{2}k^2b + \int\left(e^{iky} - 1 - \frac{iky}{1+y^2}\right)\phi(dy) \tag{3.4}$$

for some $a \in \mathbb{R}$, $b \geq 0$, and some Lévy measure $\phi(dy)$. This Lévy representation $\mu \simeq [a, b, \phi]$ is unique.

Proof. The proof is based on a compound Poisson approximation, see Meerschaert and Scheffler [146, Theorem 3.1.11]. □

Example 3.5. If $Y \simeq \mathcal{N}(a, \sigma^2)$ then Theorem 3.4 holds with $b = \sigma^2$ and $\phi = 0$.

Example 3.6. If Y is compound Poisson, then Theorem 3.4 holds with $b = 0$, $\phi(dy) = \lambda\,\omega(dy)$, and

$$a = \lambda \int \frac{y}{1+y^2}\,\omega(dy).$$

To check this, write

$$\begin{aligned}\psi(k) &= ik\lambda \int \frac{y}{1+y^2}\,\omega(dy) + \int \left(e^{iky} - 1 - \frac{iky}{1+y^2}\right)\lambda\,\omega(dy)\\ &= \lambda \int e^{iky}\omega(dy) - 1 = \lambda\left[\hat{\omega}(k) - 1\right].\end{aligned}$$

Note that the integral expression for a exists, since the integrand is bounded.

The next result shows that every infinitely divisible law is essentially compound Poisson. Suppose that Y_n is a random variable with cdf $F_n(x)$ and probability measure μ_n for each positive integer n. We say that $Y_n \Rightarrow Y$ (convergence in distribution, sometimes called convergence in law, or weak convergence) if $F_n(x) \to F(x)$ for all $x \in \mathbb{R}$ such that $F(x+) = F(x-)$. In view of the continuity theorem for FT (see Theorem 1.3), this is equivalent to $\hat{\mu}_n(k) \to \hat{\mu}(k)$ for every $k \in \mathbb{R}$.

Proposition 3.7. *Every infinitely divisible law is the weak limit of compound Poisson laws.*

Proof. Use the Lévy Representation Theorem 3.4 to write $\hat{\mu}(k) = e^{\psi(k)}$ where (3.4) holds. Then $\hat{\mu}(k) = [\hat{\mu}_n(k)]^n$ where $\hat{\mu}_n(k) = e^{\psi(k)/n}$. This shows that $Y \simeq X_1 + \cdots + X_n$ where the iid summands $(X_n) \simeq \mu_n$. Now define $Y_n = X_1 + \cdots + X_N$ where N is Poisson with mean n. Then Y_n is compound Poisson with characteristic function $\hat{\nu}_n(k) = \exp(n[\hat{\mu}_n(k) - 1])$. Fix $k \in \mathbb{R}$ and write

$$\hat{\mu}_n(k) - 1 = \left(1 + \frac{1}{n}\psi(k) + \frac{1}{2!}\left(\frac{1}{n}\psi(k)\right)^2 + \cdots\right) - 1 = \frac{1}{n}\psi(k) + O(n^{-2})$$

so that $n[\hat{\mu}_n(k) - 1] = \psi(k) + O(n^{-1})$. Then $\hat{\nu}_n(k) = \exp(\psi(k) + O(n^{-1})) \to \exp(\psi(k)) = \hat{\mu}(k)$ for all $k \in \mathbb{R}$, so $\nu_n \Rightarrow \mu$. □

The compound Poisson approximation gives a concrete interpretation of the Lévy measure. Suppose that $\mu \simeq [0, 0, \phi]$ so that

$$\hat{\mu}(k) = \exp\left[\int \left(e^{iky} - 1 - \frac{iky}{1+y^2}\right)\phi(dy)\right].$$

Define

$$\begin{aligned}\hat{\nu}_n(k) &= \exp\left[\int_{|y|>1/n} \left(e^{iky} - 1 - \frac{iky}{1+y^2}\right)\phi(dy)\right]\\ &= \exp\left(\lambda_n \int \left(e^{iky} - 1\right)\omega_n(dy) - ika_n\right)\end{aligned}$$

where

$$\lambda_n = \int_{|y|>1/n} \phi(dy) = \phi\{y : |y| > n^{-1}\}$$
$$\omega_n(B) = \lambda_n^{-1} \int_{|y|>1/n, y\in B} \phi(dy) = \lambda_n^{-1}\phi(B \cap \{y : |y| > n^{-1}\})$$
$$a_n = \int_{|y|>1/n} \frac{y}{1+y^2}\,\phi(dy).$$

Then $\nu_n \simeq Y_n + a_n$ a shifted compound Poisson where $Y_n \simeq W_1 + \cdots + W_N$, $(W_n) \simeq \omega_n$ is iid, and N is Poisson with mean λ_n independent of (W_n). The Lévy Representation Theorem 3.4 implies that $\hat{\nu}_n(k) \to e^{\psi(k)} = \hat{\mu}(k)$, so $\nu_n \Rightarrow \mu$. Every infinitely divisible law with no normal component can be approximated by such a compound Poisson, the sum of a random number of jumps. The Lévy measure controls both the number and size of the jumps.

Details

The Lebesgue-Stieltjes integral and the distributional derivative were introduced briefly in Remark 2.13. Here we provide more detail, with an emphasis on probability distributions. The cumulative distribution function $F(x) = \mathbb{P}[Y \le x]$ is monotone nondecreasing and continuous from the right, and it follows that there exists a Borel measure μ such that $\mu(a, b] = F(b) - F(a)$ for all $a < b$ in $\mathbb{R}$. If the pdf $f(x) = F'(x)$ exists, then we can define the probability measure

$$\mu(a, b] = \mathbb{P}[a < Y \le b] = \int_a^b f(x)\,dx,$$

and the characteristic function

$$\hat{\mu}(k) = \int e^{ikx} f(x)\,dx.$$

If the random variable Y has atoms, i.e., if $\mathbb{P}[Y = x_k] > 0$ for some real numbers x_k, then $F(x_k) > F(x_k-)$ and the cumulative distribution function is not continuous, so it is certainly not differentiable. Then the pdf cannot exist at every $x \in \mathbb{R}$. In this case, the characteristic function

$$\hat{\mu}(k) = \mathbb{E}[e^{ikY}] = \int e^{ikx}\mu(dx) = \int e^{ikx} F(dx)$$

is defined using the Lebesgue integral with respect to the probability measure μ, or equivalently, the Lebesgue-Stieltjes integral with respect to the cumulative distribution function $F(x)$. If the atoms of Y (i.e., the discontinuity points of the cumulative distribution function $F(x)$) are isolated, then we may also write the pdf of Y using physics notation. For example, if Y is a Poisson random variable with mean λ, then $\mathbb{P}[Y = j] = \mu\{j\} = e^{-\lambda}\lambda^j/j!$ for $j = 0, 1, 2, \ldots$, and we can use physics notation to write

$$f(x) = \partial_x F(x) = \sum_{j=0}^{\infty} e^{-\lambda}\frac{\lambda^j}{j!}\,\delta(x - j)$$

where $\partial_x F(x)$ is the weak or distributional derivative of $F(x)$. This is a completely rigorous alternative notation for the pdf. Readers of this book who are more familiar with the physics notation may consider the Lebesgue integral $\int e^{ikx}\mu(dx)$ or the Lebesgue-Stieltjes integral $\int e^{ikx}F(dx)$ as an alternative notation for $\int e^{ikx}f(x)\,dx$, with the understanding that the pdf $f(x)$ may contain Dirac delta function terms to represent atoms of the probability distribution. In a similar manner, readers who are more familiar with the physics notation may interpret the Lévy measure as $\phi(dy) = \phi(y)dy$ where the function $\phi(y)$ is integrable over $\{y : |y| > R\}$ and the function $y^2\phi(y)$ is integrable over $\{0 < |y| \le R\}$. It is possible that the Lévy measure $\phi(dy)$ contains atoms. For example, a Poisson random variable with mean λ has Lévy measure $\phi(dy) = \lambda\,\delta(y - 1)\,dy$. For readers who are familiar with Lebesgue integrals and Lebesgue-Stieltjes integrals, it is worth while to learn the alternative notation, since it is commonly used without explanation in the physics literature. This notation also appears frequently in the literature on partial differential equations.

If $X \simeq \mu$ and $Y \simeq \nu$ are independent then $\mathbb{P}[X \in A, Y \in B] = \mathbb{P}[X \in A]\mathbb{P}[Y \in B] = \mu(A)\nu(B)$ is the joint distribution of (X, Y), so the characteristic function of $X + Y$ is $\mathbb{E}[e^{ik(X+Y)}] = \int e^{ik(x+y)}\mu(dx)\nu(dy) = \int e^{ikx}\mu(dx)\int e^{iky}\nu(dy) = \hat{\mu}(k)\hat{\nu}(k)$.

Since the integrand in (3.4) is bounded, the integral exists over $\{y : |y| > R\}$. To show that the integral exists over $\{y : 0 < |y| \le R\}$ for any $k \in \mathbb{R}$, use (3.3) along with

$$e^{iky} - 1 - \frac{iky}{1 + y^2} = (e^{iky} - 1 - iky) + \left(iky - \frac{iky}{1 + y^2}\right) := f(y) + ikg(y)$$

where $f(y) = O(y^2)$ as $y \to 0$ by a Taylor series approximation, and

$$g(y) = y - \frac{y}{1 + y^2} = \frac{y(1 + y^2) - y}{1 + y^2} = \frac{y^3}{1 + y^2} = O(y^3) \quad \text{as } y \to 0.$$

Since ϕ is a Lévy measure,

$$\int_{0<|y|\le R} y^2\phi(dy) < \infty,$$

and

$$\int_{0<|y|\le R} |y|^3\phi(dy) \le R \int_{0<|y|\le R} y^2\phi(dy) < \infty.$$

3.2 Stable characteristic functions

Here we compute the characteristic function of a stable law, using the Lévy representation. First we need to develop some alternative forms.

Theorem 3.8. *Suppose $Y \simeq \mu$ is infinitely divisible with characteristic function $\hat{\mu}(k) = e^{\psi(k)}$ and (3.4) holds. Then we can also write $\hat{\mu}(k) = e^{\psi_0(k)}$ where*

$$\psi_0(k) = ika_0 - \tfrac{1}{2}k^2 b + \int \left(e^{iky} - 1 - ikyI(|y| \le R)\right) \phi(dy) \tag{3.5}$$

for any $R > 0$, for some unique a_0 depending on R and a. Furthermore:

(a) If

$$\int_{0<|y|\le R} |y|\phi(dy) < \infty \tag{3.6}$$

then we can also write $\hat{\mu}(k) = e^{\psi_1(k)}$ where

$$\psi_1(k) = ika_1 - \tfrac{1}{2}k^2 b + \int \left(e^{iky} - 1\right) \phi(dy) \tag{3.7}$$

for some unique a_1 depending on a_0; and

(b) If

$$\int_{|y|>R} |y|\phi(dy) < \infty \tag{3.8}$$

then we can also write $\hat{\mu}(k) = e^{\psi_2(k)}$ where

$$\psi_2(k) = ika_2 - \tfrac{1}{2}k^2 b + \int \left(e^{iky} - 1 - iky\right) \phi(dy) \tag{3.9}$$

for some unique a_2 depending on a_0.

Proof. The integral

$$\delta_0 = \int \left(\frac{y}{1+y^2} - yI(|y| \le R)\right) \phi(dy)$$

exists, since the integrand is bounded and $O(y^3)$ as $y \to 0$. If we take $a_0 = a - \delta_0$, then $\psi(k) = \psi_0(k)$. If (3.7) holds, then $\psi_0(k) = \psi_1(k)$, where

$$a_1 = a_0 - \int_{0<|y|\le R} y\phi(dy).$$

If (3.9) holds, then $\psi_0(k) = \psi_2(k)$, where

$$a_2 = a_0 + \int_{|y|>R} y\phi(dy).$$

Uniqueness follows from Theorem 3.4. □

We now define a *one-sided stable law* μ to be an infinitely divisible law with Lévy representation $[a, 0, \phi]$ where $a \in \mathbb{R}$ and

$$\phi(dy) = \begin{cases} C\alpha y^{-\alpha-1}dy & \text{for } y > 0 \\ 0 & \text{for } y < 0 \end{cases} \tag{3.10}$$

for some $0 < \alpha < 2$. We call α the *index* of that stable law. Note that (3.10) is a Lévy measure since

$$\phi\{y : |y| > R\} = \int_R^\infty \phi(dy) = \int_R^\infty C\alpha y^{-\alpha-1}dy = CR^{-\alpha}$$

and

$$\int_{0<|y|\le R} y^2\phi(dy) = \int_0^R C\alpha y^{1-\alpha}dy = \frac{C\alpha}{2-\alpha}R^{2-\alpha}$$

are both finite for any $R > 0$.

Example 3.9. Suppose $Y \simeq \mu$ is a one-sided stable law stable with index $0 < \alpha < 1$. Since

$$\int_{0<|y|\le R} |y|\phi(dy) = \int_0^R C\alpha y^{-\alpha}dy = \frac{C\alpha}{1-\alpha}R^{1-\alpha}$$

is finite, we can use Theorem 3.8 (a) to write

$$\hat{\mu}(k) = e^{\psi_1(k)} = \exp\left[ika_1 + \int_0^\infty \left(e^{iky} - 1\right) C\alpha y^{-\alpha-1}dy\right]. \tag{3.11}$$

We want to evaluate this integral.

Proposition 3.10. *When* $0 < \alpha < 1$, *the stable characteristic function* (3.11) *with* $a_1 = 0$ *can be written in the form*

$$\hat{\mu}(k) = \mathbb{E}[e^{ikY}] = \exp\left[-C\Gamma(1-\alpha)(-ik)^\alpha\right]. \tag{3.12}$$

Proof. We follow the proof in Feller [68], see also [146, Lemma 7.3.7]. We will approximate the integral

$$I(\alpha) = \int_0^\infty \left(e^{iky} - 1\right)\alpha y^{-\alpha-1}dy$$

by another integral

$$I_s(\alpha) = \int_0^\infty \left(e^{(ik-s)y} - 1\right)\alpha y^{-\alpha-1}dy$$

for $s > 0$. Integrate by parts with $u = e^{(ik-s)y} - 1$ to see that

$$I_s(\alpha) = \left[\left(e^{(ik-s)y} - 1\right)(-y^{-\alpha})\right]_0^\infty + (ik - s)\int_0^\infty e^{(ik-s)y}y^{-\alpha}dy \tag{3.13}$$

and note that the boundary terms vanish, since $e^{(ik-s)y} - 1 = O(y)$ as $y \to 0$. The characteristic function of a gamma pdf is

$$\int_0^\infty e^{iky}\frac{b^a}{\Gamma(a)}y^{a-1}e^{-by}dy = \left(1 - \frac{ik}{b}\right)^{-a}$$

for $a > 0$ and $b > 0$. Set $a - 1 = -\alpha$ and $b = s$ to see that

$$I_s(\alpha) = (ik - s)\frac{\Gamma(1-\alpha)}{s^{1-\alpha}}\left(1 - \frac{ik}{s}\right)^{\alpha-1} = -\Gamma(1-\alpha)(s - ik)^\alpha$$

for $s > 0$. Apply the dominated convergence theorem to see that $I_s(\alpha) \to I(\alpha)$ as $s \to 0$. This shows that

$$I(\alpha) = -\Gamma(1-\alpha)(-ik)^\alpha \tag{3.14}$$

and then (3.12) follows. □

The FT of this stable law is $\mathbb{E}[e^{-ikY}] = \hat{\mu}(-k) = \exp[-C\Gamma(1-\alpha)(ik)^\alpha]$. Given any infinitely divisible law μ with characteristic function $\hat{\mu}(k) = e^{\psi(k)}$, we can define a *Lévy process* Z_t such that $\mathbb{E}[e^{ikZ_t}] = e^{t\psi(k)}$ for all $t \geq 0$. A Lévy process Z_t is infinitely divisible, with $Z_0 = 0$, $Z_{t+s} - Z_t \simeq Z_s$ for all $s, t > 0$ (stationary increments), and Z_t independent of $Z_{t+s} - Z_t$ for all $s, t > 0$ (independent increments). See Section 4.3 for more details. Note that $Z_t \simeq [ta, tb, t\phi]$ since

$$t\psi(k) = ikta - \tfrac{1}{2}k^2tb + \int\left(e^{iky} - 1 - \frac{iky}{1+y^2}\right)t\phi(dy).$$

Taking μ as above, the stable Lévy process Z_t has FT $\hat{p}(k,t) = \mathbb{E}[e^{-ikZ_t}] = e^{-Dt(ik)^\alpha}$ where $D = C\Gamma(1-\alpha) > 0$. Then

$$\frac{d}{dt}\hat{p}(k,t) = -D(ik)^\alpha\hat{p}(k,t).$$

Invert the FT to see that $p(x,t)$ solves the fractional diffusion equation

$$\frac{\partial}{\partial t}p(x,t) = -D\frac{\partial^\alpha}{\partial x^\alpha}p(x,t).$$

Note that in this case $(0 < \alpha < 1)$ there is a minus sign on the right-hand side.

Example 3.11. Now suppose that $Y \simeq \mu$ is a one-sided stable law stable with index $1 < \alpha < 2$. Since

$$\int_{|y|>R}|y|\phi(dy) = \int_R^\infty C\alpha y^{-\alpha}dy = \frac{C\alpha}{\alpha-1}R^{1-\alpha}$$

is finite, we can use Theorem 3.8 (b) to write

$$\hat{\mu}(k) = e^{\psi_2(k)} = \exp\left[ika_2 + \int_0^\infty \left(e^{iky} - 1 - iky\right) C\alpha y^{-\alpha-1} dy\right]. \tag{3.15}$$

Proposition 3.12. *When $1 < \alpha < 2$, the stable characteristic function* (3.15) *with $a_2 = 0$ can be written in the form*

$$\hat{\mu}(k) = \mathbb{E}[e^{ikY}] = \exp\left[C\frac{\Gamma(2-\alpha)}{\alpha - 1}(-ik)^\alpha\right]. \tag{3.16}$$

Proof. The proof is similar to Proposition 3.10. Write

$$J(\alpha) = \int_0^\infty \left(e^{iky} - 1 - iky\right) \alpha y^{-\alpha-1} dy$$

and

$$J_s(\alpha) = \int_0^\infty \left(e^{(ik-s)y} - 1 - (ik-s)y\right) \alpha y^{-\alpha-1} dy$$

for $s > 0$. Integrate by parts with $u = e^{(ik-s)y} - 1 - (ik-s)y$ to see that the boundary terms vanish (see details) and

$$\begin{aligned} J_s(\alpha) &= (ik-s)\int_0^\infty \left(e^{(ik-s)y} - 1\right) y^{-\alpha} dy \\ &= \frac{ik-s}{\alpha-1}\int_0^\infty \left(e^{(ik-s)y} - 1\right)(\alpha-1) y^{-(\alpha-1)-1} dy \end{aligned} \tag{3.17}$$

where $0 < \alpha - 1 < 1$. Then we can apply the calculation in the proof of Proposition 3.10 to see that

$$\begin{aligned} J_s(\alpha) &= \frac{ik-s}{\alpha-1} I_s(\alpha-1) \\ &= \frac{ik-s}{\alpha-1}\left[-\Gamma(1-(\alpha-1))(s-ik)^{\alpha-1}\right] = \frac{\Gamma(2-\alpha)}{\alpha-1}(s-ik)^\alpha \end{aligned}$$

for $s > 0$. Then dominated convergence theorem implies

$$J_s(\alpha) \to J(\alpha) = \frac{\Gamma(2-\alpha)}{\alpha-1}(-ik)^\alpha$$

as $s \to 0$. □

Taking μ as above, the stable Lévy process Z_t with $Z_1 \simeq \mu$ has FT $\hat{p}(k,t) = \mathbb{E}[e^{-ikZ_t}] = e^{Dt(ik)^\alpha}$ where $D = C\Gamma(2-\alpha)/(\alpha-1) > 0$. Then

$$\frac{d}{dt}\hat{p}(k,t) = D(ik)^\alpha \hat{p}(k,t)$$

which leads to the fractional diffusion equation

$$\frac{\partial}{\partial t}p(x,t) = D\frac{\partial^\alpha}{\partial x^\alpha}p(x,t).$$

Note that in this case ($1 < \alpha < 2$) there is no minus sign on the right-hand side.

Details

The Dominated Convergence Theorem (DCT) (e.g., see Rudin [181, Theorem 11.32]) states that if $f_n(y) \to f(y)$ for all y and if $|f_n(y)| \le g(y)$ for all n and all y, where $\int g(y)\,dy$ exists, then $\int f_n(y)\,dy \to \int f(y)\,dy$ and these integrals exist. Write

$$\begin{aligned} I_s(\alpha) &= \int_0^\infty \left(e^{(ik-s)y} - 1\right)\alpha y^{-\alpha-1}dy \\ &= \int_0^\infty (e^{-sy}\cos(ky) - 1)\,\alpha y^{-\alpha-1}dy + i\int_0^\infty (e^{-sy}\sin(ky))\,\alpha y^{-\alpha-1}dy. \end{aligned}$$

Since $|e^{(ik-s)y} - 1| \le 2$, both integrands are bounded by $C_1 y^{-\alpha-1}$ for all $y > 0$. To establish an integrable bound near zero, apply the mean value theorem on $[0, y]$ for $0 < y < 1$ to get

$$|e^{-sy}\cos(ky) - 1| \le e^{-sy}|s\cos(ky) + k\sin(ky)|y \le (|k| + s)y$$

Since $s \to 0$, eventually $s < 1$, and with $C_2 = |k| + 1$, $|e^{-sy}\cos(ky) - 1| \le C_2 y$. Note that k is fixed in this argument. Similarly

$$|e^{-sy}\sin(ky)| \le C_2 y,$$

so both integrands are also bounded by $C_2 y\,\alpha y^{-\alpha-1} = C_3 y^{-\alpha}$ for $0 < y < 1$. Define

$$g(y) = \begin{cases} C_3 y^{-\alpha} & \text{for } 0 < y < 1, \text{ and} \\ C_1 y^{-\alpha-1} & \text{for } y \ge 1. \end{cases}$$

Then $\int_0^\infty g(y)\,dy$ exists, and the dominated convergence theorem applies to the real and imaginary parts of the integral, which shows that $I_s(\alpha) \to I(\alpha)$. It is also possible to apply the DCT directly to the complex-valued integrand.

A similar bound shows that the boundary terms in (3.13) vanish, since:

$$\begin{aligned} \left|\left(e^{(ik-s)y} - 1\right)(-y^{-\alpha})\right| &\le 2y^{-\alpha} \to 0 \quad \text{as } y \to \infty; \text{ and} \\ \left|\left(e^{(ik-s)y} - 1\right)(-y^{-\alpha})\right| &\le 2C_2 y^{1-\alpha} \to 0 \quad \text{as } y \to 0. \end{aligned}$$

The boundary terms in (3.17) vanish since, for fixed $s > 0$ and $k \in \mathbb{R}$,

$$\begin{aligned} \left|\left(e^{(ik-s)y} - 1 - (ik-s)y\right)(-y^{-\alpha})\right| &\le C_4 y^{1-\alpha} \to 0 \quad \text{as } y \to \infty; \text{ and} \\ \left|\left(e^{(ik-s)y} - 1 - (ik-s)y\right)(-y^{-\alpha})\right| &\le C_5 y^{2-\alpha} \to 0 \quad \text{as } y \to 0. \end{aligned}$$

3.3 Semigroups

The theory of semigroups allows an elegant treatment of fractional diffusion equations as ordinary differential equations on a space of functions. It also explains the generator form of the fractional derivative. A semigroup is a family of linear operators on a Banach space. A *Banach space* $\mathbb{B}$ is a complete normed vector space. That is, if $f_n \in \mathbb{B}$ is a Cauchy sequence in this vector space, such that $\|f_n - f_m\| \to 0$ as $m, n \to \infty$, then there exists some $f \in \mathbb{B}$ such that $\|f_n - f\| \to 0$ as $n \to \infty$ in the Banach space norm. In this section, we will use some basic results on semigroups. For more on the general theory of semigroups, see [8, 90, 165].

Example 3.13. The Banach space $\mathbb{B} = C(\mathbb{R})$ consists of bounded continuous functions $f : \mathbb{R} \to \mathbb{R}$ with the norm $\|f\| = \sup\{|f(x)| : x \in \mathbb{R}\}$. The space $\mathbb{B} = C_0(\mathbb{R})$ consists of continuous functions with $f(x) \to 0$ as $|x| \to \infty$, with the same norm.

Example 3.14. The Banach space L^2 consists of finite variance random variables X with the norm $\|X\| = \sqrt{\mathbb{E}[X^2]}$. We will use this space in the proofs of Section 7.9. Some authors write $L^2(\Omega, P)$ to emphasize that this is a space of random variables on the sample space Ω with probability measure P.

Example 3.15. The Banach space $L^p(\mathbb{R})$ consists of functions $f : \mathbb{R} \to \mathbb{R}$ such that $\int |f(x)|^p\,dx < \infty$, with the norm $\|f\|_p = (\int |f(x)|^p\,dx)^{1/p}$ for $0 < p < \infty$. The most common choices are $p = 1$ and $p = 2$. The Sobolev space $W^{k,p}(\mathbb{R})$ consists of all functions such that f and all of its derivatives $f^{(j)}$ up to order k exist and are in $L^p(\mathbb{R})$, with the norm

$$\|f\|_{k,p} = \left(\sum_{j=0}^{k} \|f^{(j)}\|_p^p\right)^{1/p}.$$

A family of linear operators $\{T_t : t \geq 0\}$ on a Banach space $\mathbb{B}$ is called a *semigroup* if $T_0 f = f$ for all $f \in \mathbb{B}$, and $T_{t+s} = T_t T_s$ (the composition of these two operators). We say that T_t is *bounded* if, for each $t \geq 0$, there exists some $M_t > 0$ such that $\|T_t f\| \leq M_t \|f\|$ for all $f \in \mathbb{B}$. We say that T_t is *strongly continuous* if $\|T_t f - f\| \to 0$ for all $f \in \mathbb{B}$. A strongly continuous, bounded semigroup is also called a C_0 semigroup.

The *generator* of the semigroup T_t is a linear operator defined by

$$Lf(x) = \lim_{t \to 0} \frac{T_t f(x) - T_0 f(x)}{t - 0}. \tag{3.18}$$

This is the abstract derivative of the semigroup evaluated at $t = 0$. Note that the limit in (3.18) is taken in the Banach space norm. For example, when $\mathbb{B} = C_0(\mathbb{R})$ we require that

$$\sup_{x \in \mathbb{R}} \left| \frac{T_t f(x) - T_0 f(x)}{t - 0} - Lf(x) \right| \to 0 \quad \text{as } t \to 0, \tag{3.19}$$

and then (3.18) also holds for each $x \in \mathbb{R}$. If T_t is a C_0 semigroup, then the generator (3.18) exists, and its domain

$$\mathrm{Dom}(L) = \{f \in \mathbb{B} : Lf \text{ exists}\}$$

is a dense subset of $\mathbb{B}$, i.e., for any $f \in \mathbb{B}$ there exists a sequence $f_n \in \mathrm{Dom}(L)$ such that $\|f_n - f\| \to 0$, see Pazy [165, Corollary I.2.5].

Theorem 3.16. *If T_t is a C_0 semigroup on the Banach space $\mathbb{B}$, then the function $q(t) = T_t f$ solves the Cauchy problem*

$$\frac{d}{dt}q = Lq; \; q(0) = f \tag{3.20}$$

for any $f \in \mathrm{Dom}(L)$.

Proof. See, for example, Pazy [165, Theorem I.2.4]. □

In our applications, the Banach space $\mathbb{B}$ is typically a space of functions, like $C_0(\mathbb{R})$ or $L^p(\mathbb{R})$. Then we can write Theorem 3.16 in a more concrete form: If T_t is a C_0 semigroup on the Banach space of functions $\mathbb{B}$, then $q(x, t) = T_t f(x)$ solves the Cauchy problem

$$\frac{\partial}{\partial t}q(x,t) = Lq(x,t); \quad q(x,0) = f(x), \tag{3.21}$$

for any $f \in \mathrm{Dom}(L)$. If $L = \partial^2/\partial x^2$, then (3.21) is the diffusion equation, and (3.20) represents this partial differential equation as an ordinary differential equation on some suitable space of functions.

Given a Lévy process $\{Z_t : t \geq 0\}$, we define a family of linear operators

$$T_t f(x) = \mathbb{E}[f(x - Z_t)] \tag{3.22}$$

for $t \geq 0$, for suitable functions $f(x)$. The next result shows that (3.22) defines a C_0 semigroup on the Banach space $C_0(\mathbb{R})$, and gives an explicit form of the generator in terms of the Lévy representation.

Theorem 3.17. *Suppose that Z_t is a Lévy process, and that $\mathbb{E}[e^{ikZ_1}] = e^{\psi(k)}$ where $\psi(k)$ is given by (3.4). Then (3.22) defines a C_0 semigroup on $C_0(\mathbb{R})$ with generator*

$$Lf(x) = -af'(x) + \tfrac{1}{2}bf''(x) + \int \left(f(x-y) - f(x) + \frac{yf'(x)}{1+y^2} \right) \phi(dy). \tag{3.23}$$

The domain $\mathrm{Dom}(L)$ contains all f such that $f, f', f'' \in C_0(\mathbb{R})$. If we also have $f, f', f'' \in L^1(\mathbb{R})$, then $\psi(-k)\hat{f}(k)$ is the FT of $Lf(x)$.

Proof. See Sato [187, Theorem 31.5] for the proof that (3.22) defines a C_0 semigroup on $C_0(\mathbb{R})$ with generator (3.23). Hille and Phillips [90, Theorem 23.14.2] proved that $Lf(x)$ has FT $\psi(-k)\hat{f}(k)$ when $f, f', f'' \in L^1(\mathbb{R})$. □

Remark 3.18. In this remark, we sketch the main ideas in the proof of Theorem 3.17. Strong continuity of the semigroup (3.22) on $C_0(\mathbb{R})$ follows from the fact that $Z_t \Rightarrow Z_0 = 0$. The *semigroup property* $T_{t+s} = T_t T_s$ follows from the fact that Z_t has stationary independent increments (see details). The generator formula (3.23) comes from a FT inversion: Suppose $\{Z_t : t \geq 0\}$ is a Lévy process with FT $\hat{p}(k, t) = \mathbb{E}[e^{-ikZ_t}] = \exp(t\psi(-k))$. If Z_t has pdf $p(x, t)$, then we have

$$T_t f(x) = \int f(x - y)p(y, t)\, dy \tag{3.24}$$

a convolution of the two functions. We define the convolution

$$f * g(x) = \int f(x - y)g(y)\, dy$$

and we note that the FT converts convolutions to products: The FT of $f*g$ is $\mathcal{F}[f*g](k) = \hat{f}(k)\hat{g}(k)$ (see details). If $f(x)$ is a probability density, and if $X \simeq f(x)$ is independent of Z_t, then $X + Z_t \simeq T_t f(x)$, since the pdf of a sum of independent random variables is a convolution of their respective densities. We can think of X as the initial particle location, with pdf $f(x)$. Then $T_t f(x)$ is the pdf of particle location at time $t \geq 0$, with $T_0 f(x) = f(x)$. Since the FT of a convolution is a product, it follows from (3.22) that $T_t f(x)$ has FT $e^{t\psi(-k)}\hat{f}(k)$. Then for suitable functions f we can pass the FT inside the limit and write

$$\begin{aligned}\mathcal{F}[Lf](k) &= \lim_{t\to 0} \frac{e^{t\psi(-k)}\hat{f}(k) - \hat{f}(k)}{t - 0} \\ &= \left[\lim_{t\to 0} \frac{[1 + t\psi(-k) + \frac{1}{2}t^2\psi(-k)^2 + \cdots] - 1}{t}\right]\hat{f}(k) = \psi(-k)\hat{f}(k).\end{aligned}$$

We call $\psi(-k)$ the *Fourier symbol* of the generator L. Use the Lévy representation (3.5) to write

$$\psi(-k)\hat{f}(k) = -a(ik)\hat{f}(k) + \tfrac{1}{2}(ik)^2 b\hat{f}(k) + \int\left(e^{-iky} - 1 + \frac{iky}{1 + y^2}\right)\hat{f}(k)\phi(dy).$$

Then invert this FT using the fact that

$$\int e^{-ikx} f(x - y)\, dx = e^{-iky}\hat{f}(k) \tag{3.25}$$

to arrive at (3.23). The condition $f, f', f'' \in L^1(\mathbb{R})$ is required to show that the FT of $Lf(x)$ exists.

Remark 3.19. In this remark, we outline the main idea behind the proof of Theorem 3.16, for the special case of an infinitely divisible semigroup. Take FT in (3.22) to get

$$\hat{q}(k, t) = e^{t\psi(-k)}\hat{f}(k); \quad \hat{q}(k, 0) = \hat{f}(k).$$

Compute

$$\frac{\partial}{\partial t}\hat{q}(k,t) = \psi(-k)\hat{q}(k,t)$$

and invert the FT to arrive at (3.21). Note that the domain $\mathrm{Dom}(L)$ of the generator (3.23) on the space $L^1(\mathbb{R})$ consists of all functions $f \in L^1(\mathbb{R})$ such that $\hat{h}(k) = \psi(-k)\hat{f}(k)$ is the FT of some function $h \in L^1(\mathbb{R})$, see Baeumer and Meerschaert [18, Theorem 2.2].

Now we illustrate the semigroup machinery with some familiar examples.

Example 3.20. If $Z_t \simeq \mathcal{N}(0, 2Dt)$ then

$$\hat{\mu}_t(k) = e^{-tDk^2} = e^{t\psi(k)}$$

with Fourier symbol $\psi(-k) = D(ik)^2$. The generator can be obtained by inverting $\psi(-k)\hat{f}(k) = D(ik)^2\hat{f}(k)$, so that $L = D\partial^2/\partial x^2$ in this case. The Cauchy problem is:

$$\frac{\partial}{\partial t}q(x,t) = D\frac{\partial^2}{\partial x^2}q(x,t); \quad q(x,0) = f(x).$$

Its solution is

$$q(x,t) = T_t f(x) = \int_{-\infty}^{\infty} f(x-y)p(y,t)\,dy$$

where

$$p(y,t) = \frac{1}{\sqrt{4\pi Dt}}\exp\left(-\frac{y^2}{4Dt}\right).$$

If the initial particle location is a random variable X with pdf $f(x)$, independent of Z_t, then the Brownian motion with a random initial location $X + Z_t$ has pdf $T_t f(x)$. This is a *Markov process*: The pdf of the displacement $(X + Z_{t+s}) - (X + Z_t) = Z_{t+s} - Z_t$ is independent of the past history of the process $\{Z_u : 0 \le u \le t\}$.

Example 3.21. If $Z_t = tv$ for some constant velocity v then

$$T_t f(x) = \mathbb{E}[f(x - Z_t)] = f(x - vt),$$

the *shift semigroup*. Its generator is

$$Lf(x) = \lim_{t\to 0}\frac{f(x-vt) - f(x)}{t-0}\frac{v}{v} = -vf'(x).$$

Here $\hat{\mu}_t(k) = \mathbb{E}[e^{ikvt}] = e^{t\psi(k)}$ so that $\psi(k) = ikv$, and then $\psi(-k) = -v(ik)$, so that $L = -v\,\partial/\partial x$. It is easy to check that $q(x,t) = f(x - vt)$ solves $\partial q/\partial t = -v\partial q/\partial x$.

Example 3.22. If $Z_t \simeq \mathcal{N}(vt, \sigma^2 t)$ is a Brownian motion with drift, take $a = v$, $b = \sigma^2$, and $\phi = 0$ in (3.23) to see that the density $q(x,t)$ of $X + Z_t$ solves

$$\frac{\partial}{\partial t}q(x,t) = -v\frac{\partial}{\partial x}q(x,t) + \frac{1}{2}\sigma^2\frac{\partial^2}{\partial x^2}q(x,t) = Lq(x,t)$$

with initial condition $q(x, 0) = f(x)$. This diffusion equation with drift comes from the sum of two semigroups. The semigroups commute, so the generators add.

Theorem 3.17 gives an explicit form for the generator of an infinitely divisible semigroup. Now we apply this result to explain the generator form of the fractional derivative. In order to apply Theorem 3.17 to stable semigroups, it is convenient to develop some alternative forms of the generator. The next result is the semigroup analogue of Theorem 3.8.

Theorem 3.23. *Suppose that Z_t is a Lévy process, and that $\mathbb{E}[e^{ikZ_1}] = e^{\psi(k)}$ where $\psi(k)$ is given by* (3.4)*. Then we can also write the generator* (3.23) *in the form*

$$Lf(x) = -a_0 f'(x) + \tfrac{1}{2} b f''(x) + \int \Big(f(x-y) - f(x) + yf'(x) I(|y| \le R)\Big)\, \phi(dy) \tag{3.26}$$

for any $R > 0$, for some unique a_0 depending on R and a. Furthermore:

(a) If (3.6) *holds, then we can also write*

$$Lf(x) = -a_1 f'(x) + \tfrac{1}{2} b f''(x) + \int (f(x-y) - f(x))\, \phi(dy) \tag{3.27}$$

for some unique a_1 depending on a_0; and

(b) If (3.8) *holds, then we can also write*

$$Lf(x) = -a_2 f'(x) + \tfrac{1}{2} b f''(x) + \int \Big(f(x-y) - f(x) + yf'(x)\Big)\, \phi(dy) \tag{3.28}$$

for some unique a_2 depending on a_0.

Proof. The proof is very similar to Theorem 3.8. Since the integral

$$\delta_0 = \int \left(\frac{y}{1+y^2} - yI(|y| \le R) \right) \phi(dy)$$

exists, we can take $a_0 = a - \delta_0$. If (3.6) holds, take

$$a_1 = a_0 - \int_{0<|y|\le R} y\phi(dy).$$

If (3.8) holds, take

$$a_2 = a_0 + \int_{|y|>R} y\phi(dy).$$

□

Example 3.24. Let Z_t be a stable Lévy process with index $0 < \alpha < 1$, such that Z_1 has the one-sided stable characteristic function (3.11) with $a_1 = 0$. Then it follows from (3.27) that the generator of this semigroup is

$$Lf(x) = \int_0^\infty (f(x-y) - f(x))\, C\alpha y^{-\alpha-1} dy.$$

Proposition 3.10 shows that $\psi(-k) = -C\Gamma(1-\alpha)(ik)^\alpha$ is the Fourier symbol of this one-sided stable semigroup. If we take $C = 1/\Gamma(1-\alpha)$, then this shows that $L = -\partial^\alpha/\partial x^\alpha$, using the generator form (2.17) of the fractional derivative for $0 < \alpha < 1$. Note the minus sign in the generator in this case. A result of Hille and Phillips [90, Theorem 23.15.2] implies that this generator exists for all $f \in L^1(0,\infty)$ such that $f(0) = 0$ and $f' \in L^1(0,\infty)$. This strengthens the result in Proposition 2.1, since it implies that the fractional derivative $d^\alpha f/dx^\alpha$ of order $0 < \alpha < 1$ exists whenever the first derivative f' exists.

Example 3.25. Let Z_t be a stable Lévy process with index $1 < \alpha < 2$, such that Z_1 has the one-sided stable characteristic function (3.15) with $a_2 = 0$. Then it follows from (3.28) that the generator of this semigroup is

$$Lf(x) = \int_0^\infty \left(f(x-y) - f(x) + yf'(x)\right) C\alpha y^{-\alpha-1}\,dy.$$

Proposition 3.12 shows that

$$\psi(-k) = C\frac{\Gamma(2-\alpha)}{\alpha-1}(ik)^\alpha$$

is the Fourier symbol of this one-sided stable semigroup. If we take $C = (\alpha-1)/\Gamma(2-\alpha)$, then this shows that $L = \partial^\alpha/\partial x^\alpha$, using the generator form (2.20) of the fractional derivative of order $1 < \alpha < 2$. Note the positive sign in the generator in this case. Theorem 3.17 shows that this fractional derivative exists when $f, f', f'' \in C_0(\mathbb{R})$, which strengthens the result in Proposition 2.1, since it implies that the fractional derivative $d^\alpha f/dx^\alpha$ of order $1 < \alpha < 2$ exists whenever the second derivative f'' exists.

Details

A substitution $z = x - y$ shows that the FT of $f * g$ is

$$\begin{aligned}\int_{-\infty}^\infty e^{-ikx}\left(\int_{-\infty}^\infty f(x-y)g(y)\,dy\right)dx &= \int_{-\infty}^\infty e^{-ik(z+y)}\int_{-\infty}^\infty f(z)g(y)\,dy\,dz\\ &= \int_{-\infty}^\infty e^{-ikz}f(z)\,dz\int_{-\infty}^\infty e^{-iky}g(y)\,dy\\ &= \hat{f}(k)\hat{g}(k).\end{aligned}$$

The proof of the semigroup property $T_{t+s} = T_tT_s$ for (3.22) uses a conditioning argument. Since Z_t is a Lévy process, $(Z_{t+s} - Z_t) \simeq Z_s$, and $(Z_{t+s} - Z_t)$ is independent of Z_t. Then we can write

$$\begin{aligned} T_{t+s}f(x) &= \mathbb{E}[f(x - Z_{t+s})] \\ &= \mathbb{E}\big[\mathbb{E}[f(x - \{(Z_{t+s} - Z_t) + Z_t\})|Z_t]\big] \\ &= \int \mathbb{E}[f(x - \{Z_s + y\})]p(y,t)\,dy \\ &= \int \mathbb{E}[f(x - y - Z_s)]p(y,t)\,dy \\ &= \int T_s f(x-y)p(y,t)\,dy \\ &= \mathbb{E}[T_s f(x - Z_t)] = T_t T_s f(x). \end{aligned} \tag{3.29}$$

This is a special case of the *Chapman-Kolmogorov equation* for Markov processes.

3.4 Poisson approximation

In order to motivate the proof of the extended central limit theorem, by the method of triangular arrays, we show here how the stable laws emerge as weak limits of compound Poisson random variables with Pareto (power law) jumps.

Example 3.26. Suppose that Y is a one-sided stable random variable with characteristic function $\hat{\mu}(k) = e^{\psi(k)}$, where

$$\psi(k) = -C\Gamma(1-\alpha)(-ik)^\alpha = \int_0^\infty \left(e^{iky} - 1\right) C\alpha y^{-\alpha-1}\,dy$$

for some $0 < \alpha < 1$, using Proposition 3.10. We will approximate Y by an infinitely divisible random variable Y_n with characteristic function $\mathbb{E}[e^{ikY_n}] = e^{\psi_n(k)}$ where

$$\psi_n(k) = \int_{1/n}^\infty \left(e^{iky} - 1\right) C\alpha y^{-\alpha-1}\,dy.$$

Define

$$\lambda_n = \int_{1/n}^\infty C\alpha y^{-\alpha-1}\,dy = \Big[-Cy^{-\alpha}\Big]_{1/n}^\infty = Cn^\alpha$$

and write

$$\psi_n(k) = \lambda_n \int \left(e^{iky} - 1\right) \omega_n(dy)$$

where

$$\omega_n(dy) = \lambda_n^{-1} C\alpha y^{-\alpha-1} I(y > 1/n)\,dy = n^{-\alpha}\alpha y^{-\alpha-1} I(y > 1/n)\,dy$$

is a probability measure. This is a special case of the *Pareto distribution*, originally invented to model the distribution of incomes. The general Pareto distribution can be defined by setting $\mathbb{P}[X > x] = Cx^{-\alpha}$ for $x > C^{1/\alpha}$ where C, α are positive constants.

Take (W_n) iid with distribution ω_n so that

$$\mathbb{P}[W_n > x] = \int_x^\infty n^{-\alpha}\alpha y^{-\alpha-1}dy = \left[n^{-\alpha}(-y^{-\alpha})\right]_x^\infty = Ax^{-\alpha}$$

for all $x > A^{1/\alpha} = 1/n$. Write

$$\psi_n(k) = \lambda_n \int \left(e^{iky} - 1\right)\omega_n(dy) = \lambda_n[\hat{\omega}_n(dy) - 1]$$

to see that Y_n is compound Poisson, in view of Example 3.3. In fact $Y_n \simeq W_1 + \cdots + W_N$ where (W_n) are iid Pareto with $\mathbb{P}[W_n > x] = Ax^{-\alpha}$ and N is Poisson with mean $\lambda_n = Cn^\alpha$ independent of (W_n). Since the integral $\psi(k)$ exists, we certainly have $\psi_n(k) \to \psi(k)$ for each fixed $k \in \mathbb{R}$, and then $\hat{\mu}_n(k) = e^{\psi_n(k)} \to e^{\psi(k)} = \hat{\mu}(k)$ as $n \to \infty$. This proves that $Y_n \Rightarrow Y$.

Hence a stable law is essentially a compound Poisson with power law jumps. The mean number of jumps $\lambda_n = Cn^\alpha \to \infty$ as the minimum jump size $1/n \to 0$, so that the jump intensity $\phi_n(dy) = \lambda_n\omega_n(dy)$ increases without bound to the Lévy measure ϕ of the stable law. This means that the stable law represents the accumulation of an infinite number of power law jumps. For any n, it combines a finite number of jumps of size greater than $1/n$ with an infinite number of jumps of size less than $1/n$.

We now define the general *two-sided stable law* μ with *index* $0 < \alpha < 2$ to be an infinitely divisible law with Lévy representation $[a, 0, \phi]$, where $a \in \mathbb{R}$ and

$$\phi(dy) = \begin{cases} pC\alpha y^{-\alpha-1}dy & \text{for } y > 0\text{, and} \\ qC\alpha|y|^{-\alpha-1}dy & \text{for } y < 0. \end{cases} \tag{3.30}$$

where $p, q \geq 0$ with $p + q = 1$. This is a Lévy measure since

$$\phi\{y : |y| > R\} = CR^{-\alpha} \quad \text{and} \quad \int_{0<|y|\leq R} y^2\phi(dy) = \frac{C\alpha}{2-\alpha}R^{2-\alpha}$$

are both finite for any $R > 0$.

Example 3.27. Consider a two-sided stable random variable Y with index $0 < \alpha < 1$. Since

$$\int_{0<|y|\leq R} |y|\phi(dy) = \int_0^R C\alpha y^{-\alpha}dy = \frac{C\alpha}{1-\alpha}R^{1-\alpha}$$

is finite, we can apply Theorem 3.8 (a). Suppose that Y is *centered* so that $a_1 = 0$ in (3.7). Then we can write $\hat{\mu}(k) = \mathbb{E}[e^{ikY}] = e^{\psi(k)}$ where

$$\begin{aligned}\psi(k) &= \int \left(e^{iky} - 1\right) \phi(dy)\\ &= p \int_0^\infty \left(e^{iky} - 1\right) C\alpha y^{-\alpha-1} dy + q \int_{-\infty}^0 \left(e^{iky} - 1\right) C\alpha(-y)^{-\alpha-1} dy\\ &= -pC\Gamma(1-\alpha)(-ik)^\alpha + q \int_0^\infty \left(e^{-ikx} - 1\right) C\alpha x^{-\alpha-1} dx\\ &= -pC\Gamma(1-\alpha)(-ik)^\alpha - qC\Gamma(1-\alpha)(ik)^\alpha \end{aligned} \tag{3.31}$$

using a substitution $x = -y$ and Proposition 3.10.

Define $Y_n \simeq \mu_n$ where $\hat{\mu}_n(k) = e^{\psi_n(k)}$ with

$$\psi_n(k) = \int_{|y|>1/n} \left(e^{iky} - 1\right) \phi(dy).$$

Let

$$\lambda_n = \int_{|y|>1/n} \phi(dy) = Cn^\alpha$$

and $\omega_n(dy) = \lambda_n^{-1} I(|y| > 1/n)\, \phi(dy)$. Take $(W_n) \simeq \omega_n$ iid so that

$$\mathbb{P}[W_n > x] = pAx^{-\alpha} \quad \text{and} \quad \mathbb{P}[W_n < -x] = qAx^{-\alpha}$$

for all $x > A^{1/\alpha} = 1/n$. Then $Y_n \simeq W_1 + \cdots + W_N$ where N is Poisson with mean $\lambda_n = Cn^\alpha$ independent of (W_n). Again we have $Y_n \Rightarrow Y$, which shows that the two-sided stable is also the accumulation of power law jumps, including a finite number of jumps larger than $1/n$ and an infinite number of very small jumps. The constants p and q balance the positive and negative jumps.

The two-sided stable law decomposes into independent positive and negative parts: Use (3.31) to write $\psi(k) = p\psi_+(k) + q\psi_-(k)$ where

$$\begin{aligned}\psi_+(k) &= \int_0^\infty \left(e^{iky} - 1\right) C\alpha y^{-\alpha-1} dy = -C\Gamma(1-\alpha)(-ik)^\alpha,\\ \psi_-(k) &= \int_{-\infty}^0 \left(e^{iky} - 1\right) C\alpha |y|^{-\alpha-1} dy = -C\Gamma(1-\alpha)(ik)^\alpha.\end{aligned}$$

Then $\hat{\mu}(k) = e^{\psi(k)} = e^{p\psi_+(k)} e^{q\psi_-(k)}$ which shows that $Y \simeq Y^+ + Y^-$ a sum of two independent stable laws. We can also write $Y_n \simeq Y_n^+ + Y_n^-$ a sum of two independent compound Poisson, the first with only positive jumps, and the second with only negative jumps.

The generator form of the negative fractional derivative comes inverting the FT for the symbol $\psi_-(-k)$: Use the fact that $\int e^{-ikx}f(x+y)\,dx = e^{iky}\hat{f}(k)$ to see that

$$\psi_-(-k)\hat{f}(k) = \int_0^\infty \left(e^{iky}\hat{f}(k) - \hat{f}(k)\right) C\alpha y^{-\alpha-1}dy$$

is the FT of

$$\int_0^\infty (f(x+y) - f(x))\, C\alpha y^{-\alpha-1}dy.$$

Take $C = 1/\Gamma(1-\alpha)$ to get

$$\begin{aligned}\frac{d^\alpha f(x)}{d(-x)^\alpha} &= \mathcal{F}^{-1}[(-ik)^\alpha \hat{f}(k)]\\ &= \mathcal{F}^{-1}[-\psi_-(-k)\hat{f}(k)]\\ &= \frac{\alpha}{\Gamma(1-\alpha)}\int_0^\infty (f(x) - f(x+y))\, y^{-\alpha-1}dy \end{aligned} \tag{3.32}$$

for $0 < \alpha < 1$. Formula (3.32) also follows from (3.27) and a simple change of variables.

Suppose that Z_t is a two-sided stable Lévy motion with $Z_1 \simeq Y$. Then $\hat{p}(k,t) = \mathbb{E}[e^{-ikZ_t}] = e^{t\psi(-k)}$ with $\psi(-k) = -pD(ik)^\alpha - qD(-ik)^\alpha$ and $D = C\Gamma(1-\alpha) > 0$. Then

$$\frac{d\hat{p}(k,t)}{dt} = \psi(-k)\hat{p}(k,t) = -pD(ik)^\alpha \hat{p}(k,t) - qD(-ik)^\alpha \hat{p}(k,t)$$

which inverts to the two-sided fractional diffusion equation

$$\frac{\partial p(x,t)}{\partial t} = -pD\frac{\partial^\alpha p(x,t)}{\partial x^\alpha} - qD\frac{\partial^\alpha p(x,t)}{\partial(-x)^\alpha}$$

for $0 < \alpha < 1$. The positive fractional derivative codes positive power law jumps, and the negative fractional derivative corresponds to the negative power law jumps.

3.5 Shifted Poisson approximation

Here we develop the Poisson approximation for stable laws with index $1 < \alpha < 2$. In this case, the Poisson approximation involves a shift.

Example 3.28. Suppose that Y is one-sided stable with characteristic function $\hat{\mu}(k) = \mathbb{E}[e^{ikY}] = e^{\psi(k)}$ where

$$\begin{aligned}\psi(k) &= \int \left(e^{iky} - 1 - iky\right)\phi(dy)\\ &= \int_0^\infty \left(e^{iky} - 1 - iky\right) C\alpha y^{-\alpha-1}dy\\ &= C\frac{\Gamma(2-\alpha)}{\alpha-1}(-ik)^\alpha\end{aligned}$$

for $1 < \alpha < 2$, using Proposition 3.12. Let Y_n be infinitely divisible with characteristic function $\hat{\mu}_n(k) = e^{\psi_n(k)}$ where

$$\psi_n(k) = \int_{1/n}^{\infty} \left(e^{iky} - 1 - iky\right) C\alpha y^{-\alpha-1} dy$$
$$= \int \left(e^{iky} - 1 - iky\right) \phi_n(dy)$$

and $\phi_n(dy) = I(y > 1/n)\, \phi(dy)$ is the Lévy measure of this infinitely divisible law. Define

$$\lambda_n = \int \phi_n(dy) = \int_{1/n}^{\infty} C\alpha y^{-\alpha-1} dy = Cn^{\alpha}$$

so that $\omega_n(dy) = \lambda_n^{-1} \phi_n(dy)$ is a probability measure. Take (W_j) iid with distribution ω_n and write

$$\psi_n(k) = \lambda_n \int \left(e^{iky} - 1 - iky\right) \omega_n(dy) = \lambda_n \int \left(e^{iky} - 1\right) \omega_n(dy) - ika_n$$

where $a_n = \lambda_n \int y\, \omega_n(dy) = \lambda_n \mathbb{E}[W_j]$. Here $\mathbb{P}[W_j > x] = Ax^{-\alpha}$ with $A = n^{-\alpha}$ so that

$$\mathbb{E}[W_j] = \int_{1/n}^{\infty} y A\alpha y^{-\alpha-1} dy = \left[A\alpha \frac{y^{1-\alpha}}{1-\alpha}\right]_{1/n}^{\infty} = \frac{A\alpha}{\alpha - 1} n^{\alpha-1}$$

is finite for all n for $1 < \alpha < 2$. Then $\psi_n(k) = \lambda_n[\hat{\omega}(k) - 1] - ika_n$ so Y_n is shifted compound Poisson: Take N Poisson with mean λ_n, independent of (W_j), and note that

$$\exp(\psi_n(k)) = \exp(\lambda_n[\hat{\omega}(k) - 1] - ika_n) = \mathbb{E}\left[\exp\left(ik[W_1 + \cdots + W_N - a_n]\right)\right]$$

so that $Y_n \simeq W_1 + \cdots + W_N - a_n$. Note that

$$\lambda_n = Cn^{\alpha} \to \infty \quad \text{and} \quad a_n = \lambda_n \mathbb{E}[W_j] = \frac{\alpha C}{\alpha - 1} n^{\alpha-1} \to \infty$$

so that both the mean number of jumps and the shift tend to infinity as the truncation threshold $1/n \to 0$. Since $\mathbb{P}[W_j > x] = Ax^{-\alpha}$ the stable random variable Y is essentially the (compensated) sum of power law jumps. The compensator adjusts the random sum of power law jumps to mean zero. As the threshold shrinks to zero, the number of jumps increases to infinity, and their accumulated mean a_n also increases to infinity, but the compensated sum (the shifted compound Poisson) converges to an α-stable limit.

Let Z_t be a stable Lévy process with $Z_1 \simeq Y$. Then $Z_t \simeq [0, 0, t\phi]$ in the alternative Lévy representation (3.9). The Lévy process $Z_t^n \simeq [0, 0, t\phi_n]$ with $Z_1^n \simeq Y_n$ is a compound Poisson process with power law jumps, centered to mean zero. In fact we can write

$$Z_t^n = W_1 + \cdots + W_{N(t)} - ta_n$$

where $N(t)$ is a Poisson process with rate λ_n. The Poisson process $N(t)$ is a Lévy process whose Lévy measure is a point mass $\phi_n\{1\} = \lambda_n$. Then $\mathbb{E}[N(t)] = \lambda_n t$ and a standard conditioning argument shows that the compound Poisson process (a random sum) has mean

$$\mathbb{E}[W_1 + \cdots + W_{N(t)}] = \mathbb{E}[N(t)]\mathbb{E}[W_j] = ta_n.$$

Example 3.29. A general two-sided stable random variable Y with $1 < \alpha < 2$ has Lévy measure (3.30). Then it follows from Proposition 3.12 and a change of variables that $E[e^{ikY}] = e^{\psi(k)}$ where

$$\begin{aligned}\psi(k) &= \int \left(e^{iky} - 1 - iky\right)\phi(dy)\\ &= pC\frac{\Gamma(2-\alpha)}{\alpha-1}(-ik)^\alpha + qC\frac{\Gamma(2-\alpha)}{\alpha-1}(+ik)^\alpha. \end{aligned} \tag{3.33}$$

If Z_t is a stable Lévy motion with $Z_1 \simeq Y$ then $\hat{p}(k,t) = E[e^{-ikZ_t}] = e^{t\psi(-k)}$. Take $D = C\Gamma(2-\alpha)/(\alpha-1)$ and write $\hat{p}(k,t) = \exp[pDt(ik)^\alpha + qDt(-ik)^\alpha]$. Then

$$\frac{d\hat{p}(k,t)}{dt} = \psi(-k)\hat{p}(k,t) = pD(ik)^\alpha\hat{p}(k,t) + qD(-ik)^\alpha\hat{p}(k,t)$$

which inverts to the two-sided fractional diffusion equation

$$\frac{\partial p(x,t)}{\partial t} = pD\frac{\partial^\alpha p(x,t)}{\partial x^\alpha} + qD\frac{\partial^\alpha p(x,t)}{\partial(-x)^\alpha}$$

for $1 < \alpha < 2$. As in the case $0 < \alpha < 1$, the positive fractional derivative comes from the positive power law jumps, and the negative fractional derivative corresponds to the negative jumps.

Example 3.29 illustrates the reason for the positive coefficients in the fractional diffusion equation for $1 < \alpha < 2$, and the negative coefficients for $0 < \alpha < 1$. This comes from the change of sign in the stable characteristic function. One can also note that the log-characteristic function $\psi(k)$ should have a negative real part, since the characteristic function $e^{\psi(k)}$ remains bounded for all real k. Since $(\pm ik)^\alpha$ has a positive real part for $0 < \alpha < 1$, and a negative real part for $1 < \alpha < 2$, the negative sign in the case $0 < \alpha < 1$ is necessary to make the real part of $\psi(k)$ negative.

We have now connected the coefficients α and D in the fractional diffusion equation with the parameters of the Pareto law. The order of the fractional derivative equals the power law index α, and the fractional dispersivity

$$D = \begin{cases} C\Gamma(1-\alpha) & \text{for } 0 < \alpha < 1, \text{ and}\\ C\dfrac{\Gamma(2-\alpha)}{\alpha-1} & \text{for } 1 < \alpha < 2.\end{cases}$$

These relations can be useful for simulating sample paths of a stable Lévy process using the compound Poisson approximation. A histogram of particle locations at time

$t > 0$ will approximate the solution to the corresponding fractional diffusion equation. This is the method of particle tracking, see for example Zhang, Benson, Meerschaert and Scheffler [223].

The two-sided stable law is a sum of independent components, segregating the positive and negative jumps. Write $\psi(k) = p\psi_+(k) + q\psi_-(k)$ where

$$\psi_+(k) = \int_0^\infty \left(e^{iky} - 1 - iky\right) C\alpha y^{-\alpha-1} dy = C\frac{\Gamma(2-\alpha)}{\alpha - 1}(-ik)^\alpha,$$

$$\psi_-(k) = \int_{-\infty}^0 \left(e^{iky} - 1 - iky\right) C\alpha |y|^{-\alpha-1} dy = C\frac{\Gamma(2-\alpha)}{\alpha - 1}(ik)^\alpha.$$

Then $Y \simeq Y^+ + Y^-$ a sum of two independent stable laws. Take $C = (\alpha - 1)/\Gamma(2-\alpha)$ to get $\psi_-(-k) = (-ik)^\alpha$ the Fourier symbol of the negative fractional derivative. Invert the FT to obtain the generator form of the negative fractional derivative in the case $1 < \alpha < 2$: A change of variables shows that

$$\psi_-(k) = \int_0^\infty \left(e^{-iky} - 1 + iky\right) C\alpha y^{-\alpha-1} dy.$$

Use $\int e^{ikx} f(x+y)\, dx = e^{iky}\hat{f}(k)$ to get

$$\begin{aligned}\frac{d^\alpha f(x)}{d(-x)^\alpha} &= \mathcal{F}^{-1}[(-ik)^\alpha \hat{f}(k)] \\ &= \mathcal{F}^{-1}\left[\int_0^\infty \left(e^{iky}\hat{f}(k) - \hat{f}(k) - iky\hat{f}(k)\right) C\alpha y^{-\alpha-1} dy\right] \\ &= \frac{\alpha(\alpha-1)}{\Gamma(2-\alpha)} \int_0^\infty (f(x+y) - f(x) - yf'(x)) y^{-\alpha-1} dy. \end{aligned} \tag{3.34}$$

Note that $f(x+y) = f(x) + yf'(x) + O(y^2)$ by a Taylor series expansion, so that this integral converges at $y = 0$ if f is sufficiently smooth. Formula (3.34) can also be derived from (3.28) by a change of variables.

Details

In the one-sided case, the shifted compound Poisson process $Z_t^n = S(N(t)) - ta_n$ where $S(n) = W_1 + \cdots + W_n$ is a random walk. Since $N(t)$ is Poisson with mean $t\lambda_n$ independent of $S(n)$, the random sum $S(N(t))$ has mean

$$\begin{aligned}\mathbb{E}[S(N(t))] &= \sum_{j=0}^\infty \mathbb{E}[S(j)|N(t) = j]\, \mathbb{P}[N(t) = j] \\ &= \sum_{j=0}^\infty j\,\mathbb{E}[W]\, \mathbb{P}[N(t) = j] \\ &= \mathbb{E}[W] t\lambda_n = ta_n\end{aligned}$$

so that $\mathbb{E}[Z_t^n] = 0$.

In the two-sided case we have

$$\lambda_n = \int \phi_n(dy) = \int_{|y|>1/n} \phi(dy)$$
$$= \int_{1/n}^{\infty} pC\alpha y^{-\alpha-1}dy + \int_{-\infty}^{-1/n} qC\alpha |y|^{-\alpha-1}dy$$
$$= (p+q)\int_{1/n}^{\infty} C\alpha y^{-\alpha-1}dy = Cn^{\alpha}$$

and the probability measure

$$\omega_n(dy) = \lambda_n^{-1}\phi_n(dy) = \begin{cases} n^{-\alpha}p\alpha y^{-\alpha-1}dy & \text{for } y > 1/n, \text{ and} \\ n^{-\alpha}q\alpha|y|^{-\alpha-1}dy & \text{for } y < -1/n. \end{cases}$$

Then

$$\mathbb{P}[W_n > x] = \omega_n(x,\infty) = \int_x^{\infty} n^{-\alpha}p\alpha y^{-\alpha-1}dy = pAx^{-\alpha}$$
$$\mathbb{P}[W_n < -x] = \int_{-\infty}^{-x} n^{-\alpha}q\alpha|y|^{-\alpha-1}dy = \int_x^{\infty} n^{-\alpha}q\alpha y^{-\alpha-1}dy = qAx^{-\alpha}$$

where $A = n^{-\alpha}$ for all n. Again $Y_n \Rightarrow Y$ since $\int \left(e^{iky} - 1 - iky\right)\phi(dy)$ exists. Here

$$\mathbb{E}[W_j] = \int_{1/n}^{\infty} y\,pA\alpha y^{-\alpha-1}dy + \int_{-\infty}^{-1/n} y\,qA\alpha|y|^{-\alpha-1}dy = (p-q)\frac{A\alpha}{\alpha-1}n^{\alpha-1}$$

so that $\mathbb{E}[W_j] = 0$ if $p = q$. In this case, the compensator $a_n = 0$, and the compound Poisson approximation converges without centering.

3.6 Triangular arrays

This section develops the general theory of triangular arrays, which is the fundamental tool used to prove the extended central limit theorem for stable laws. Recall that Y is infinitely divisible if for every positive integer n we can write $Y \simeq X_{n1} + \cdots + X_{nn}$ a sum of iid random variables. A *triangular array* of random variables is a set

$$\{X_{nj} : j = 1, \ldots, k_n; n = 1, 2, 3, \ldots\} \tag{3.35}$$

where $X_{n1}, \ldots, X_{nk_n}$ are independent for each $n \geq 1$, and $k_n \to \infty$ as $n \to \infty$. Then the *row sum*

$$S_n = X_{n1} + \cdots + X_{nk_n}$$

is a sum of independent random variables. We will make the usual assumption that

$$\lim_{n\to\infty} \sup_{1\le j\le k_n} \mathbb{P}[|X_{nj}| > \varepsilon] = 0 \quad \text{for all } \varepsilon > 0. \tag{3.36}$$

This condition ensures that every summand is asymptotically negligible. A general result [146, Theorem 3.2.14] states that Y is infinitely divisible if and only if $S_n - a_n \Rightarrow Y$ for some triangular array that satisfies (3.36) and some sequence (a_n).

Example 3.30. Take (W_n) iid with $\mathbb{E}[W_n] = 0$ and $\mathbb{E}[W_n^2] = \sigma^2 < \infty$. Then

$$X_{nj} = \frac{1}{\sigma\sqrt{n}} W_j : \quad 1 \le j \le n$$

forms a triangular array with $k_n = n$, and the row sums $S_n \Rightarrow Y \simeq \mathcal{N}(0, 1)$. Note that X_{nj} are iid for $1 \le j \le n$, but the distribution of X_{nj} depends on n.

Example 3.31. Take (W_n) iid with $\mathbb{E}[W_n] = 0$ and $\mathbb{E}[W_n^2] = \sigma^2 < \infty$. Then

$$X_{nj} = n^{-1/2} W_j : \quad 1 \le j \le [nt]$$

forms a triangular array with $k_n = [nt]$, and the row sums $S_n \Rightarrow Y \simeq \mathcal{N}(0, \sigma^2 t)$. In other words, $S_n \Rightarrow B(t)$ for any single $t \ge 0$, where $B(t)$ is a Brownian motion.

Example 3.32. Take (W_n) iid with $\mu = \mathbb{E}[W_n] \ne 0$ and $\mathbb{E}[(W_n - \mu)^2] = \sigma^2 < \infty$. Then

$$X_{nj} = \frac{1}{\sqrt{n}}(W_j - \mu) + \frac{1}{n}\mu : \quad 1 \le j \le [nt]$$

forms a triangular array with $k_n = [nt]$, and the row sums

$$S_n = \sum_{j=1}^{k_n} X_{nj} = \frac{1}{\sqrt{n}} \sum_{j=1}^{[nt]} (W_j - \mu) + \frac{[nt]}{n}\mu \Rightarrow B(t) + \mu t$$

a Brownian motion with drift, where $B(t) \simeq \mathcal{N}(0, \sigma^2 t)$. Note that two scales are necessary here: We must divide the mean by n and the deviation from the mean by $\sqrt{n}$ to represent both terms in the limit.

The proof the extended central limit theorem with normal or stable limits depends on the convergence theory for triangular arrays. Define the truncated random variables

$$X_{nj}^R = X_{nj} I(|X_{nj}| \le R) = \begin{cases} X_{nj} & \text{if } |X_{nj}| \le R; \text{ and} \\ 0 & \text{if } |X_{nj}| > R. \end{cases}$$

We say that a sequence of σ-finite Borel measures $\phi_n(dy) \to \phi(dy)$ on $\{y : y \ne 0\}$ if $\phi_n(B) \to \phi(B)$ for any Borel set B that is bounded away from zero, and such that $\phi(\partial B) = 0$. This is called *vague convergence*. In Section 3.4 we defined a sequence of compound Poisson random variables whose Lévy measures ϕ_n converged vaguely to the Lévy measure ϕ of a stable law. See the details at the end of this section for more discussion.

Theorem 3.33 (Triangular array convergence). *Given a triangular array (3.35) such that (3.36) holds, there exists a random variable Y and a sequence (a_n) such that $S_n - a_n \Rightarrow Y$ if and only if:*

(i) $\sum_{j=1}^{k_n} \mathbb{P}[X_{nj} \in dy] \to \phi(dy)$ *for some σ-finite Borel measure on* $\{y : y \neq 0\}$; *and*

(ii) $\lim_{\varepsilon \to 0} \limsup_{n\to\infty} \sum_{j=1}^{k_n} \mathrm{Var}[X_{nj}^{\varepsilon}] = \lim_{\varepsilon \to 0} \liminf_{n\to\infty} \sum_{j=1}^{k_n} \mathrm{Var}[X_{nj}^{\varepsilon}] = b \geq 0.$

In this case, Y is infinitely divisible with Lévy representation $[a, b, \phi]$, where a depends on the choice of centering constants (a_n). We can take

$$a_n = \sum_{j=1}^{k_n} \mathbb{E}[X_{nj}^R] \tag{3.37}$$

for any $R > 0$ such that $\phi\{y : |y| = R\} = 0$, and then $\mathbb{E}[e^{ikY}] = e^{\psi_0(k)}$ where

$$\psi_0(k) = -\tfrac{1}{2}k^2 b + \int \left(e^{iky} - 1 - ikyI(|y| \leq R)\right) \phi(dy). \tag{3.38}$$

That is, (3.5) holds with $a_0 = 0$.

Proof. This is a special case of [146, Theorem 3.2.2]. □

Remark 3.34. To establish vague convergence condition (*i*), it suffices to show

$$\sum_{j=1}^{k_n} \mathbb{P}[X_{nj} > y] \to \phi(y, \infty) \quad \text{and} \quad \sum_{j=1}^{k_n} \mathbb{P}[X_{nj} < -y] \to \phi(-\infty, -y) \tag{3.39}$$

for every $y > 0$ such that $\phi\{y\} = \phi\{-y\} = 0$. The centering constants a_n in (3.37) and the log characteristic function $\psi_0(k)$ both depend on the choice of $R > 0$. If the Lévy measure has a density, as is the case for stable laws, then any $R > 0$ may be used, since we always have $\phi\{R\} = \phi\{-R\} = 0$. To establish the truncated variance condition (*ii*), it is of course sufficient to show that

$$\lim_{\varepsilon \to 0} \lim_{n\to\infty} \sum_{j=1}^{k_n} \mathrm{Var}[X_{nj}^{\varepsilon}] = b. \tag{3.40}$$

Remark 3.35. The proof of Theorem 3.33 is based on a Poisson approximation. First we approximate $S_n \approx S_N$ where N is Poisson with mean k_n, independent from the triangular array elements. Then we use the converge criteria for infinitely divisible laws. Suppose $Y_n \simeq [a_n, b_n, \phi_n]$ and $Y \simeq [a, b, \phi]$ in terms of the Lévy representation. Then $Y_n \Rightarrow Y$ if and only if $\psi_n(k) \to \psi(k)$ for each k, i.e., the log characteristic functions converge [146, Lemma 3.1.10]. Write

$$f(y, k) = e^{iky} - 1 - \frac{iky}{1 + y^2}$$

and note that $y \mapsto f(y,k)$ is a bounded continuous function such that

$$f(y,k) = -\tfrac{1}{2}k^2y^2 + O(y^2) \quad \text{as } y \to 0$$

for any fixed k. Now it is not hard to show that

$$\int_{|y|>\varepsilon} f(y,k)\phi_n(dy) \to \int_{|y|>\varepsilon} f(y,k)\phi(dy)$$

whenever $\phi\{|y| = \varepsilon\} = 0$, which must be true for almost every $\varepsilon > 0$. Then

$$\lim_{\varepsilon\to 0}\lim_{n\to\infty} \int_{|y|>\varepsilon} f(y,k)\phi_n(dy) = \lim_{\varepsilon\to 0} \int_{|y|>\varepsilon} f(y,k)\phi(dy) = \int f(y,k)\phi(dy)$$

since $\int y^2 I(0 < |y| \le \varepsilon)\,\phi(dy)$ exists for a Lévy measure. To handle the the remaining part of the integral term in the Lévy representation for $\psi_n(k)$ we write

$$\begin{aligned}
&\lim_{\varepsilon\to 0}\lim_{n\to\infty}\left[-\tfrac{1}{2}k^2 b_n + \int_{0<|y|\le\varepsilon} f(y,k)\phi_n(dy)\right] \\
&= \lim_{\varepsilon\to 0}\lim_{n\to\infty}\left[-\tfrac{1}{2}k^2 b_n - \tfrac{1}{2}k^2 \int_{0<|y|\le\varepsilon} y^2\phi_n(dy)\right] = -\tfrac{1}{2}k^2 b
\end{aligned}$$

provided that

$$\lim_{\varepsilon\to 0}\lim_{n\to\infty}\left[b_n + \int_{0<|y|\le\varepsilon} y^2\phi_n(dy)\right] = b. \tag{3.41}$$

Then it can be shown that $Y_n \Rightarrow Y$ if and only if (3.41) holds along with $a_n \to a$ and $\phi_n \to \phi$, see [146, Theorem 3.1.16]. The proof of Theorem 3.33 uses these ideas, along with some delicate centering arguments.

Here we prove the traditional central limit theorem with iid summands, to illustrate the use of Theorem 3.33.

Theorem 3.36 (Central Limit Theorem). *Suppose that (W_n) are iid and that $\mu_1 = \mathbb{E}[W_n]$ and $\mu_2 = \mathbb{E}[W_n^2]$ exist. Then*

$$\frac{W_1 + \cdots + W_n - n\mu_1}{n^{1/2}} \Rightarrow Y \simeq \mathcal{N}(0,\sigma^2) \tag{3.42}$$

where $\sigma^2 = \mu_2 - \mu_1^2$.

Proof. Define a triangular array with row elements $X_{nj} = n^{-1/2}W_j$ for $j = 1,\dots,n$. Then condition (3.36) holds (see details), and then in order to prove that $S_n - a_n \Rightarrow Y$

normal, it suffices to check conditions (i) and (ii) in Theorem 3.33. For condition (i) we have for each $\varepsilon > 0$ that

$$\begin{aligned}\sum_{j=1}^{k_n} \mathbb{P}[|X_{nj}| > \varepsilon] &= n\mathbb{P}[|n^{-1/2}W_j| > \varepsilon] \\ &= n\mathbb{P}[|W_j| > n^{1/2}\varepsilon] \\ &= n\mathbb{E}[I(|W_j| > n^{1/2}\varepsilon)] \\ &\le n\mathbb{E}\left[\left(\frac{W_j}{n^{1/2}\varepsilon}\right)^2 I(|W_j| > n^{1/2}\varepsilon)\right] \\ &= \varepsilon^{-2}\mathbb{E}\left[W_j^2 I(|W_j| > n^{1/2}\varepsilon)\right] \to 0\end{aligned}$$

as $n \to \infty$, since $\mu_2 = \mathbb{E}[W_n^2]$ exists. Then (i) holds with $\phi = 0$.

Condition (ii) in this case is a form of the *Lindeberg Condition*. Write

$$\begin{aligned}&\sum_{j=1}^{k_n} \mathrm{Var}[X_{nj}^\varepsilon] = n\left\{E\left[(X_{nj}^\varepsilon)^2\right] - \mathbb{E}\left[X_{nj}^\varepsilon\right]^2\right\} \\ &= nE\left[(n^{-1/2}W_j)^2 I(|n^{-1/2}W_j| \le \varepsilon)\right] - nE\left[n^{-1/2}W_j I(|n^{-1/2}W_j| \le \varepsilon)\right]^2 \\ &= E\left[W_j^2 I(|W_j| \le n^{1/2}\varepsilon)\right] - E\left[W_j I(|W_j| \le n^{1/2}\varepsilon)\right]^2 \to \mu_2 - \mu_1^2\end{aligned}$$

since the first and second moments exist. Then Theorem 3.33 shows that $S_n - a_n \Rightarrow Y \simeq [a, b, 0]$ where $b = \mu_2 - \mu_1^2 = \sigma^2 = \mathrm{Var}(Y)$. This shows that Y is normal. From (3.37) we get

$$\begin{aligned}a_n = \sum_{j=1}^{k_n} \mathbb{E}[X_{nj}^R] &= n\mathbb{E}\left[n^{-1/2}W_j I(|W_j| \le n^{1/2}R)\right] \\ &= n^{1/2}\left\{\mu_1 - \mathbb{E}\left[W_j I(|W_j| > n^{1/2}R)\right]\right\}\end{aligned}$$

where

$$\begin{aligned}\left|n^{1/2}\mathbb{E}\left[W_j I(|W_j| > n^{1/2}R)\right]\right| &\le n^{1/2}\mathbb{E}\left[|W_j| I(|W_j| > n^{1/2}R)\right] \\ &\le n^{1/2}\mathbb{E}\left[|W_j|\left(\frac{|W_j|}{n^{1/2}R}\right) I(|W_j| > n^{1/2}R)\right] \\ &= R^{-1}\mathbb{E}\left[W_j^2 I(|W_j| > n^{1/2}R)\right] \to 0\end{aligned}$$

since μ_2 exists. This shows that $a_n - n^{1/2}\mu_1 \to 0$ and then we have $S_n - n^{1/2}\mu_1 = S_n - a_n + (a_n - n^{1/2}\mu_1) \Rightarrow Y$. Then (3.42) follows. □

Details

Theorem 3.33 uses the concept of *vague convergence*: We say that a sequence of σ-finite Borel measures $\phi_n(dy) \to \phi(dy)$ on $\{y : y \neq 0\}$ if $\phi_n(B) \to \phi(B)$ for any Borel set B

that is bounded away from zero, and such that $\phi(\partial B) = 0$. Here ∂B is the topological boundary of the set B, defined as the closure of B (the intersection of all closed sets that contain B) minus the interior of B (the union of all open sets contained in B). The Borel measure is a standard tool in real analysis and probability (e.g., see [35, 62, 180]). In the physics notation introduced in the details at the end of Section 3.1, we noted that a Lévy density can often be interpreted in terms of generalized functions, with Dirac delta function terms to represents atoms in the Lévy measure. Readers who are more comfortable with the physics notation may interpret the vague convergence $\phi_n \to \phi$ to mean that, if $\phi_n(dy) = \phi_n(y)\,dy$ and $\phi(dy) = \phi(y)\,dy$, then

$$\phi_n(a,b) = \int_a^b \phi_n(y)dy \to \int_a^b \phi(y)dy = \phi(a,b)$$

for all $0 < a < b$ or $a < b < 0$ such that $\phi(y)$ has no Dirac delta function terms at the points a, b, i.e., $\phi\{a\} = \phi\{b\} = 0$. Stable distributions all have Lévy densities $\phi(y)$ with no Dirac delta function terms. However, these Lévy measures are not finite, since $\int_0^\infty \phi(y)\,dy = \infty$ or $\int_{-\infty}^0 \phi(y)\,dy = \infty$. In this case, the Lévy measure is called σ-finite because it assigns finite measure to the sets $\{y : |y| > 1/n\}$, and the set $\{y : y \neq 0\}$ is the countable union of these.

If X is any random variable, then the distribution of X is *tight*, meaning that

$$\mathbb{P}[|X| > r] \to 0 \quad \text{as } r \to \infty. \tag{3.43}$$

Equation (3.43) follows by a simple application of the dominated convergence theorem. It follows that

$$\mathbb{P}[|X_{nj}| > \varepsilon] = \mathbb{P}[|W_j| > n^{1/2}\varepsilon] \to 0$$

as $n \to \infty$, so that condition (3.36) holds.

3.7 One-sided stable limits

Here we prove that one-sided stable laws with Lévy measure (3.10) are the limits for sums of iid Pareto jumps. We also specify a convenient centering.

Theorem 3.37. *Suppose that (W_n) are iid and positive with $\mathbb{P}[W_n > x] = Cx^{-\alpha}$ for all $x > C^{1/\alpha}$ for some $C > 0$ and $0 < \alpha < 2$. Then*

$$n^{-1/\alpha}(W_1 + \cdots + W_n) - a_n \Rightarrow Y \tag{3.44}$$

for some sequence (a_n), where Y is a one-sided stable law with Lévy representation $[a, 0, \phi]$, and the Lévy measure is given by (3.10). If $0 < \alpha < 1$, we can choose $a_n = 0$, and then (3.12) holds. If $1 < \alpha < 2$, we can choose $a_n = n^{1-1/\alpha}\mu_1$ where $\mu_1 = \mathbb{E}[W_n]$, and then (3.16) holds.

Proof. Define a triangular array $X_{nj} = n^{-1/\alpha} W_j$ for $j = 1, \dots, n$. Then condition (3.36) holds (see details), and we just need to check the convergence criteria (*i*) and (*ii*) from Theorem 3.33. For $y > 0$ we have

$$\begin{aligned}\sum_{j=1}^{k_n} \mathbb{P}[X_{nj} > y] &= n\mathbb{P}[n^{-1/\alpha} W_j > y] \\ &= n\mathbb{P}[W_j > n^{1/\alpha} y] \\ &= nC\left(n^{1/\alpha} y\right)^{-\alpha} = Cy^{-\alpha}\end{aligned}$$

whenever $n^{1/\alpha} y > C^{1/\alpha}$, as well as

$$\sum_{j=1}^{k_n} \mathbb{P}[X_{nj} < -y] = 0.$$

Then (*i*) holds with $\phi(y, \infty) = Cy^{-\alpha}$ for all $y > 0$, and $\phi(-\infty, 0) = 0$. This is equivalent to (3.10). Note that $0 < \alpha < 2$ is required here, so that $\phi(dy)$ is a Lévy measure:

$$\int_{|y|\le R} y^2 \phi(dy) = \int_0^R y^2 C\alpha y^{-\alpha-1} dy = \left[\frac{C\alpha}{2-\alpha} y^{2-\alpha}\right]_0^R = \frac{C\alpha}{2-\alpha} R^{2-\alpha} < \infty.$$

For any $\varepsilon > 0$ we have, whenever n is sufficiently large to make $n^{1/\alpha}\varepsilon > C^{1/\alpha}$, that

$$\begin{aligned}0 \le \sum_{j=1}^{k_n} \operatorname{Var}[X_{nj}^\varepsilon] &= n\left\{E\left[(X_{nj}^\varepsilon)^2\right] - \mathbb{E}\left[X_{nj}^\varepsilon\right]^2\right\} \le nE\left[(X_{nj}^\varepsilon)^2\right] \\ &= nE\left[(n^{-1/\alpha} W_j)^2 I(|n^{-1/\alpha} W_j| \le \varepsilon)\right] \\ &= n^{1-2/\alpha} E\left[W_j^2 I(|W_j| \le n^{1/\alpha}\varepsilon)\right] \\ &= n^{1-2/\alpha} \int_{C^{1/\alpha}}^{n^{1/\alpha}\varepsilon} y^2 C\alpha y^{-1-\alpha} dy \\ &= n^{1-2/\alpha} \left[C\alpha \frac{y^{2-\alpha}}{2-\alpha}\right]_{C^{1/\alpha}}^{n^{1/\alpha}\varepsilon} \\ &= n^{1-2/\alpha} C\alpha \left[\frac{(n^{1/\alpha}\varepsilon)^{2-\alpha}}{2-\alpha} - \frac{(C^{1/\alpha})^{2-\alpha}}{2-\alpha}\right] \\ &= n^{1-2/\alpha} \frac{C\alpha}{2-\alpha} \left[n^{2/\alpha-1}\varepsilon^{2-\alpha} - C^{2/\alpha-1}\right] \\ &= \varepsilon^{2-\alpha}\frac{C\alpha}{2-\alpha} - n^{1-2/\alpha}\frac{\alpha}{2-\alpha} C^{2/\alpha} \sim \varepsilon^{2-\alpha}\frac{C\alpha}{2-\alpha}\end{aligned} \tag{3.45}$$

as $n \to \infty$, since $1 - 2/\alpha < 0$. Then we have

$$\lim_{\varepsilon\to 0} \limsup_{n\to\infty} \sum_{j=1}^{k_n} \operatorname{Var}[X_{nj}^\varepsilon] \le \lim_{\varepsilon\to 0} \varepsilon^{2-\alpha}\frac{C\alpha}{2-\alpha} = 0$$

since $2 - \alpha > 0$, so that (ii) holds with $b = 0$. This proves that $S_n - a_n \Rightarrow Y_0$ holds for some sequence (a_n), where $S_n = X_{n1} + \cdots + X_{nn} = n^{-1/\alpha}(W_1 + \cdots + W_n)$ is the row sum of this triangular array, and Y_0 is infinitely divisible with Lévy measure ϕ and no normal component.

Suppose that $0 < \alpha < 1$. Theorem 3.33 shows that, if we choose (a_n) according to (3.37), then $\mathbb{E}[e^{ikY_0}] = e^{\psi_0(k)}$ where

$$\begin{aligned}\psi_0(k) &= \int \left(e^{iky} - 1 - ikyI(|y| \le R)\right) \phi(dy)\\ &= \int \left(e^{iky} - 1\right) \phi(dy) - ik \int yI(|y| \le R)\phi(dy)\\ &= -C\Gamma(1-\alpha)(-ik)^\alpha - ika \end{aligned} \tag{3.46}$$

by Proposition 3.10, where we can take

$$a = \int_0^R y\, C\alpha y^{-\alpha-1} dy = C\alpha \left[\frac{y^{1-\alpha}}{1-\alpha}\right]_0^R = \frac{C\alpha}{1-\alpha} R^{1-\alpha} \tag{3.47}$$

for any $R > 0$, since ϕ has a density. Write

$$\begin{aligned} a_n = \sum_{j=1}^{k_n} \mathbb{E}\left[X_{nj}^R\right] &= n\mathbb{E}\left[n^{-1/\alpha} W_j I(|n^{-1/\alpha} W_j| \le R)\right]\\ &= n^{1-1/\alpha} \int_{C^{1/\alpha}}^{n^{1/\alpha}R} y\, C\alpha y^{-1-\alpha} dy\\ &= n^{1-1/\alpha} C\alpha \left[\frac{(n^{1/\alpha}R)^{1-\alpha}}{1-\alpha} - \frac{(C^{1/\alpha})^{1-\alpha}}{1-\alpha}\right]\\ &= \frac{C\alpha}{1-\alpha} R^{1-\alpha} - n^{1-1/\alpha} \frac{\alpha}{1-\alpha} C^{1/\alpha} \to \frac{C\alpha}{1-\alpha} R^{1-\alpha} = a \end{aligned} \tag{3.48}$$

as $n \to \infty$, since $1 - 1/\alpha < 0$ in this case.

Let Y be a one-sided stable law with characteristic function $\exp[-C\Gamma(1-\alpha)(-ik)^\alpha]$, so that $Y - a = Y_0$ in view of (3.46). Since $a_n - a \to 0$ we also have $S_n - a = S_n - a_n + (a_n - a) \Rightarrow Y_0$, and then we also have $S_n = S_n - a + a \Rightarrow Y_0 + a = Y$. Hence we can take $a_n = 0$ in this case, and then the limit has characteristic function (3.12).

Suppose that $1 < \alpha < 2$. Theorem 3.33 shows that, if we choose (a_n) according to (3.37), then $\mathbb{E}[e^{ikY_0}] = e^{\psi_0(k)}$ where

$$\begin{aligned}\psi_0(k) &= \int \left(e^{iky} - 1 - ikyI(|y| \le R)\right) \phi(dy)\\ &= \int \left(e^{iky} - 1 - iky\right) \phi(dy) + ik \int yI(|y| > R)\phi(dy)\\ &= C\frac{\Gamma(2-\alpha)}{\alpha - 1}(-ik)^\alpha + ika \end{aligned} \tag{3.49}$$

by Proposition 3.12, where we can choose

$$a = \int_R^\infty y\, C\alpha y^{-\alpha-1} dy = C\alpha \left[\frac{y^{1-\alpha}}{1-\alpha}\right]_R^\infty = \frac{C\alpha}{\alpha-1} R^{1-\alpha} \tag{3.50}$$

for any $R > 0$, since ϕ has a density. Using (3.48) we have

$$a_n = \frac{C\alpha}{1-\alpha} R^{1-\alpha} - n^{1-1/\alpha} \frac{\alpha}{1-\alpha} C^{1/\alpha} = -a + n^{1-1/\alpha} \mu_1 \tag{3.51}$$

since

$$\begin{aligned} \mu_1 = \mathbb{E}[W_n] = \int_{C^{1/\alpha}}^\infty y\, C\alpha y^{-1-\alpha} dy &= \left[C\alpha \frac{y^{1-\alpha}}{1-\alpha}\right]_{C^{1/\alpha}}^\infty \\ &= \frac{C\alpha}{\alpha-1} (C^{1/\alpha})^{1-\alpha} = \frac{\alpha}{\alpha-1} C^{1/\alpha} \end{aligned} \tag{3.52}$$

exists in this case.

Let $Y = Y_0 - a$, so that Y has characteristic function (3.16). Since $S_n - a_n \Rightarrow Y_0$ and $a_n + a = n^{1-1/\alpha}\mu_1$, it follows that $S_n - n^{1-1/\alpha}\mu_1 = S_n - a_n - a \Rightarrow Y_0 - a = Y$. □

Remark 3.38. Theorem 3.37 shows that no centering is needed to get convergence when $0 < \alpha < 1$, and when $1 < \alpha < 2$ we can center to zero expectation. The stable limits in this case will be called *centered stable*. When $1 < \alpha < 2$, it is not hard to check that a centered stable law has mean zero, by differentiating the characteristic function. See the details at the end of this section.

Details

Since W_j is tight for any fixed j, so that (3.43) holds with $X = W_j$, it follows that

$$\mathbb{P}[|X_{nj}| > \varepsilon] = \mathbb{P}[|W_j| > n^{1/\alpha}\varepsilon] \to 0$$

as $n \to \infty$, so that condition (3.36) holds.

If Y is centered stable with index $1 < \alpha < 2$, then $\hat{\mu}(k) = e^{\psi(k)}$ where $\psi(k) = \int \left(e^{iky} - 1 - iky\right) \phi(dy)$. Then

$$\frac{d}{dk}\hat{\mu}(k) = e^{\psi(k)} \int iy \left(e^{iky} - 1\right) \phi(dy)$$

where the integrand is $O(y^2)$ as $y \to 0$, and $O(y)$ as $y \to \infty$, so that the integral exists. Using the general fact that $\frac{d}{dk}\hat{\mu}(0) = i\,\mathbb{E}[Y]$ if $\mathbb{E}[|Y|] < \infty$ (see Proposition 1.1), it follows that $\mathbb{E}[Y] = 0$ in this case. The same argument shows that $\mathbb{E}[Y] = a_2$ for any infinitely divisible law that satisfies condition (3.8) in Theorem 3.8, see [146, Remark 3.1.15].

3.8 Two-sided stable limits

We prove that general two-sided stable laws are the limits for Pareto random walks that allow both positive and negative jumps. Centering is unnecessary when $0 < \alpha < 1$, and we can center to mean zero when $1 < \alpha < 2$.

Theorem 3.39. *Suppose (W_n) are iid with $\mathbb{P}[W_n > x] = pCx^{-\alpha}$ and $\mathbb{P}[W_n < -x] = qCx^{-\alpha}$ for all $x > C^{1/\alpha}$ for some $C > 0$ and $0 < \alpha < 2$, and some $p, q \geq 0$ such that $p + q = 1$. Then*

$$n^{-1/\alpha}(W_1 + \cdots + W_n) - a_n \Rightarrow Y \tag{3.53}$$

for some sequence (a_n), where Y is a stable law with Lévy representation $[a, 0, \phi]$, and the Lévy measure ϕ is given by (3.30). If $0 < \alpha < 1$, we can choose $a_n = 0$, and then Y has characteristic function

$$\hat{\mu}(k) = \mathbb{E}[e^{ikY}] = \exp\left[- pC\Gamma(1-\alpha)(-ik)^{\alpha} - qC\Gamma(1-\alpha)(ik)^{\alpha}\right]. \tag{3.54}$$

If $1 < \alpha < 2$, we can choose $a_n = n^{1-1/\alpha}\mu_1$ where $\mu_1 = \mathbb{E}[W_n]$, and then Y has characteristic function

$$\hat{\mu}(k) = \mathbb{E}[e^{ikY}] = \exp\left[pC\frac{\Gamma(2-\alpha)}{\alpha-1}(-ik)^{\alpha} + qC\frac{\Gamma(2-\alpha)}{\alpha-1}(ik)^{\alpha}\right]. \tag{3.55}$$

Proof. The proof is similar to Theorem 3.37. Use the triangular array $X_{nj} = n^{-1/\alpha}W_j$ for $j = 1, \ldots, n$, so that condition (3.36) holds. For any $y > 0$ we have

$$\sum_{j=1}^{k_n} \mathbb{P}[X_{nj} > y] = n\mathbb{P}[W_j > n^{1/\alpha}y] = npC\left(n^{1/\alpha}y\right)^{-\alpha} = pCy^{-\alpha}$$

and

$$\sum_{j=1}^{k_n} \mathbb{P}[X_{nj} < -y] = n\mathbb{P}[W_j < -n^{1/\alpha}y] = qCy^{-\alpha}$$

whenever $n^{1/\alpha}y > C^{1/\alpha}$. Then condition (i) from Theorem 3.33 holds with $\phi(y, \infty) = Cpy^{-\alpha}$ and $\phi(-\infty, -y) = Cqy^{-\alpha}$ for all $y > 0$. This is equivalent to (3.30). Note that the condition (3.3) for a Lévy measure requires $0 < \alpha < 2$.

For any $\varepsilon > 0$, for all n is sufficiently large, we have

$$\begin{aligned}
0 \leq \sum_{j=1}^{k_n} \mathrm{Var}[X_{nj}^{\varepsilon}] &\leq nE\left[(X_{nj}^{\varepsilon})^2\right] \\
&= n^{1-2/\alpha}E\left[W_j^2 I(|W_j| \leq n^{1/\alpha}\varepsilon)\right] \\
&= \varepsilon^{2-\alpha}\frac{C\alpha}{2-\alpha} - n^{1-2/\alpha}\frac{\alpha}{2-\alpha}C^{2/\alpha} \sim \varepsilon^{2-\alpha}\frac{C\alpha}{2-\alpha}
\end{aligned}$$

as $n \to \infty$, by exactly the same argument as the one-sided case (see the proof of Theorem 3.37), since the distribution of W_n^2 is the same. It follows that condition (ii) from Theorem 3.33 holds with $b = 0$. Then the centered row sums $S_n - a_n \Rightarrow Y_0$ where Y_0 is infinitely divisible with no normal component, and Lévy measure (3.30).

Suppose that $0 < \alpha < 1$. Theorem 3.33 implies that, if the norming sequence (a_n) is chosen according to (3.37), then $\mathbb{E}[e^{ikY_0}] = e^{\psi_0(k)}$ where

$$\begin{aligned}\psi_0(k) &= \int \left(e^{iky} - 1 - ikyI(|y| \le R)\right)\phi(dy)\\ &= \int \left(e^{iky} - 1\right)\phi(dy) - ik\int yI(|y| \le R)\phi(dy)\\ &= -pC\Gamma(1-\alpha)(-ik)^\alpha - qC\Gamma(1-\alpha)(ik)^\alpha - ika\end{aligned}$$

by (3.31). Here

$$\begin{aligned}a &= \int yI(|y| \le R)\phi(dy)\\ &= \int_0^R y\,pC\alpha y^{-\alpha-1}\,dy + \int_{-R}^0 y\,qC\alpha(-y)^{-\alpha-1}\,dy = \frac{C\alpha}{1-\alpha}(p-q)R^{1-\alpha}\end{aligned} \tag{3.56}$$

which reduces to (3.47) if $p = 1$. Theorem 3.33 shows that we can choose

$$\begin{aligned}a_n &= \sum_{j=1}^{k_n} \mathbb{E}\left[X_{nj}^R\right] = n\mathbb{E}\left[n^{-1/\alpha}W_jI(|W_j| \le n^{1/\alpha}R)\right]\\ &= n^{1-1/\alpha}\left[\int_{C^{1/\alpha}}^{n^{1/\alpha}R} y\,pC\alpha y^{-1-\alpha}dy + \int_{-n^{1/\alpha}R}^{-C^{1/\alpha}} y\,qC\alpha|y|^{-1-\alpha}dy\right]\\ &= n^{1-1/\alpha}C\alpha\Bigg[p\left(\frac{(n^{1/\alpha}R)^{1-\alpha}}{1-\alpha} - \frac{(C^{1/\alpha})^{1-\alpha}}{1-\alpha}\right)\\ &\qquad - q\left(\frac{(n^{1/\alpha}R)^{1-\alpha}}{1-\alpha} - \frac{(C^{1/\alpha})^{1-\alpha}}{1-\alpha}\right)\Bigg]\\ &= \frac{C\alpha}{1-\alpha}(p-q)R^{1-\alpha} - n^{1-1/\alpha}(p-q)\frac{\alpha}{1-\alpha}C^{1/\alpha}\\ &\to \frac{C\alpha}{1-\alpha}(p-q)R^{1-\alpha} = a\end{aligned} \tag{3.57}$$

as $n \to \infty$, since $1 - 1/\alpha < 0$ in this case.

Define $Y = Y_0 + a$. Since $a_n \to a$ it follows that $S_n \Rightarrow Y_0 + a = Y$. Hence we can choose $a_n = 0$ in this case, and then the limit has characteristic function (3.54).

Suppose that $1 < \alpha < 2$. If (3.37) holds, then $\mathbb{E}[e^{ikY_0}] = e^{\psi_0(k)}$ where

$$\begin{aligned}\psi_0(k) &= \int \left(e^{iky} - 1 - ikyI(|y| \le R)\right)\phi(dy)\\ &= \int \left(e^{iky} - 1 - iky\right)\phi(dy) + ik\int yI(|y| > R)\phi(dy)\\ &= pC\frac{\Gamma(2-\alpha)}{\alpha-1}(-ik)^\alpha + qC\frac{\Gamma(2-\alpha)}{\alpha-1}(ik)^\alpha + ika\end{aligned} \tag{3.58}$$

by (3.33), where

$$a = \int yI(|y| > R)\phi(dy)$$
$$= \int_R^\infty y\,pC\alpha y^{-\alpha-1}dy + \int_{-\infty}^{-R} y\,qC\alpha(-y)^{-\alpha-1}dy = \frac{C\alpha}{\alpha-1}(p-q)R^{1-\alpha}$$

for some arbitrary fixed $R > 0$. Using (3.57) we have

$$a_n = \frac{C\alpha}{1-\alpha}(p-q)R^{1-\alpha} - n^{1-1/\alpha}(p-q)\frac{\alpha}{1-\alpha}C^{1/\alpha} = -a + n^{1-1/\alpha}\mu_1 \tag{3.59}$$

since

$$\mu_1 = \mathbb{E}[W_n] = p\int_{C^{1/\alpha}}^\infty y\,C\alpha y^{-1-\alpha}dy + q\int_{-\infty}^{-C^{1/\alpha}} y\,C\alpha(-y)^{-1-\alpha}dy$$
$$= \frac{C\alpha}{\alpha-1}(p-q)(C^{1/\alpha})^{1-\alpha} = (p-q)\frac{\alpha}{\alpha-1}C^{1/\alpha}$$

exists in this case.

Define $Y = Y_0 - a$. Since $a_n + a = n^{1-1/\alpha}\mu_1$ it follows that $S_n - n^{1-1/\alpha}\mu_1 \Rightarrow Y_0 - a = Y$, and the limit Y has characteristic function (3.55). □

Remark 3.40. Theorem 3.39 shows that no centering is needed to get convergence when $0 < \alpha < 1$, and when $1 < \alpha < 2$ we can center to zero expectation (see the details at the end of Section 3.7). The stable limits in this case will be called *centered stable.*

Now we extend the convergence in Theorem 3.39 to process limits. The next result shows that a random walk with power law jumps, suitably centered, converges to an α-stable Lévy motion. If $0 < \alpha < 1$, then no centering is needed. If $1 < \alpha < 2$, we can center to zero expectation.

Theorem 3.41. *Suppose (W_n) are iid with $\mathbb{P}[W_n > x] = pCx^{-\alpha}$ and $\mathbb{P}[W_n < -x] = qCx^{-\alpha}$ for all $x > C^{1/\alpha}$ for some $C > 0$ and $0 < \alpha < 2$, and some $p, q \geq 0$ such that $p + q = 1$.*
(a) If $0 < \alpha < 1$, then

$$n^{-1/\alpha}\sum_{j=1}^{[nt]} W_j \Rightarrow Z_t \tag{3.60}$$

for all $t > 0$, where

$$\mathbb{E}[e^{ikZ_t}] = \exp\left[-tpC\Gamma(1-\alpha)(-ik)^\alpha - tqC\Gamma(1-\alpha)(ik)^\alpha\right]; \tag{3.61}$$

(b) If $1 < \alpha < 2$, then $\mu_1 = \mathbb{E}[W_n]$ exists and

$$n^{-1/\alpha}\sum_{j=1}^{[nt]} (W_j - \mu_1) \Rightarrow Z_t \tag{3.62}$$

for all $t > 0$, *where*

$$\mathbb{E}[e^{ikZ_t}] = \exp\left[tpC\frac{\Gamma(2-\alpha)}{\alpha-1}(-ik)^\alpha + tqC\frac{\Gamma(2-\alpha)}{\alpha-1}(ik)^\alpha\right]. \tag{3.63}$$

Proof. If $0 < \alpha < 1$, then Theorem 3.39 shows that $n^{-1/\alpha}S(n) \Rightarrow Y$, where the random walk $S(n) = W_1 + \cdots + W_n$, and the limit Y has characteristic function (3.54). Let $\hat{\mu}_n(k)$ be the characteristic function of $n^{-1/\alpha}W_j$, so that $\hat{\mu}_n(k)^n \to \hat{\mu}(k)$ for all $k \in \mathbb{R}$. Then we have

$$\hat{\mu}_n(k)^{[nt]} = (\hat{\mu}_n(k)^n)^{[nt]/n} \to \hat{\mu}(k)^t \tag{3.64}$$

for any $t > 0$, and (3.60) follows, where the limit Z_t has characteristic function $\mu(k)^t$, so that (3.61) also holds.

If $1 < \alpha < 2$, then Theorem 3.39 shows that $n^{-1/\alpha}S(n) - n^{1-1/\alpha}\mu_1 \Rightarrow Y$, where the limit Y has characteristic function (3.55). Letting $\hat{\mu}_n(k)$ be the characteristic function of $n^{-1/\alpha}(W_j - \mu_1)$, it follows that $\hat{\mu}_n(k)^n \to \hat{\mu}(k)$ for all $k \in \mathbb{R}$. Again (3.64) holds, and then (3.62) follows, where the limit Z_t has characteristic function (3.63). □

Theorem 3.41 relates the parameters of a Pareto random walk to the FT of the limit process, an α-stable Lévy motion Z_t. For example, in the case $1 < \alpha < 2$ we have $\hat{p}(k,t) = \mathbb{E}[e^{-ikZ_t}] = \exp[tpD(ik)^\alpha + tqD(-ik)^\alpha]$, where $D = C\Gamma(2-\alpha)/(\alpha-1)$. Then

$$\frac{\partial p(x,t)}{\partial t} = pD\frac{\partial^\alpha p(x,t)}{\partial x^\alpha} + qD\frac{\partial^\alpha p(x,t)}{\partial(-x)^\alpha}$$

and we can see that the weights p, q on the positive and negative fractional derivatives come from the relative probability of large jumps in the positive or negative directions. This is consistent with our earlier conclusions, based on the Poisson approximation.

Remark 3.42. It is also possible to prove Theorem 3.41 directly. We illustrate the proof in the case $0 < \alpha < 1$ and $p = 1$. Define a triangular array $X_{nj} = n^{-1/\alpha}W_j$ for $j = 1, \ldots, [nt]$. Then condition (i) from Theorem 3.33 holds since:

$$\sum_{j=1}^{k_n} \mathbb{P}[X_{nj} > y] = [nt]\mathbb{P}[n^{-1/\alpha}W_j > y] = \frac{[nt]}{n} n\mathbb{P}[W_j > n^{1/\alpha}y] \to tCy^{-\alpha}; \text{ and}$$

$$\sum_{j=1}^{k_n} \mathbb{P}[X_{nj} < -y] = [nt]\mathbb{P}[n^{-1/\alpha}W_j < -y] \to 0.$$

Condition (ii) holds since

$$0 \le \sum_{j=1}^{k_n} \mathrm{Var}[X_{nj}^\varepsilon] \le [nt]E\left[(X_{nj}^\varepsilon)^2\right] = \frac{[nt]}{n} nE\left[(X_{nj}^\varepsilon)^2\right] \to 0.$$

Then $S_n - a_n \Rightarrow Z_t^0$, where the limit Z_t^0 is infinitely divisible with no normal component and Lévy measure $t\phi(dy)$, with ϕ given by (3.10). If (3.37) holds, then

$$\begin{aligned}\mathbb{E}[e^{ikZ_t^0}] &= \exp\left[\int \left(e^{iky} - 1 - ikyI(|y| \le R)\right) t\phi(dy)\right] \\ &= \exp\left[\int \left(e^{iky} - 1\right) t\phi(dy) - ikta\right]\end{aligned}$$

where a is given by (3.47). Since

$$a_n = \sum_{j=1}^{k_n} \mathbb{E}\left[X_{nj}^R\right] = [nt]\mathbb{E}\left[X_{nj}^R\right] = \frac{[nt]}{n}\left[\frac{C\alpha}{1-\alpha}R^{1-\alpha} - n^{1-1/\alpha}\frac{\alpha}{1-\alpha}C^{1/\alpha}\right] \to ta$$

as $n \to \infty$, we can let $Z_t = Z_t^0 - ta$, and it follows that (3.60) holds, where $\mathbb{E}[e^{ikZ_t}] = \exp\left[\int \left(e^{iky} - 1\right) t\phi(dy)\right]$. Then (3.61) follows from Proposition 3.10.

Remark 3.43. The convergence arguments in Theorem 3.41 shed some light on the structure of the limit process Z_t. This topic will be covered systematically in Chapter 4. Under the assumptions of this theorem, suppose that (3.60) holds for some $0 < \alpha < 1$, or some $1 < \alpha < 2$ with $E[W_n] = 0$. Given $s, t > 0$, write $S_n = n^{-1/\alpha}(W_1 + \cdots + W_n)$ and note that $S_{[n(t+s)]} \Rightarrow Z_{t+s}$. We also have

$$\begin{aligned}S_{[n(t+s)]} &= S_{[nt]} + \left(S_{[n(t+s)]} - S_{[nt]}\right) \\ &= n^{-1/\alpha}\sum_{j=1}^{[nt]} W_j + n^{-1/\alpha}\sum_{j=[nt]+1}^{[n(t+s)]} W_j \\ &\Rightarrow Z_t + \left(Z_{t+s} - Z_t\right)\end{aligned}$$

since the two sums are independent. This shows that $Z_{t+s} = Z_t + (Z_{t+s} - Z_t)$ a sum of two independent increments. Since the second sum is identically distributed with $S_{[ns]}$, it also shows that $Z_{t+s} - Z_t \simeq Z_s$, i.e., the distribution of the increments is stationary. In general, we define a Lévy process Z_t as an infinitely divisible process with stationary independent increments. Assuming Z_t has characteristic function $\hat{\mu}(k)^t = e^{t\psi(k)}$ *is not sufficient* to make Z_t a Lévy process. For example, take $Z \simeq \mathcal{N}(0, 2)$ and define $Z_t = t^{1/2}Z$. Then Z_t has characteristic function e^{-tk^2}, but Z_t does not have independent increments.

4 Continuous Time Random Walks

We begin this chapter by refining the stable limit theory from Chapter 3. We introduce regular variation as a technical tool to describe the full range of random walks attracted to a normal or stable limit. This shows that fractional diffusion is a robust model. Then we extend to the continuous time random walk (CTRW) by imposing a random waiting time between random walk jumps. The CTRW is studied as a random walk in space-time, which is then reduced to a time-changed process in space, using the fundamental ideas of Skorokhod. Finally, we develop the space-time fractional diffusion equations that govern CTRW scaling limits.

4.1 Regular variation

Regular variation is a technical tool that formalizes the idea of power law asymptotics. The necessary and sufficient conditions for the central limit to hold, even in the case of a normal limit, are written in terms of regular variation. Suppose that (W_n) are iid random variables, and Y is a random variable that is not degenerate (i.e., there is no constant y such that $\mathbb{P}[Y = y] = 1$). We want to know when

$$a_n(W_1 + \cdots + W_n) - b_n \Rightarrow Y \tag{4.1}$$

for some $a_n > 0$ and $b_n \in \mathbb{R}$.

Suppose that $R : [A, \infty) \to (0, \infty)$ is Borel measurable, for some $A > 0$. We say that $R(x)$ varies regularly with index ρ, and we write $R \in \mathrm{RV}(\rho)$, if

$$\lim_{x\to\infty} \frac{R(\lambda x)}{R(x)} = \lambda^\rho \quad \text{for all } \lambda > 0. \tag{4.2}$$

Then $R(\lambda x) \approx \lambda^\rho R(x)$, so that $R(x)$ behaves like a power law as $x \to \infty$. If $\rho = 0$ we also say that $R(x)$ is slowly varying. We say that a sequence of positive real numbers (a_n) is regularly varying with index ρ, and we write $(a_n) \in \mathrm{RV}(\rho)$, if

$$\lim_{n\to\infty} \frac{a_{[\lambda n]}}{a_n} = \lambda^\rho \quad \text{for all } \lambda > 0.$$

Example 4.1. The function $R(x) = x^\rho \log x$ is regularly varying with index ρ. The function $R(x) = x^{-\alpha}[2 + \cos x]$ is not regularly varying, because $R(\lambda x)/R(x)$ oscillates too fast to approach a limit as $x \to \infty$. The function $R(x) = \log x$ is slowly varying. If $R(x) \in \mathrm{RV}(\rho)$, then $L(x) = x^{-\rho}R(x)$ is slowly varying.

Remark 4.2. If a sequence of positive real numbers (a_n) is regularly varying with index ρ, then the function $R(x) = a_{[x]}$ varies regularly with the same index. Conversely, if a function $R(x)$ varies regularly with index ρ, then the sequence $a_n = R(n)$ varies regularly with the same index. The proof is surprisingly delicate, see Meerschaert and Scheffler [146, Theorem 4.2.9].

Let W be identically distributed with W_n and define

$$U_2(x) = \mathbb{E}[W^2 I(|W| \le x)] \quad \text{and} \quad V_0(x) = \mathbb{P}[|W| > x] \tag{4.3}$$

the truncated second moment and tail of W.

Example 4.3. Suppose that $V_0(x) = \mathbb{P}[W > x] = x^{-\alpha}$ for some $\alpha > 0$, for all $x \ge 1$. Then W has cdf $F(x) = \mathbb{P}[W \le x] = 1 - x^{-\alpha}$ and pdf $f(x) = \alpha x^{-\alpha-1}$ for $x \ge 1$. For $\zeta > \alpha$ we define the truncated moment

$$\begin{aligned} U_\zeta(x) = \mathbb{E}[W^\zeta I(W \le x)] &= \int_1^x y^\zeta f(y)\, dy \\ &= \int_1^x \alpha y^{\zeta-\alpha-1} dy = \frac{\alpha}{\zeta - \alpha}\left[y^{\zeta-\alpha}\right]_1^x \\ &= \frac{\alpha}{\zeta-\alpha}\left[x^{\zeta-\alpha} - 1\right] \sim \frac{\alpha}{\zeta-\alpha} x^{\zeta-\alpha} \end{aligned}$$

as $x \to \infty$. Then $U_\zeta(x) \to \infty$ as $x \to \infty$, i.e., $\mathbb{E}[W^\zeta]$ does not exist. For $0 \le \eta < \alpha$ we define the tail moment

$$\begin{aligned} V_\eta(x) = \mathbb{E}[W^\eta I(W > x)] &= \int_x^\infty y^\eta f(y)\, dy \\ &= \int_x^\infty \alpha y^{\eta-\alpha-1} dy = \frac{\alpha}{\eta-\alpha}\left[y^{\eta-\alpha}\right]_x^\infty = \frac{\alpha}{\alpha-\eta} x^{\eta-\alpha} \end{aligned}$$

so that $V_\eta(x) \to 0$ as $x \to \infty$. Combine to obtain the Karamata relation:

$$\frac{x^{\zeta-\eta} V_\eta(x)}{U_\zeta(x)} \to \frac{\zeta-\alpha}{\alpha-\eta} \quad \text{as } x \to \infty. \tag{4.4}$$

Theorem 4.4 (Karamata Theorem). *Suppose W is a random variable such that $U_\zeta(x) = \mathbb{E}[|W|^\zeta I(|W| \le x)]$ and $V_\eta(x) = \mathbb{E}[|W|^\eta I(|W| > x)]$ exist.*
(a) If $U_\zeta(x)$ is RV(ρ), then $\rho = \zeta - \alpha \ge 0$ for some α, and (4.4) holds;
(b) If $V_\eta(x)$ is RV(ρ), then $\rho = \eta - \alpha \le 0$ for some α, and (4.4) holds;
(c) If (4.4) holds for some $\alpha \in (\eta, \zeta]$, then $U_\zeta(x)$ is RV$(\zeta - \alpha)$;
(d) If (4.4) holds for some $\alpha \in [\eta, \zeta)$, then $V_\eta(x)$ is RV$(\eta - \alpha)$.

Proof. This is a special case of [146, Theorem 5.3.11]. See also Feller [68, VIII.8]. The proof uses integration by parts to relate U_ζ to V_η, along with some hard analysis. □

We say that W belongs to the *domain of attraction* of Y, and we write $W \in \text{DOA}(Y)$, if (4.1) holds for some $a_n > 0$ and $b_n \in \mathbb{R}$, where (W_n) are iid with W, and Y is nondegenerate. The following theorem gives necessary and sufficient conditions for $W \in \text{DOA}(Y)$ in terms of regular variation. It also proves that normal and stable laws are the only possible limits. The proof is based on Theorem 3.33, the convergence criteria for triangular arrays. It uses regular variation together with the Karamata Theorem 4.4 to compare the tail (condition (i) of Theorem 3.33) and the truncated second moment (condition (ii) of Theorem 3.33). The first part of the theorem, regarding normal limits, will be proved in this section. The second part, regarding stable limits, will be proven in Section 4.2.

Theorem 4.5 (Extended Central Limit Theorem). *If $W \in \text{DOA}(Y)$ then Y is either normal, or stable with index $0 < \alpha < 2$, and:*
(a) If Y is normal, then $W \in \text{DOA}(Y)$ if and only if $U_2(x)$ is slowly varying;
(b) If Y is stable with index $0 < \alpha < 2$, then $W \in \text{DOA}(Y)$ if and only if $V_0(x)$ is regularly varying with index $-\alpha$ and

$$\lim_{x\to\infty} \frac{\mathbb{P}[W > x]}{V_0(x)} = p \quad \textit{for some } 0 \le p \le 1. \tag{4.5}$$

Proof of Theorem 4.5 (a). Suppose that (W_j) are iid with W and that $U_2(x)$ is slowly varying. Then $\mu_1 = \mathbb{E}[W]$ exists (see Proposition 4.14 in the details at the end of this section). If $\mathbb{E}[W^2] < \infty$, we have already proven in Theorem 3.36 that (4.1) holds. Otherwise if $\mathbb{E}[W^2] = \infty$, then $U_2(x) \to \infty$ as $x \to \infty$. Choose $a_n \to 0$ such that

$$na_n^2 U_2(a_n^{-1}) \to \sigma^2 > 0 \tag{4.6}$$

(see Corollary 4.13 at the end of this section for an explicit construction). Define a triangular array with row elements $X_{nj} = a_n W_j$ for $j = 1, \dots, n$. Then condition (3.36) holds (see details), and so it suffices to check conditions (i) and (ii) from Theorem 3.33.

Condition (i): Apply Theorem 4.4 (a) with $\zeta = 2$, $\eta = 0$, and $\rho = 0$ to see that the Karamata equation (4.4) holds with $\alpha = 2$. Then

$$\begin{aligned}
\sum_{j=1}^{k_n} \mathbb{P}[|X_{nj}| > \varepsilon] &= n\mathbb{P}[|a_n W_j| > \varepsilon] \\
&= nV_0(a_n^{-1}\varepsilon) \\
&= \frac{(a_n^{-1}\varepsilon)^2 V_0(a_n^{-1}\varepsilon)}{U_2(a_n^{-1}\varepsilon)} \cdot \varepsilon^{-2} \cdot na_n^2 U_2(a_n^{-1}\varepsilon) \\
&\to 0 \cdot \varepsilon^{-2} \cdot \sigma^2
\end{aligned}$$

since $x^2 V_0(x)/U_2(x) \to (2-\alpha)/\alpha = 0$ by (4.4), and

$$na_n^2 U_2(a_n^{-1}\varepsilon) = na_n^2 U_2(a_n^{-1}) \cdot \frac{U_2(a_n^{-1}\varepsilon)}{U_2(a_n^{-1})} \to \sigma^2 \cdot 1$$

by (4.6), and the fact that $U_2(x\varepsilon)/U_2(x) \to 1$ as $x \to \infty$. This shows that (i) holds with $\phi\{x : |x| > \varepsilon\} = 0$ for all $\varepsilon > 0$, i.e., $\phi = 0$, the zero measure.

Condition(ii): Since $U_2(a_n^{-1}) \to \infty$ it follows from (4.6) that $na_n^2 \to 0$. Then with $X_{nj}^\varepsilon = X_{nj}I[|X_{nj}| \le \varepsilon]$ we have

$$\begin{aligned}\sum_{j=1}^{k_n} \mathrm{Var}[X_{nj}^\varepsilon] &= n\mathbb{E}[(X_{nj}^\varepsilon)^2] - n\mathbb{E}[X_{nj}^\varepsilon]^2 \\ &= n\mathbb{E}[(a_n W)^2 I(|a_n W| \le \varepsilon)] - n\mathbb{E}[a_n W I(|a_n W| \le \varepsilon)]^2 \\ &= na_n^2 U_2(a_n^{-1}\varepsilon) - na_n^2\mathbb{E}[WI(|W| \le a_n^{-1}\varepsilon)]^2 \\ &\sim na_n^2 U_2(a_n^{-1}\varepsilon) \to \sigma^2 \end{aligned} \tag{4.7}$$

since $\mathbb{E}[WI(|W| \le a_n^{-1}\varepsilon)] \to \mu_1$ by the dominated convergence theorem, and $na_n^2 \to 0$. Then it follows from Theorem 3.33 that (4.1) holds with Y normal.

Since the direct half of Theorem 4.5(a) is our main interest, we only sketch the proof of the converse, highlighting the role of regular variation arguments. Suppose that (4.1) holds with Y normal. Assume for now that $\mu_1 = \mathbb{E}[W] = 0$. Then conditions (i) and (ii) hold from Theorem 3.33. Writing (ii) as in (4.7) it follows that

$$na_n^2 U_2(a_n^{-1}\varepsilon) - na_n^2 U_1(a_n^{-1}\varepsilon)^2 \to \sigma^2 = \mathrm{Var}(Y). \tag{4.8}$$

If (4.1) holds with Y nondegenerate, a simple argument using characteristic functions [146, Lemma 3.3.3] shows that $a_n \to 0$. Then a dominated convergence argument yields $U_1(a_n^{-1}\varepsilon) = \mathbb{E}[WI(|W| \le a_n^{-1}\varepsilon)] \to \mu_1 = 0$. A similar argument shows that $U_2(a_n^{-1}\varepsilon) = \mathbb{E}[W^2 I(|W| \le a_n^{-1}\varepsilon)] \to E[W^2]$, where $0 < E[W^2] \le \infty$ since W is not degenerate. It follows that $U_1(a_n^{-1}\varepsilon)^2 = o(U_2(a_n^{-1}\varepsilon))$ as $n \to \infty$, and then (4.8) yields $na_n^2 U_2(a_n^{-1}\varepsilon) \to \sigma^2$ for all $\varepsilon > 0$. Then an argument similar to the first part of the proof of Proposition 4.15 in the next section shows that $x^{-2}U_2(x)$ varies regularly with index -2, and it follows that U_2 is slowly varying. See Feller [68, XVII.5] or [146, Theorem 8.1.11] for complete details.

Finally, if (4.1) holds with Y normal, then a convergence of types argument [146, Theorem 8.1.5] shows that (a_n) is RV$(-1/2)$, and then it follows from condition (ii) and a regular variation estimate [146, Proposition 8.1.6] that $\mu_1 = \mathbb{E}[W]$ exists, so the assumption $\mu_1 = 0$ entails no loss of generality: Simply replace W_j by $W_j - \mu_1$, which changes the shift b_n. □

Corollary 4.6. *We can choose $b_n = na_n\mu_1$ in (4.1) when Y is normal.*

Proof. This was already proven in Theorem 3.36, in the case $\mathbb{E}[W^2] < \infty$. In the general case, Theorem 3.33 implies that we can take

$$b_n = \sum_{j=1}^{k_n} \mathbb{E}[X_{nj}^R] = n\mathbb{E}[a_n W I(|a_n W| \le R)] = na_n\mathbb{E}[WI(|W| \le a_n^{-1}R)] \sim na_n\mu_1$$

since $\mathbb{E}[WI(|W| \le a_n^{-1}R)] \to \mu_1$ as $n \to \infty$. □

Remark 4.7. For finite variance jumps, we can take $a_n = n^{-1/2}$ in (4.1). For infinite variance jumps and Y normal, Corollary 4.13 in the details at the end of this section shows that the sequence (a_n) is RV$(-1/2)$. Then we can write $a_n = n^{-1/2}\ell_n$ where (ℓ_n) is slowly varying. Now Proposition 4.9 together with Remark 4.2 show that for any $\varepsilon > 0$, for some $n_0 > 0$ we have

$$n^{-\varepsilon} < \ell_n < n^{\varepsilon} \tag{4.9}$$

for all $n \geq n_0$. In other words, the norming constants $a_n \to 0$ about as fast as $n^{-1/2}$ when Y is normal.

Details

Since W_j is tight for any fixed j, so that (3.43) holds with $X = W_j$, it follows that

$$\mathbb{P}[|X_{nj}| > \varepsilon] = \mathbb{P}[|W_j| > a_n^{-1}\varepsilon] \to 0,$$

since $a_n \to 0$ as $n \to \infty$, so that condition (3.36) holds.

The theory of regular variation is simpler for monotone functions. We will restrict to this case, since it suffices for all our applications. The next four results remain true if we remove the assumption that $R(x)$ is monotone, but the proofs are significantly harder, see Seneta [197, Theorem 1.1 and Section 1.5].

Proposition 4.8. *If $R(x)$ is monotone and $R(x) \in$ RV(ρ) for some $\rho > 0$, then $R(x) \to \infty$ as $x \to \infty$.*

Proof. Fix some $\lambda > 1$ large and note that for all $\delta > 0$ small, there exists some $x_0 > 0$ such that

$$\frac{R(\lambda x)}{R(x)} \geq \lambda^{\rho}(1-\delta) > 1$$

for all $x \geq x_0$. Given $x > x_0$, we can write $x = \zeta\lambda^n x_0$ for some unique nonnegative integer $n = n_x$ and some unique real number $\zeta = \zeta_x \in [1, \lambda)$. Then

$$\frac{R(x)}{R(x_0)} = \frac{R(\zeta\lambda^n x_0)}{R(x_0)} = \frac{R(\zeta\lambda^n x_0)}{R(\lambda^n x_0)}\frac{R(\lambda^n x_0)}{R(\lambda^{n-1} x_0)}\cdots\frac{R(\lambda x_0)}{R(x_0)} \geq [\lambda^{\rho}(1-\delta)]^n$$

tends to infinity as $x \to \infty$. □

Proposition 4.9. *If $R(x)$ is monotone and $R(x) \in$ RV(ρ), then for any $\varepsilon > 0$, for some $x_0 > 0$ we have*

$$x^{\rho-\varepsilon} < R(x) < x^{\rho+\varepsilon} \tag{4.10}$$

for all $x \geq x_0$.

Proof. The function $x^{-\rho+\varepsilon}R(x)$ is RV(ε), so it tends to infinity as $x \to \infty$ by Proposition 4.8. This proves that $x^{\rho-\varepsilon} < R(x)$ for all large x. The proof of the upper bound is similar. □

Theorem 4.10 (Uniform Convergence Theorem). *Suppose $R(x)$ is monotone and $R(x) \in \mathrm{RV}(\rho)$. Then for any sequence $\lambda_n \to \lambda > 0$, and any sequence $x_n \to \infty$, we have*

$$\frac{R(\lambda_n x_n)}{R(x_n)} \to \lambda^\rho \tag{4.11}$$

as $n \to \infty$.

Proof. Since $\lambda_n \to \lambda$ as $n \to \infty$, for any $\delta > 0$ such that $\lambda - \delta > 0$, there exists some $n_0 > 0$ such that $\lambda - \delta < \lambda_n < \lambda + \delta$ for all $n \geq n_0$. If R is monotone nondecreasing, write

$$\frac{R(x_n(\lambda - \delta))}{R(x_n)} \leq \frac{R(x_n\lambda_n)}{R(x_n)} \leq \frac{R(x_n(\lambda + \delta))}{R(x_n)}$$

The left-hand side of the above inequality converges to $(\lambda - \delta)^\rho$, and the right-hand side converges to $(\lambda + \delta)^\rho$. Since $\delta > 0$ can be made arbitrarily small, it follows that (4.11) holds. The proof for R monotone nonincreasing is similar. □

Remark 4.11. The condition that (4.11) holds for all $\lambda_n \to \lambda > 0$ and all $x_n \to \infty$ is equivalent to the condition that (4.2) holds uniformly on compact subsets of $\lambda > 0$. Theorem 4.10 is usually stated in terms of uniform convergence on compact sets, e.g., see Seneta [197, Theorem 1.1]. The proof is much harder when R is not monotone.

Proposition 4.12. *If $R(x)$ is monotone and $R(x) \in \mathrm{RV}(\rho)$ for some $\rho > 0$, then there exists a sequence $r_n \to \infty$ such that $R(r_n) \sim n$ as $n \to \infty$. In that case, the sequence (r_n) varies regularly with index $1/\rho$.*

Proof. Define $r_n = \inf\{x > 0 : R(x) > n\}$, which exists because $R(x) \to \infty$ by Proposition 4.8. Since R is monotone, $r_n \leq r_{n+1}$, so the limit of r_n as $n \to \infty$ exists. This limit cannot be finite: If $r_n \to r < \infty$, then $r_n \leq r$ for all n, so $R(r + 1) \geq R(r_n + 1) > n$ by definition of r_n. Then $R(r + 1) = \infty$, which is a contradiction. Therefore $r_n \to \infty$ as $n \to \infty$. Since R is monotone, $R(r_n + \varepsilon_n) > n$ for any $\varepsilon_n \downarrow 0$, and $R(r_n - \varepsilon_n) \leq n$. Write

$$\frac{R(r_n)}{R(r_n + \varepsilon_n)} < \frac{R(r_n)}{n} \leq \frac{R(r_n)}{R(r_n - \varepsilon_n)}$$

and apply Theorem 4.10 with $\lambda_n = (r_n + \varepsilon_n)/r_n \to 1$ to see that

$$\frac{R(r_n)}{R(r_n + \varepsilon_n)} = \frac{R(r_n)}{R(\lambda_n r_n)} \to 1$$

as $n \to \infty$. A similar argument shows that the right-hand side tends to the same limit, and then it follows that $R(r_n) \sim n$.

It remains to show that the sequence (r_n) varies regularly with index $1/\rho$. Since $R(r_n) \sim n$, and since R is $\mathrm{RV}(\rho)$ and monotone, it follows from Theorem 4.10 that

$$\frac{R(r_n x_n)}{n} = \frac{R(r_n)}{n}\frac{R(r_n x_n)}{R(r_n)} \to x^\rho \quad \text{whenever } x_n \to x > 0. \tag{4.12}$$

Define $x_n = r_{[\lambda n]} r_n^{-1}$ for some fixed $\lambda > 0$, and write

$$\frac{R(r_n x_n)}{n} = \frac{[\lambda n]}{n} \frac{R(r_{[\lambda n]})}{[\lambda n]} \to \lambda.$$

Then a simple proof by contradiction shows that $x_n \to \lambda^{1/\rho}$: If any subsequence $(x_{n'})$ of (x_n) satisfies $x_{n'} \to 0$, then (4.12) implies that $R(r_{n'} x_{n'})/n' \to 0$; if $x_{n'} \to \infty$, then $R(r_{n'} x_{n'})/n' \to \infty$; and if $x_{n'} \to b \neq \lambda^{1/\rho}$, then $R(r_{n'} x_{n'})/n' \to b^\rho \neq \lambda$. □

Corollary 4.13. *If $U_2(x)$ is slowly varying, then* (4.6) *holds for some $a_n \to 0$, and (a_n) is* RV(−1/2).

Proof. If U_2 is slowly varying, then $R(x) = \sigma^2 x^2 / U_2(x)$ is RV(2). Apply Proposition 4.12 to obtain a sequence $r_n = a_n^{-1}$ in RV(1/2) such that $\sigma^2 a_n^2 U_2(a_n^{-1}) \sim n$, which is equivalent to (4.6). □

Proposition 4.14. *If $U_2(x)$ is slowly varying, then $\mathbb{E}[W]$ exists.*

Proof. Apply Karamata (4.4) to see that $x^2 V_0(x)/U_2(x) \to (2 - \alpha)/\alpha = 0$. Then for some $x_0 > 0$ we have $V_0(x) \le x^{-2} U_2(x)$ for all $x \ge x_0$. Given any $\varepsilon > 0$, Proposition 4.9 implies that $V_0(x) \le x^{\varepsilon - 2}$ for all $x \ge x_0$. Write

$$\mathbb{E}[|W|] = \int_0^\infty \mathbb{P}[|W| > x]\, dx = \int_0^\infty V_0(x)\, dx \le x_0 + \int_{x_0}^\infty x^{\varepsilon-2} dx < \infty$$

for any $0 < \varepsilon < 1$. □

4.2 Stable Central Limit Theorem

In this section, we will prove part (b) of Theorem 4.5, the necessary and sufficient conditions for the central limit theorem (4.1) to hold when Y is not normal. We say W is regularly varying if

$$n\mathbb{P}[a_n W \in dy] \to \phi(dy) \quad \text{as } n \to \infty \tag{4.13}$$

for some $a_n \to 0$ and some σ-finite Borel measure ϕ on $\{y \neq 0\}$ which is not the zero measure. The vague convergence in (4.13) is the same as for condition (i) in Theorem 3.33, the convergence criteria for triangular arrays.

Proposition 4.15. *Suppose that W is regularly varying and* (4.13) *holds. Then:*
(a) For some $\alpha > 0$ we have

$$\phi(dy) = \begin{cases} pC\alpha y^{-\alpha-1} dy & \text{for } y > 0 \\ qC\alpha |y|^{-\alpha-1} dy & \text{for } y < 0 \end{cases} \tag{4.14}$$

for some $C > 0$ and some $p, q \geq 0$ with $p + q = 1$;

(b) The sequence (a_n) is RV$(-1/\alpha)$, that is,

$$\frac{a_{[\lambda n]}}{a_n} \to \lambda^{-1/\alpha} \quad \text{as } n \to \infty \tag{4.15}$$

for all $\lambda > 0$;

(c) The tail $V_0(x) = \mathbb{P}[|W| > x]$ is RV$(-\alpha)$ and the tail balance condition (4.5) holds. Conversely, these two conditions imply W is regularly varying and (4.13) holds.

See details at the end of this section for proof. When Proposition 4.15 holds, we will also say that W is RV$(-\alpha)$.

Proof of Theorem 4.5 (b). In view of Proposition 4.15 (c), it suffices to show that (4.1) holds with Y nonnormal if and only if W is RV$(-\alpha)$. Suppose that (W_j) are iid with W, and that W is RV$(-\alpha)$ for some $0 < \alpha < 2$. Define a triangular array with row elements $X_{nj} = a_n W_j$ for $j = 1, \ldots, n$. Then condition (3.36) holds (see details), and so in order to show that (4.1) holds, it suffices to check the convergence criteria (i) and (ii) for triangular arrays in Theorem 3.33. Proposition 4.15 (a) along with (4.13) shows that (i) holds, where ϕ is given by the formula (4.14). Since $0 < \alpha < 2$, it is not hard to check that ϕ is a Lévy measure. For condition (ii) we apply the Karamata Theorem 4.4 to see that

$$\frac{x^2 V_0(x)}{U_2(x)} \to \frac{2-\alpha}{\alpha} \quad \text{as } x \to \infty$$

so that $U_2(x) \sim \alpha x^2 V_0(x)/(2-\alpha)$ as $x \to \infty$. Then

$$\begin{aligned}
0 \leq \sum_{j=1}^{k_n} \operatorname{Var}[X_{nj}^{\varepsilon}] &\leq n\mathbb{E}[(X_{nj}^{\varepsilon})^2] \\
&= n a_n^2 \mathbb{E}[W^2 I(|a_n W| \leq \varepsilon)] \\
&= n a_n^2 U_2(a_n^{-1}\varepsilon) \\
&\sim n a_n^2 \frac{\alpha}{2-\alpha} (a_n^{-1}\varepsilon)^2 V_0(a_n^{-1}\varepsilon) \\
&= \frac{\alpha}{2-\alpha} \varepsilon^2 n V_0(a_n^{-1}\varepsilon) \\
&= \frac{\varepsilon^2 \alpha}{2-\alpha} n\mathbb{P}[|a_n W| > \varepsilon] \\
&\to \frac{\varepsilon^2 \alpha}{2-\alpha} \phi\{y : |y| > \varepsilon\} = \frac{\varepsilon^2 \alpha}{2-\alpha} C\varepsilon^{-\alpha}
\end{aligned}$$

so that

$$\lim_{\varepsilon \to 0} \limsup_{n \to \infty} \sum_{j=1}^{k_n} \operatorname{Var}[X_{nj}^{\varepsilon}] \leq \lim_{\varepsilon \to 0} \frac{\alpha}{2-\alpha} C\varepsilon^{2-\alpha} = 0.$$

This proves that (4.1) holds, where Y has Lévy representation $[0, 0, \phi]$. Then it follows from Proposition 4.15 (a) that Y is stable with index α.

Conversely, if (4.1) holds where Y is nonnormal, the triangular array convergence condition (*i*) shows that (4.13) holds, where ϕ is not the zero measure. Then W is RV$(-\alpha)$, and since ϕ is a Lévy measure, it is easy to check that $0 < \alpha < 2$. □

The next result provides specific details about the centering constants and limit distribution in the stable case.

Proposition 4.16. *Suppose that* (4.1) *holds, where Y is stable with index $0 < \alpha < 2$ and Lévy measure* (4.14).

(a) If $0 < \alpha < 1$, we can take $b_n = 0$, and then the limit Y is centered stable with characteristic function

$$\mathbb{E}[e^{ikY}] = \exp\left(-C\Gamma(1-\alpha)[p(-ik)^\alpha + q(ik)^\alpha]\right); \tag{4.16}$$

(b) If $1 < \alpha < 2$, we can take $b_n = na_n\mathbb{E}[W]$, and then the limit Y is centered stable with mean zero and characteristic function

$$\mathbb{E}[e^{ikY}] = \exp\left(C\frac{\Gamma(2-\alpha)}{\alpha-1}[p(-ik)^\alpha + q(ik)^\alpha]\right). \tag{4.17}$$

Proof. We illustrate the proof in the special case where $W > 0$, so that $p = 1$. For the general case, see [146, Theorem 8.2.7]. Suppose that $a_n(W_1 + \cdots + W_n) - b_n \Rightarrow Y_1$. In case (a), by exactly the same argument as for the Pareto (see Proposition 3.10), we get

$$\begin{aligned}\mathbb{E}[e^{ikY_1}] &= \exp\left[\int (e^{iky} - 1)\phi(dy) - ik\int yI(|y| \le R)\phi(dy)\right]\\ &= \exp\left[-C\Gamma(1-\alpha)(-ik)^\alpha - ikb\right]\end{aligned}$$

where

$$b = \int_{|y|\le R} y\phi(dy) = \frac{C\alpha}{1-\alpha}R^{1-\alpha}.$$

By Karamata (4.4) we have $U_1(x) \sim \alpha x V_0(x)/(1-\alpha)$. The centering constants are given by

$$\begin{aligned}b_n &= \sum_{j=1}^{k_n}\mathbb{E}[X_{nj}^R] = \sum_{j=1}^{k_n}\mathbb{E}[X_{nj}I(|X_{nj}| \le R)]\\ &= na_n\mathbb{E}[WI(|W| \le a_n^{-1}R)]\\ &= na_nU_1(a_n^{-1}R)\\ &\sim na_n\frac{\alpha}{1-\alpha}a_n^{-1}RV_0(a_n^{-1}R)\\ &= \frac{R\alpha}{1-\alpha}n\mathbb{P}[|a_nW| > R] \to \frac{R\alpha}{1-\alpha}\phi\{|y| > R\} = b\end{aligned}$$

since $\phi\{|y| > R\} = CR^{-\alpha}$. Define $Y = Y_1 + b$. Then $a_n(W_1 + \cdots + W_n) \Rightarrow Y$, and the limit is centered stable, i.e., stable with characteristic function given by (4.16).

In case (b), by exactly the same argument as for the Pareto (see Proposition 3.12), we get

$$\mathbb{E}[e^{ikY_1}] = \exp\left[\int (e^{iky} - 1 - iky)\phi(dy) + ik\int yI(|y| > R)\phi(dy)\right]$$
$$= \exp\left[C\frac{\Gamma(2-\alpha)}{\alpha - 1}(-ik)^\alpha + ikb\right]$$

where

$$b = \int\limits_{|y|>R} y\phi(dy) = \frac{C\alpha}{\alpha - 1}R^{1-\alpha}.$$

From Karamata (4.4) we get

$$\frac{xV_1(x)}{U_2(x)} \to \frac{2-\alpha}{\alpha-1} \quad \text{and} \quad \frac{x^2V_0(x)}{U_2(x)} \to \frac{2-\alpha}{\alpha}$$

so that

$$V_1(x) \sim \frac{2-\alpha}{\alpha-1}x^{-1}U_2(x) \sim \frac{\alpha}{\alpha-1}xV_0(x).$$

The centering constants are

$$\begin{aligned} b_n &= na_n\mathbb{E}[WI(|a_nW| \le R)] \\ &= na_n\left\{\mu_1 - \mathbb{E}[WI(|W|| > a_n^{-1}R)]\right\} \\ &= na_n\mu_1 - na_nV_1(a_n^{-1}R) \\ &\sim na_n\mu_1 - na_n\frac{\alpha}{\alpha-1}a_n^{-1}RV_0(a_n^{-1}R) \\ &= na_n\mu_1 - \frac{R\alpha}{\alpha-1}n\mathbb{P}[|a_nW| > R] \\ &\sim na_n\mu_1 - \frac{R\alpha}{\alpha-1}\phi\{|y| > R\} = na_n\mu_1 - b \end{aligned}$$

since $\phi\{|y| > R\} = CR^{-\alpha}$. Define $Y = Y_1 - b$. Then

$$a_n(W_1 + \cdots + W_n) - na_n\mu_1 = a_n\sum_{j=1}^{n}(W_j - \mathbb{E}[W_j]) \Rightarrow Y.$$

This limit Y is centered stable, with characteristic function (4.17), and Remark 3.40 shows that $\mathbb{E}[Y] = 0$. □

Remark 4.17. The convergence (4.1) extends to random walk limits. If

$$a_n(W_1 + \cdots + W_n) - b_n \Rightarrow Z_1$$

where Z_1 is normal or stable, then we also have convergence of the characteristic functions

$$\hat{\mu}_n(k)^n \to \hat{\mu}(k) = e^{\psi(k)}$$

where μ_n is the distribution of $a_n W - n^{-1} b_n$, and μ is the distribution of the infinitely divisible random variable Z_1. It follows easily that

$$\hat{\mu}_n(k)^{[nt]} = (\hat{\mu}_n(k)^n)^{[nt]/n} \to e^{t\psi(k)}$$

for any $t \geq 0$, which means that

$$a_n(W_1 + \cdots + W_{[nt]}) - \frac{[nt]}{n} b_n \Rightarrow Z_t \tag{4.18}$$

for any $t \geq 0$. The limit Z_t is a *Lévy process*, see Section 4.3 for more details. If $0 < \alpha < 1$, then Proposition 4.16 (a) shows that we can take $b_n = 0$. If $1 < \alpha < 2$, then Proposition 4.16 (b) shows that we can take $b_n = na_n\mathbb{E}[W]$ where $a_n \to 0$, and Corollary 4.6 shows that the same is true when $\alpha = 2$. In the case $1 < \alpha \leq 2$, equation (4.18) can also be written in the form

$$a_n \sum_{j=1}^{[nt]} (W_j - v) \Rightarrow Z_t$$

where $v = \mathbb{E}[W]$. Using two scales leads to a Lévy process with drift:

$$a_n \sum_{j=1}^{[nt]} (W_j - v) + n^{-1} \sum_{j=1}^{[nt]} v \Rightarrow Z_t + vt \tag{4.19}$$

since $[nt]/n \to t$. Two different scales are required here since $a_n \to 0$ at a different rate than n^{-1} when $\alpha \neq 1$.

Remark 4.18. Some authors use a different centering in Remark 4.17. Suppose that (4.18) holds where Z_t is either normal, or stable with index $1 < \alpha < 2$, so that $\mathbb{E}[W]$ exists. Then

$$\left|\frac{[nt]}{n} b_n - tb_n\right| = \left(\frac{nt - [nt]}{n}\right) b_n \leq \left(\frac{1}{n}\right) na_n\mathbb{E}[W] = a_n\mathbb{E}[W] \to 0.$$

Now it follows from (4.18) that

$$a_n(W_1 + \cdots + W_{[nt]}) - tb_n \Rightarrow Z_t \tag{4.20}$$

for any $t \geq 0$.

Details

Proof of Proposition 4.15. First we will prove part (c). Suppose that W is regularly varying and (4.13) holds. Define $B = \{y : |y| > x\}$ and $G(x) = \phi(B)$, and apply (4.13) to see that

$$nV_0(a_n^{-1}x) = n\mathbb{P}[a_n W \in B] \to \phi(B) = G(x) \tag{4.21}$$

as $n \to \infty$ for all x such that $G(x+) = G(x-)$. Since ϕ is not the zero measure, we have $G(x) > 0$ for some $x > 0$. Since $G(x)$ is monotone, it has at most a countable number of discontinuities. Without loss of generality, we may assume that $x = 1$ is a continuity point, with $C = G(1) > 0$. Define $n = n(x) = \inf\{n > 0 : a_{n+1}^{-1} > x\}$ so that $a_n^{-1} \le x < a_{n+1}^{-1}$. Then

$$\frac{nV_0(a_n^{-1}r)}{nV_0(a_{n+1}^{-1})} \le \frac{V_0(rx)}{V_0(x)} \le \frac{nV_0(a_{n+1}^{-1}r)}{nV_0(a_n^{-1})}$$

where

$$nV_0(a_{n+1}^{-1}r) = \frac{n}{n+1}(n+1)V_0(a_{n+1}^{-1}r) \to G(r)$$

if r is a continuity point of G. Define $\varphi(r) = G(r)/G(1)$. Let $n \to \infty$ to see that

$$\lim_{x\to\infty} \frac{V_0(rx)}{V_0(x)} = \varphi(r) \tag{4.22}$$

if r is a continuity point. If r, λ, and $r\lambda$ are continuity points, we can take the limit as $x \to \infty$ on both sides of the equation

$$\frac{V_0(r\lambda x)}{V_0(x)} = \frac{V_0(r\lambda x)}{V_0(rx)}\frac{V_0(rx)}{V_0(x)}$$

to see that $\varphi(r\lambda) = \varphi(r)\varphi(\lambda)$. It follows that $\varphi(r) = r^\rho$ for some $\rho \in \mathbb{R}$ (see Seneta [197, Lemma 1.6]), and then $G(r) = Cr^\rho$. Hence every $r > 0$ is a continuity point, so (4.22) holds for every $r > 0$. Since $G(r) \to 0$ as $r \to \infty$, $\rho < 0$. Then V_0 varies regularly with index $\rho = -\alpha$ for some $\alpha > 0$. Now write

$$\frac{nV_+(a_n^{-1})}{nV_0(a_{n+1}^{-1})} \le \frac{V_+(x)}{V_0(x)} \le \frac{nV_+(a_{n+1}^{-1})}{nV_0(a_n^{-1})}$$

and let $x \to \infty$ (which means that $n = n(x) \to \infty$ as well) to see that the tail balance condition (4.5) holds with $p = \phi\{y : y > 1\}/\phi\{y : |y| > 1\}$, so that $0 \le p \le 1$.

Conversely, suppose that $V_0(x)$ is $RV(-\alpha)$ and (4.5) holds. Apply Proposition 4.12 with $R(x) = C/V_0(x)$ to obtain a sequence r_n such that $R(r_n) \sim n$. Define $a_n = r_n^{-1}$ so that $nV_0(a_n^{-1}) = n\mathbb{P}[|a_n W| > 1] \to C > 0$. Since $a_n^{-1} = r_n \to \infty$, it follows from (4.2) that

$$n\mathbb{P}[|a_n W| > x] = nV_0(a_n^{-1}x) = nV_0(a_n^{-1})\frac{nV_0(a_n^{-1}x)}{nV_0(a_n^{-1})} \to Cx^{-\alpha}$$

for all $x > 0$. Using (4.5) it follows that $n\mathbb{P}[a_n W > x] \sim np\mathbb{P}[|a_n W| > x] \to pCx^{-\alpha}$ and similarly for the left tail. This is sufficient to prove that (4.13) holds with ϕ given by (4.14), which proves part (c) and also part (a). Proposition 4.12 also implies that (r_n) varies regularly with index $1/\alpha$. Then (a_n) varies regularly with index $-1/\alpha$ so that (4.15) holds, which proves part (b). This concludes the proof. □

If W is RV$(-\alpha)$ for some $0 < \alpha < 2$, then (a_n) is RV$(-1/\alpha)$. Then Proposition 4.9 together with Remark 4.2 imply that $a_n \to 0$. Since W_j is tight for any fixed j, so that (3.43) holds with $X = W_j$, it follows that

$$\mathbb{P}[|X_{nj}| > \varepsilon] = \mathbb{P}[|W_j| > a_n^{-1}\varepsilon] \to 0,$$

so that condition (3.36) holds.

4.3 Continuous time random walks

In a continuous time random walk (CTRW), we assume a random waiting time between particle jumps. Let $S(n) = Y_1 + \cdots + Y_n$ be a random walk with iid particle jumps. Define another random walk $T(n) = J_1 + \cdots + J_n$ where $J_n \geq 0$ are iid waiting times between particle jumps, so that a particle arrives at location $S(n)$ at time $T(n)$. Here we also suppose that (Y_n) are independent of (J_n), so the CTRW is uncoupled. Let

$$N(t) = \max\{n \geq 0 : T(n) \leq t\}$$

denote the number of particle jumps by time $t \geq 0$, where $T(0) = 0$. Then the CTRW $S(N(t))$ is the particle location at time $t \geq 0$. Our goal is to determine the limit process for this CTRW. Then in Section 4.5, we will derive the governing equation of the CTRW limit.

Since $T(n)$ is a random walk, its limit distribution can be obtained as we did for $S(n)$. Suppose that Y_n are iid with Y, and that $Y \in \text{DOA}(A)$ where A is either normal, or stable with index $0 < \alpha < 2$. Then

$$a_n S(n) - b_n \Rightarrow A \tag{4.23}$$

for some $a_n > 0$ and b_n real. Suppose that $b_n = 0$, e.g., assume that $\mathbb{E}[Y] = 0$ in the case $1 < \alpha \leq 2$. Then Remark 4.17 shows that we also get random walk convergence

$$a_n S([nt]) \Rightarrow A(t) \tag{4.24}$$

where the limit $A(t)$ is a Brownian motion, or an α-stable Lévy motion. Suppose J_n are iid with J, and $J \in \text{DOA}(D)$. If $\mathbb{E}[J]$ exists, then the renewal theorem (e.g., see Durrett [62, Theorem 2.4.6]) shows that $N(t)/t \to \lambda = 1/\mathbb{E}[J]$ with probability one as $t \to \infty$. That is, $N(t) \approx \lambda t$ for t large. The proof of this fact is a simple application of the strong law of large numbers. Then

$$a_n S(N(nt)) \Rightarrow A(\lambda t)$$

and the effect of the waiting times is just a change of scale (see details). However, if $\mathbb{E}[J] = \infty$, the CTRW behaves quite differently.

Suppose that $J \in \text{DOA}(D)$ where D is β-stable with $0 < \beta < 1$. For example, we could take $\mathbb{P}[J > t] = Bt^{-\beta}$ for some $B > 0$. Then Proposition 4.16 (a) shows that

$c_n T_n \Rightarrow D$ for some $c_n \to 0$, and Remark 4.17 shows that the random walk converges:

$$c_n T([nt]) \Rightarrow D(t) \tag{4.25}$$

where $D(t)$ is called a β-stable *subordinator*. Since every $J_n \geq 0$, $D(t)$ is a one-sided stable with $p = 1$ and $q = 0$. Also, if $0 < t_1 < t_2$, then $c_n T([nt_1]) \leq c_n T([nt_2])$ for all n, which shows that the limit $D(t_1) \leq D(t_2)$, i.e., the process $D(t)$ is increasing. In fact, we have

$$\begin{aligned} c_n T([nt_2]) &= c_n T([nt_1]) + c_n(T([nt_2]) - T([nt_1])) \\ &= c_n \sum_{j=1}^{[nt_1]} J_j + c_n \sum_{j=[nt_1]+1}^{[nt_2]} J_j \Rightarrow D(t_1) + [D(t_2) - D(t_1)] \end{aligned}$$

and since the sums are independent, the process $D(t)$ has *independent increments*. Take weak limits on both sides of

$$c_n \sum_{j=[nt_1]+1}^{[nt_2]} J_j \simeq c_n \sum_{j=1}^{[nt_2]-[nt_1]} J_j$$

to see that $D(t_2) - D(t_1) \simeq D(t_2 - t_1)$, i.e., the process $D(t)$ has *stationary increments*. A process $\{D(t) : t \geq 0\}$ with stationary, independent increments is called a *Lévy process*. (A subordinator is a Lévy process with nondecreasing sample paths.) Usually we also assume that $D(0) = 0$ with probability one, which is certainly true here. Clearly a Lévy process is infinitely divisible, since $D(t) = D(t/n) + [D(2t/n) - D(t/n)] + \cdots + [D(t) - D((n-1)t/n)]$ is a sum of n iid random variables. Hence the FT of $D = D(1)$ can be written as $\mathbb{E}[e^{-ikD}] = e^{\psi(-k)}$ with Fourier symbol $\psi(-k)$ from the Lévy representation (3.4), and then $D(t)$ has FT $e^{t\psi(-k)}$ for all $t \geq 0$. See Sato [187] or Applebaum [7] for more information on Lévy processes.

The random walk $T(n)$ and the renewal process $N(t)$ are inverses: Obviously we have $\{N(t) \geq n\} = \{T(n) \leq t\}$, which formalizes the fact that there are at least n jumps by time t, if and only if the nth jump occurs by time t. In fact, we also have $\{N(t) \geq u\} = \{T(\lceil u \rceil) \leq t\}$ where $\lceil u \rceil$ is the smallest integer $n \geq u$. The idea of inverse processes can be used, along with the random walk limit for $T(n)$, to get the limit behavior of the renewal process $N(t)$. For ease of notation, we specialize to the case $c_n = n^{-1/\beta}$. The general argument uses the fact that $(c_n) \in \mathrm{RV}(-1/\beta)$ and asymptotic inverses. Using (4.25) we have

$$\begin{aligned} \mathbb{P}[c^{-\beta} N(ct) < u] &= \mathbb{P}[N(ct) < c^\beta u] \\ &= \mathbb{P}[T(\lceil c^\beta u \rceil) > ct] \\ &= \mathbb{P}[c^{-1} T(\lceil c^\beta u \rceil) > t] \\ &= \mathbb{P}[(c^\beta)^{-1/\beta} T(\lceil c^\beta u \rceil) > t] \\ &\to \mathbb{P}[D(u) > t] \end{aligned} \tag{4.26}$$

for all $t > 0$, since every stable law has a density (we will prove this in Section 4.5). Define the inverse stable subordinator

$$E(t) = \inf\{u > 0 : D(u) > t\} \tag{4.27}$$

which is also the first passage time of the process $D(t)$. It is not hard to check that D, E are inverses, with $\{E(t) \le u\} = \{D(u) \ge t\}$. Since the inverse stable subordinator $E(t)$ also has a density (we will also prove this in Section 4.5), we have

$$\mathbb{P}[D(u) > t] = \mathbb{P}[D(u) \ge t] = \mathbb{P}[E(t) \le u] = \mathbb{P}[E(t) < u].$$

Then it follows from (4.26) that $c^{-\beta}N(ct) \Rightarrow E(t)$. Since (J_n) is independent of (Y_n), we also have

$$(c^{-1/\alpha}S([ct]), c^{-\beta}N(ct)) \Rightarrow (A(t), E(t))$$

for each $t > 0$. To simplify notation, we assume $a_n = n^{-1/\alpha}$ here. To proceed further, we need to introduce some ideas about stochastic process convergence.

Finite dimensional convergence: Given $0 < t_1 < t_2 < \cdots < t_n < \infty$ we want to show that

$$(c^{-1/\alpha}S([ct_1]), \ldots, c^{-1/\alpha}S([ct_n])) \Rightarrow (A(t_1), \ldots, A(t_n)). \tag{4.28}$$

To check this, define $t_0 = 0$ and $S(0) = 0$ and note that

$$c^{-1/\alpha}S([ct_k]) - c^{-1/\alpha}S([ct_{k-1}]) = c^{-1/\alpha} \sum_{j=[ct_{k-1}]+1}^{[ct_k]} J_j \Rightarrow A(t_k) - A(t_{k-1})$$

for $k = 1, \ldots, n$, and since the sums are all independent, we also have

$$(c^{-1/\alpha}S([ct_k]) - c^{-1/\alpha}S([ct_{k-1}]) : k = 1, \ldots, n) \Rightarrow (A(t_k) - A(t_{k-1}) : k = 1, \ldots, n)$$

weak convergence of these n dimensional random vectors. To prove (4.28) we will use the following fundamental result on weak convergence:

Theorem 4.19 (Continuous Mapping Theorem). *If $X_c \Rightarrow X$ as $c \to \infty$ and $f(x)$ is continuous, then $f(X_c) \Rightarrow f(X)$ as $c \to \infty$.*

Proof. See for example Billingsley [36]. □

Define $f(x_1, \ldots, x_n) = (x_1, x_1 + x_2, \ldots, x_1 + \cdots + x_n)$ so that f is continuous, with

$$f(c^{-1/\alpha}S([ct_k]) - c^{-1/\alpha}S([ct_{k-1}]) : k = 1, \ldots, n) = (c^{-1/\alpha}S([ct_k]) : k = 1, \ldots, n)$$

and

$$f(A(t_k) - A(t_{k-1}) : k = 1, \ldots, n) = (A(t_k) : k = 1, \ldots, n).$$

Apply Theorem 4.19 to see that (4.28) holds in the sense of finite dimensional distributions. In Section 4.4, we will extend this result to obtain stochastic process convergence.

Next we consider the waiting times. Given $0 < t_1 < t_2 < \cdots < t_n < \infty$ and real numbers $u_1, \ldots, u_n$ we can write

$$\begin{aligned}\mathbb{P}(c^{-\beta}N(ct_k) < u_k : k = 1, \ldots, n) &= \mathbb{P}(N(ct_k) < c^{\beta}u_k : k = 1, \ldots, n) \\ &= \mathbb{P}(T(\lceil c^{\beta}u_k \rceil) > ct_k : k = 1, \ldots, n) \\ &= \mathbb{P}((c^{\beta})^{-1/\beta}T(\lceil c^{\beta}u_k \rceil) > t_k : k = 1, \ldots, n) \\ &\to \mathbb{P}(D(u_k) > t_k : k = 1, \ldots, n) \\ &= \mathbb{P}(E(t_k) < u_k : k = 1, \ldots, n)\end{aligned}$$

which proves that $c^{-\beta}N(ct) \Rightarrow E(t)$ in the sense of finite dimensional distributions. Since (J_n) is independent of (Y_n), we also have

$$(c^{-1/\alpha}S([ct]), c^{-\beta}N(ct)) \Rightarrow (A(t), E(t))$$

in the sense of finite dimensional distributions.

Remark 4.20. Proposition 4.16 (a) shows that if $0 < \alpha < 1$ we can always choose $b_n = 0$. If $\alpha = 1$, we can always choose $b_n = 0$ if the distribution of W is symmetric. Corollary 4.6 and Proposition 4.16 (b) show that we can always choose $b_n = 0$ if $1 < \alpha \leq 2$ and $\mathbb{E}[W] = 0$. Suppose that $1 < \alpha \leq 2$ and $v = \mathbb{E}[W] \neq 0$. Then Remark 4.17 shows that

$$a_n(S_{[nt]} - [nt]v) + n^{-1}[nt]v \Rightarrow A(t)$$

as $n \to \infty$ for any $t > 0$, where the limit $A(t) = Z_t + vt$ is a Brownian motion with drift in the case $\alpha = 2$, or a stable Lévy motion with drift in the case $1 < \alpha < 2$. It is not hard to show, arguing as in (4.28), that we also get convergence in the sense of finite dimensional distributions in this case.

Details

Suppose that (4.24) holds with $b_n = 0$, and that $N(nt)/n \to \lambda t$ almost surely as $n \to \infty$. Then a transfer theorem from Becker-Kern, Meerschaert and Scheffler [25, Proposition 4.1] implies that

$$a_nS(N(nt)) = a_nS(n \cdot N(nt)/n) \Rightarrow A(\lambda t) \quad \text{as } n \to \infty$$

for any $t > 0$. An alternative proof uses the Continuous Mapping Theorem 4.19: Since the waiting times J_n are independent of the jumps Y_n, we also have joint convergence $(a_nS([nt]), N(nt)/n) \Rightarrow (A(t), \lambda t)$. Extend to joint convergence in the Skorokhod space $\mathbb{D}[0, \infty)$ using (4.29) in the next section, and mimic the proof of (4.32).

4.4 Convergence in Skorokhod space

We want to understand CTRW convergence, and the limit process, in terms of sample paths. These sample paths represent particle traces in the diffusion model. Let $\mathbb{D}[0, \infty)$ denote the set of real-valued functions $x : [0, \infty) \to \mathbb{R}$ which are continuous from the right:

$$\lim_{\varepsilon \to 0+} x(t + \varepsilon) = x(t),$$

with left-hand limits:

$$\lim_{\varepsilon \to 0+} x(t - \varepsilon) = x(t-).$$

In some literature these are called càdlàg functions, an acronym for the French phrase, "continue à droite, limite à gauche," which means "continuous on the right, with limits on the left." We would like to show that $c^{-1/\alpha} S([ct]) \Rightarrow A(t)$ in the space $\mathbb{D}[0, \infty)$, and likewise for the waiting times. Then we will use the Continuous Mapping Theorem 4.19 to get the CTRW process limit.

Weak convergence theory requires a topology on the space $\mathbb{D}[0, \infty)$, suitable for stochastic process convergence. In other words, we need to say what it means for a sample path $x_n(t)$ to be close to $x(t)$. The obvious choice is to require $x_n(t) \to x(t)$ for all t, but this excludes the possibility that $x_n(t)$ has a jump at the point $t - \varepsilon_n$ for some $\varepsilon_n \to 0$ and $x(t)$ has a jump of the same size at t. For this reason, Skorokhod introduced his (J_1) topology: In this topology, $x_n(t) \to x(t)$ in $\mathbb{D}[0, T]$ if for some increasing continuous functions $\lambda_n : [0, T] \to [0, T]$ such that $\lambda_n(0) = 0$, $\lambda_n(T) = T$, and

$$\lim_{n \to \infty} \sup_{0 \le t \le T} |\lambda_n(t) - t| = 0,$$

we have

$$\lim_{n \to \infty} \sup_{0 \le t \le T} |x(t) - x_n(\lambda_n(t))| = 0.$$

Then we say that $x_n(t) \to x(t)$ in $\mathbb{D}[0, \infty)$ if $x_n(t) \to x(t)$ in $\mathbb{D}[0, T]$ for every continuity point $T > 0$ of $x(t)$. This topology is useful if the processes have isolated jumps, as in a random walk. In fact, Skorokhod [201] proved that if $Y \in \mathrm{DOA}(A)$ and (4.23) holds, then

$$a_n S([nt]) - t b_n \Rightarrow A(t) \quad \text{in } \mathbb{D}[0, \infty) \tag{4.29}$$

with this topology. This strengthens (4.20). The Skorokhod M_1 topology (see details) is a bit more flexible. It allows multiple jumps of $x_n(t)$ to coalesce into a single jump of $x(t)$ in the limit. For a beautiful description, and additional discussion, see Avram and Taqqu [12].

Theorem 3 in Bingham [38] states that if:

(a) $X_n(t) \Rightarrow X(t)$ in the sense of finite dimensional distributions;
(b) $X(t)$ is continuous in probability; and
(c) $X_n(t)$ is monotone,

then $X_n(t) \Rightarrow X(t)$ in the space $\mathbb{D}[0,\infty)$ with the Skorokhod (J_1) topology. We say that $X(t)$ is continuous in probability if

$$\mathbb{P}[|X(t_n) - X(t)| > \varepsilon] \to 0 \quad \text{as } n \to \infty$$

for all $\varepsilon > 0$, whenever $t_n \to t$. Since the sample paths of the stable subordinator $D(u)$ are strictly increasing, it follows from (4.27) that the sample paths of $E(t)$ are continuous, and hence $E(t)$ is continuous in probability. Since sample paths of the process $N(t)$ are monotone nondecreasing, it follows that

$$c^{-\beta}N(ct) \Rightarrow E(t) \quad \text{in } \mathbb{D}[0,\infty). \tag{4.30}$$

Suppose $a_n = n^{-1/\alpha}$ to ease notation, and suppose that the random walk jumps are centered so that $b_n = 0$. Since (Y_n) and (J_n) are independent, it follows from (4.29) and (4.30) that

$$(c^{-1/\alpha}S([ct]), c^{-\beta}N(ct)) \Rightarrow (A(t), E(t)) \tag{4.31}$$

in the product space in $\mathbb{D}[0,\infty) \times \mathbb{D}[0,\infty)$. From here it is hard to prove CTRW convergence in the J_1 topology. But Theorem 13.2.4 in Whitt [219] shows that $x(y(t))$ is a continuous mapping from $\mathbb{D}[0,\infty) \times \mathbb{D}[0,\infty)$ to $\mathbb{D}[0,\infty)$ with the M_1 topology, so long as $u = y(t)$ is strictly increasing whenever $x(u) \neq x(u-)$, i.e., when u is a jump point of x.

In order to apply this to the CTRW limit, we need to know that $u = E(t)$ is a point of increase whenever $A(u) \neq A(u-)$. Since the constant intervals of $u = E(t)$ correspond to the jumps of the inverse process $t = D(u)$, this is equivalent to the condition that $A(u)$ and $D(u)$ have no simultaneous jumps. This follows from the fact that $A(u)$ and $D(u)$ are independent (see details). Then, since $x_c(u) = c^{-1/\alpha}S([cu]) \Rightarrow x(u) = A(u)$ and $y_c(t) = c^{-\beta}N(ct) \Rightarrow y(t) = E(t)$ jointly in $\mathbb{D}[0,\infty)\times\mathbb{D}[0,\infty)$, it follows from (4.31) and the Continuous Mapping Theorem 4.19 that

$$c^{-\beta/\alpha}S(N([ct])) = (c^{\beta})^{-1/\alpha}S(c^{\beta}\, c^{-\beta}N([ct])) = x_{c^\beta}(y_c(t)) \Rightarrow x(y(t)) = A(E_t) \tag{4.32}$$

as $c \to \infty$, in the space $\mathbb{D}[0,\infty)$ with the M_1 topology. The convergence (4.32) also holds in the J_1 topology, but the proof is more delicate, see Straka and Henry [210, Theorem 3.6].

Recall that $A(ct) \simeq c^{1/\alpha}A(t)$ for all $c > 0$ and $t \geq 0$. It is not hard to extend to finite dimensional distributions, using the fact that $A(t)$ has independent increments. A process with this scaling property for finite dimensional distributions is called *self-similar* with index $1/\alpha$, see for example Embrechts and Maejima [64]. Since $D(t)$ is also stable, the processes $D(t)$ is self-similar with index $1/\beta$. We have noted previously that $\{E(t) \leq u\} = \{D(u) \geq t\}$. In fact, we also have

$$\{E(t_k) \leq u_k \;\forall\, k = 1, \ldots, n\} = \{D(u_k) \geq t_k \forall\, k = 1, \ldots, n\}$$

for any $0 < t_1 < t_2 < \cdots < t_n < \infty$ and real numbers $u_1, \ldots, u_n$. It follows that $E(t)$ is self-similar with index β. Then, since $A(u)$ and $E(t)$ are independent, the CTRW limit process $A(E(t))$ is self-similar with index β/α. This index codes the rate at which a plume of particles spreads away from their center of mass.

Remark 4.21. Suppose $1 < \alpha \le 2$ and that $v = \mathbb{E}[W] \ne 0$, and suppose that $a_n = n^{-1/\alpha}$ to ease notation. Then another continuous mapping argument along with (4.29) shows that

$$c^{-1/\alpha} \sum_{j=1}^{[ct]} (W_j - v) + c^{-1} \sum_{j=1}^{[ct]} v \Rightarrow A'(t) \tag{4.33}$$

as $c \to \infty$, in the space $\mathbb{D}[0, \infty)$ with the J_1 topology, where $A'(t) = A(t) + vt$ is a Brownian motion with drift in the case $\alpha = 2$, or a stable Lévy motion with drift in the case $1 < \alpha < 2$. Then (4.31) extends to

$$(c^{-1/\alpha}(S([ct]) - [ct]v) + c^{-1}[ct]v, c^{-\beta}N(ct)) \Rightarrow (A'(t), E(t)) \tag{4.34}$$

and (4.32) extends to

$$c^{-\beta/\alpha}(S(N([ct])) - N([ct])v) + c^{-1}N([ct])v \Rightarrow A'(E_t). \tag{4.35}$$

Details

Suppose that $t > r > 0$. In Section 4.5, we will prove that every stable law has a density. Since $D(t) - D(r)$ is identically distributed with $D(t - r)$, and $D(t - r)$ has a density, $D(t - r) > 0$ and $D(t) > D(r)$ with probability one, i.e., the process $D(t)$ is strictly increasing.

Since D is strictly increasing, if $D(u) \ge t$, then $D(y) > t$ for all $y > u$, so that $E(t) \le u$. Since D is right-continuous, if $D(u) < t$, then $D(y) < t$ for all $y > u$ sufficiently close to u, so $E(t) > u$. This proves that $\{E(t) \le u\} = \{D(u) \ge t\}$.

The Skorokhod M_1 topology is defined as follows: The graph of a function $x(t)$ in $\mathbb{D}[0, T]$ is the set $\{(t, x(t)) : 0 \le t \le T\}$. The completed graph also contains the points $\{px(t)+(1-p)x(t-) : 0 \le p \le 1\}$, so that it becomes a connected compact subset of $\mathbb{R}\times[0, T]$. A parametric representation $(u(s), r(s))$ is a continuous function that maps the interval $s \in [0, 1]$ onto the completed graph, such that $u(s)$ is an increasing function from $[0, 1]$ onto $[0, T]$. Then $x_n \to x$ in $\mathbb{D}[0, T]$ with the M_1 topology if and only if there exists a parametric representation $(u(s), r(s))$ of $x(t)$ and parametric representations $(u_n(s), r_n(s))$ of $x_n(t)$ such that

$$\lim_{n\to\infty} \sup_{0\le s\le 1} [|u_n(s) - u(s)| + |r_n(s) - r(s)|] = 0.$$

Also $x_n(t) \to x(t)$ in $\mathbb{D}[0, \infty)$ if $x_n(t) \to x(t)$ in $\mathbb{D}[0, T]$ for every $T > 0$ that is a point of continuity of $x(t)$. See Whitt [219] for additional discussion.

Since the stable Lévy processes $A(u)$ and $D(u)$ are independent, they have no simultaneous jumps. This follows from consideration of the two dimensional Lévy process $(A(u), D(u))$. The Lévy Representation Theorem 6.8 in dimension $d > 1$ will be discussed in Chapter 6. Remark 6.19 shows that every jump of the two dimensional Lévy process $(A(u), D(u))$ lies on one of the coordinate axes. Then it follows from the Lévy-Itô Decomposition Theorem [187, Theorem 19.2] that every jump of $(A(u), D(u))$ lies on one of the coordinate axes, i.e., $A(u)$ and $D(u)$ have no simultaneous jumps

The full proof of the CTRW limit depends on asymptotic inverses. Suppose that $a_n S(n) \Rightarrow A$ (centered jumps) and $b_n T(n) \Rightarrow D$. Define $b(c) = b_{[c]}$ for $c > 0$, and note that $1/b \in \mathrm{RV}(1/\beta)$. The asymptotic inverse $\tilde{b}(c) = \inf\{x > 0 : b(x) < c^{-1}\}$ of $1/b$ is regularly varying with index β, and $b(\tilde{b}(c)) \sim 1/c$, see Seneta [197, p. 21]. The proof is similar to Proposition 4.12. Write

$$\begin{aligned}\mathbb{P}[\tilde{b}(c)^{-1}N(ct) < u] &= \mathbb{P}[N(ct) < \tilde{b}(c)u] \\ &= \mathbb{P}[T(\lceil \tilde{b}(c)u \rceil) > ct] \\ &= \mathbb{P}[c^{-1}T(\lceil \tilde{b}(c)u \rceil) > t] \\ &\approx \mathbb{P}[b(\tilde{b}(c))T(\lceil \tilde{b}(c)u \rceil) > t] \\ &\to \mathbb{P}[D(u) > t] = \mathbb{P}[E(t) < u].\end{aligned}$$

Extend to finite dimensional convergence as before, and then to $\mathbb{D}[0, \infty)$. Use independence to get joint convergence

$$(a_n S([ct]), \tilde{b}(c)^{-1}N(ct)) \Rightarrow (A(t), E(t))$$

in $\mathbb{D}[0, \infty) \times \mathbb{D}[0, \infty)$. Define $a(c) = a_{[c]}$ and $A(c) = a(\tilde{b}(c))$, and apply Whitt [219, Theorem 13.2.4] along with continuous mapping to get

$$A(c)S(N([ct])) = a(\tilde{b}(c))S(\tilde{b}(c)\,\tilde{b}(c)^{-1}N([ct])) \Rightarrow A(E_t)$$

in the M_1 topology. For complete details, see Meerschaert and Scheffler [151, Theorem 4.2]. For J_1 convergence, see Henry and Straka [210].

4.5 CTRW governing equations

In Section 4.4, we showed that the CTRW limit is $A(E(t))$. The outer process $x = A(u)$ is an α-stable Lévy motion with index $0 < \alpha \le 2$, the long-time limit of the random walk of particle jumps. The inner process $u = E(t)$ is the inverse of a β-stable subordinator $D(t)$ with index $0 < \beta < 1$, the limit of the random walk of waiting times. If $\alpha = 2$, then $A(u)$ is a Brownian motion. In this section, we develop the fractional diffusion equation that governs the probability densities of the CTRW limit.

First note that $x = A(u)$ has a density function $p(x, u)$ for all $u > 0$. This follows by the Fourier inversion formula

$$p(x, u) = \frac{1}{2\pi} \int_{-\infty}^{\infty} e^{ikx} \hat{p}(k, u)\, dk \tag{4.36}$$

from Theorem 1.4. Since $A(u)$ is the limit of a CTRW with centered jumps, it follows from Corollary 4.6 and Proposition 4.16 that the FT of $A(u)$ is

$$\hat{p}(k,u) = \mathbb{E}[e^{-ikA(u)}] = \exp\left(Du[p(ik)^\alpha + q(-ik)^\alpha]\right), \tag{4.37}$$

where $D > 0$ if $1 < \alpha \leq 2$, and $D < 0$ if $0 < \alpha < 1$. A computation using complex exponential functions (see details) shows that $|\hat{p}(k,u)| \leq \exp(-D_0 u|k|^\alpha)$, where $D_0 > 0$ for $1 < \alpha \leq 2$, and also for $0 < \alpha < 1$. Then it follows that $\hat{p}(k,u)$ is absolutely integrable for all $u > 0$, and so (4.36) implies that $x = A(u)$ has a density.

Since $t = D(u)$ is also stable, it has a density $g(t,u)$ for all $u > 0$. Write

$$\mathbb{P}[E(t) \leq u] = \mathbb{P}[D(u) \geq t] = \int_t^\infty g(r,u)\,dr = 1 - \int_0^t g(r,u)\,dr$$

which implies that $u = E(t)$ has a density

$$h(u,t) = -\frac{d}{du}\int_0^t g(r,u)\,dr \tag{4.38}$$

for all $u > 0$ and $t > 0$. Then a conditioning argument gives the density $m(x,t)$ of $x = A(E(t))$:

$$\begin{aligned}
\mathbb{P}[A(E(t)) \leq x] &= \mathbb{E}\big[\mathbb{P}[A(E(t)) \leq x | E(t)]\big] \\
&= \int_0^\infty \mathbb{P}[A(u) \leq x | E(t) = u] P_{E(t)}(du) \\
&= \int_0^\infty \mathbb{P}[A(u) \leq x] h(u,t)\,du
\end{aligned}$$

so that

$$\begin{aligned}
m(x,t) &= \frac{d}{dx}\int_0^\infty \mathbb{P}[A(u) \leq x] h(u,t)\,du \\
&= \int_0^\infty \frac{d}{dx}\mathbb{P}[A(u) \leq x] h(u,t)\,du \\
&= \int_0^\infty p(x,u) h(u,t)\,du.
\end{aligned} \tag{4.39}$$

(In the details at the end of this section, we will prove that the derivative can be taken inside the integral in (4.39).) Heuristically, we write

$$\mathbb{P}[A(E(t)) = x] \approx \sum_u \mathbb{P}[A(u) = x]\mathbb{P}[E(t) = u].$$

Recall from (4.37) that $\hat{p}(k, u) = e^{u\psi(-k)}$ where the Fourier symbol of the stable law $x = A(u)$ is $\psi(-k) = D[p(ik)^\alpha + q(-ik)^\alpha]$. Take derivatives to get

$$\frac{d}{du}\hat{p}(k, u) = \psi(-k)e^{u\psi(-k)} = D[p(ik)^\alpha + q(-ik)^\alpha]\,\hat{p}(k, u)$$

and note that $\hat{p}(k, 0) \equiv 1$. Inverting the FT shows that the density $p(x, u)$ of the outer process $x = A(u)$ solves the space-fractional diffusion equation

$$\frac{\partial}{\partial u}p(x, u) = Dp\frac{\partial^\alpha}{\partial x^\alpha}p(x, u) + Dq\frac{\partial^\alpha}{\partial(-x)^\alpha}p(x, u) \tag{4.40}$$

with the Dirac delta function initial condition $p(x, 0) = \delta(x)$. The distribution function $P(x, u) = \mathbb{P}[A(u) \le x]$ solves the same space-fractional diffusion equation

$$\frac{\partial}{\partial u}P(x, u) = Dp\frac{\partial^\alpha}{\partial x^\alpha}P(x, u) + Dq\frac{\partial^\alpha}{\partial(-x)^\alpha}P(x, u)$$

with the Heaviside function initial condition: $P(x, 0) = H(x) = I(x \ge 0)$. This is related to the fact that $\delta(x) = \partial_x H(x)$ in terms of weak or distributional derivatives. See the details at the end of Section 3.1 for more information.

Since $t = D(u)$ is the limit of a random walk with positive jumps, it follows from Proposition 4.16 (a) that $D(u)$ is one-sided stable with characteristic function

$$\mathbb{E}[e^{ikD(u)}] = \exp\left[-Bu\Gamma(1 - \beta)(-ik)^\beta\right],$$

where $B > 0$ depends on the sequence of norming constants c_n in (4.25). If the norming constants c_n are chosen so that $B = 1/\Gamma(1 - \beta)$, then $\mathbb{E}[e^{ikD(u)}] = \exp(-u(-ik)^\beta)$ for $0 < \beta < 1$ (see details). Then the Laplace transform

$$\tilde{g}(s, u) = \int_0^\infty e^{-st}g(t, u)\,dt = \mathbb{E}[e^{-sD(u)}] = e^{-us^\beta} \tag{4.41}$$

for all $u \ge 0$ and all $s > 0$. There are two ways to make this rigorous. One is to develop the theory of positive infinitely divisible laws using Laplace transforms, see for example Sato [187]. The other is to prove the Laplace transform $\tilde{g}(s, u)$ exists for complex s, see Zolotarev [228]. See the details at the end of this section for more information.

Then the density (4.38) of $u = E(t)$ has LT

$$
\begin{aligned}
\tilde{h}(u,s) &= \int_0^\infty e^{-st} h(u,t)\,dt \\
&= -\int_0^\infty e^{-st}\left(\frac{d}{du}\int_0^t g(r,u)dr\right)dt \\
&= -\frac{d}{du}\int_0^\infty e^{-st}\int_0^t g(r,u)drdt \\
&= -\frac{d}{du}\int_0^\infty g(r,u)\left(\int_r^\infty e^{-st}dt\right)dr \\
&= -\frac{d}{du}\int_0^\infty g(r,u)s^{-1}e^{-rs}dr \\
&= -\frac{d}{du}\left[s^{-1}e^{-us^\beta}\right] \\
&= s^{-1}s^\beta e^{-us^\beta} = s^{\beta-1}e^{-us^\beta}
\end{aligned} \tag{4.42}
$$

and the density (4.39) of $x = A(E(t))$ has FLT

$$
\begin{aligned}
\bar{m}(k,s) &= \int_0^\infty \int_{-\infty}^\infty e^{-st}e^{-ikx}m(x,t)\,dx\,dt \\
&= \int_0^\infty \int_{-\infty}^\infty e^{-st}e^{-ikx}\int_0^\infty p(x,u)h(u,t)\,du\,dx\,dt \\
&= \int_0^\infty \left(\int_{-\infty}^\infty e^{-ikx}p(x,u)\,dx\right)\left(\int_0^\infty e^{-st}h(u,t)\,dt\right)du \\
&= \int_0^\infty e^{u\psi(-k)}s^{\beta-1}e^{-us^\beta}du \\
&= s^{\beta-1}\int_0^\infty e^{-u[s^\beta-\psi(-k)]}du = \frac{s^{\beta-1}}{s^\beta-\psi(-k)}
\end{aligned} \tag{4.43}
$$

by Fubini, using the fact that $|\hat{p}(k,u)| = |e^{u\psi(-k)}| \le \exp(-D_0 u|k|^\alpha)$ (see details). Rewrite (4.43) in the form

$$s^\beta \bar{m}(k,s) - s^{\beta-1} = \psi(-k)\bar{m}(k,s)$$

and note that $\hat{m}(k,0) = \mathbb{E}[e^{-ikA(E(0))}] \equiv 1$ since $A(0) = E(0) = 0$. Invert the LT to get

$$\partial_t^\beta \hat{m}(k,t) = \psi(-k)\hat{m}(k,t)$$

where ∂_t^β is the Caputo fractional derivative. Then invert the FT to see that the density $m(x, t)$ of the CTRW limit process $x = A(E(t))$ solves the space-time fractional diffusion equation

$$\partial_t^\beta m(x, t) = Dp\frac{\partial^\alpha}{\partial x^\alpha}m(x, t) + Dq\frac{\partial^\alpha}{\partial(-x)^\alpha}m(x, t). \tag{4.44}$$

If the Lévy measure of $A = A(1)$ is given by (4.14) (e.g, for Pareto jumps $\mathbb{P}[Y > y] \sim pCy^{-\alpha}$ and $\mathbb{P}[Y < -y] \sim qCy^{-\alpha}$ with $a_n = n^{-1/\alpha}$ in (4.1)), then the fractional dispersivity constant:

$$D = \begin{cases} -C\Gamma(1-\alpha) & \text{if } 0 < \alpha < 1; \\ C\dfrac{\Gamma(2-\alpha)}{\alpha - 1} & \text{if } 1 < \alpha < 2. \end{cases}$$

If $\alpha = 2$, then $A(u)$ is normal with mean zero and variance $2Du$, since

$$\hat{p}(k, u) = \mathbb{E}[e^{ikA(u)}] = \exp(Du(ik)^2) = \exp(-\tfrac{1}{2}\sigma^2 k^2)$$

with $\sigma^2 = 2Du$. We have developed the space-time fractional diffusion equation (4.44) from the extended central limit theorem, and connected the parameters of this equation to those of the continuous time random walk. The fractional derivative in space codes power law jumps, leading to anomalous super-diffusion. The fractional derivative in time models power law waiting times, leading to anomalous sub-diffusion. The CTRW combines both effects. For example, if $\alpha = 2\beta$, then the limit $A(E(t))$ has the same scaling as a Brownian motion.

For practical applications, we would like to explicitly compute solutions to the space-time fractional diffusion equation (4.44). We know that the point source solution with constant coefficients is an integral (4.39) involving the density $p(x, u)$ of a stable Lévy motion, and the density $h(u, t)$ of an inverse stable subordinator. Since we know an explicit formula for $\hat{p}(k, u)$, in principle we can use the FT inversion formula (4.36) to compute $p(x, u)$. In practice, this is a hard integral! But it does reduce to “nicer” forms that are easier to numerically integrate. Nolan [163] has developed fast and accurate computer codes to compute the stable density, see his personal web page for more information. There are also R codes, based on the same ideas. We demonstrate these codes in Section 5.1.

As for the inverse stable density, recall that $t = D(u)$ has a density $g(t, u)$ with FT

$$\hat{g}(k, u) = \int_0^\infty e^{-ikt} g(t, u)\, dt = \mathbb{E}[e^{-ikD(u)}] = e^{-u(ik)^\beta} \tag{4.45}$$

and it follows that $D(u) \simeq u^{1/\beta}D(1)$. To check this, write

$$\mathbb{E}[e^{-ik\, u^{1/\beta}D(1)}] = \mathbb{E}[e^{-i(ku^{1/\beta})D(1)}] = e^{-1(iku^{1/\beta})^\beta} = e^{-u(ik)^\beta} = \mathbb{E}[e^{-ikD(u)}].$$

Let $g_\beta(t) = g(t, 1)$ be the density of $D = D(1)$, a *standard stable subordinator*. Then $D(u)$ has density

$$g(t, u) = u^{-1/\beta} g_\beta(u^{-1/\beta} t)$$

by a simple change of variables (or just differentiate $\mathbb{P}[u^{1/\beta}D \le t]$). Write

$$\begin{aligned}\mathbb{P}[E(t) \le u] &= \mathbb{P}[D(u) \ge t]\\ &= \mathbb{P}[u^{1/\beta}D \ge t]\\ &= \mathbb{P}[D \ge tu^{-1/\beta}]\\ &= \mathbb{P}[(D/t)^{-\beta} \le u]\end{aligned} \tag{4.46}$$

which shows that $E(t) \simeq (D/t)^{-\beta}$ for all $t > 0$. Differentiate (4.46) to see that $u = E(t)$ has density (see details)

$$h(u,t) = \frac{t}{\beta}u^{-1-1/\beta}g_\beta(tu^{-1/\beta}). \tag{4.47}$$

Then (4.39) becomes

$$m(x,t) = \int_0^\infty p(x,u)\frac{t}{\beta}u^{-1-1/\beta}g_\beta(tu^{-1/\beta})\,du$$

and we can compute this explicitly using existing codes for the stable density. An alternative form can be obtained by substituting $r = tu^{-1/\beta}$, which leads to

$$m(x,t) = \int_0^\infty p(x,(t/r)^\beta)g_\beta(r)\,dr. \tag{4.48}$$

Remark 4.22. The waiting time process $t = D(u)$ has a density $g(t,u)$ with FT $\hat{g}(k,u) = e^{-u(ik)^\beta}$ and hence

$$\frac{d}{du}\hat{g}(k,u) = -(ik)^\beta\hat{g}(k,u).$$

Invert the FT to see that $g(t,u)$ solves the fractional partial differential equation

$$\frac{\partial}{\partial u}g(t,u) = -\frac{\partial^\beta}{\partial t^\beta}g(t,u)$$

using the Riemann-Liouville derivative. Note that here the roles of space and time are reversed. The inverse stable process $u = E(t)$ has a density $h(u,t)$ with LT $\tilde{h}(u,s) = s^{\beta-1}e^{-us^\beta}$ and FLT

$$\bar{h}(k,s) = \int_0^\infty e^{-iku}\tilde{h}(u,s)\,du = \int_0^\infty e^{-iku}s^{\beta-1}e^{-us^\beta}\,du = \frac{s^{\beta-1}}{s^\beta + ik}.$$

Rewrite in the form

$$s^\beta\bar{h}(k,s) - s^{\beta-1} = -ik\bar{h}(k,s)$$

and invert to see that this density solves

$$\partial_t^\beta h(u,t) = -\frac{\partial}{\partial u}h(u,t), \tag{4.49}$$

using the Caputo derivative in time. This is a degenerate case of the CTRW with $Y_n = 1$. Then $x = A(u) = u$ (the shift semigroup), $\mathbb{E}[e^{-ikA(u)}] = e^{-iku}$, and $\psi(k) = ik$. It is also possible to derive the CTRW governing equation (4.44) from (4.49), together with (4.39) and (4.40). If we wish to interpret (4.49) as a differential equation on $u \in \mathbb{R}$ then, since the function $u \mapsto h(u,t)$ has a jump at the point $u = 0$, the derivative $\partial h/\partial u$ must be interpreted as a weak derivative, as in Remark 2.13. For an alternative derivation of the governing equation for $h(u,t)$ using LT in both variables, and the explicit form of the limit $h(0+,t)$, see Hahn, Kobayashi, and Umarov [82]. An explicit formula for the moments of $E(t)$ was given by Piryatinska, Saichev and Woyczynski [168]. For a recent survey on the inverse stable subordinator, see Meerschaert and Straka [154].

Remark 4.23. In Remark 4.21 we showed that, when the random walk jumps have a finite mean in the case $1 < \alpha \le 2$, the CTRW scaling limit is $A'(E(t))$. The outer process $x = A'(u)$ is a Brownian motion with drift in the case $\alpha = 2$, or a stable Lévy motion with drift in the case $1 < \alpha < 2$. When $1 < \alpha < 2$, the probability densities $p(x,u)$ of $A'(u)$ solve the space-fractional diffusion equation with drift

$$\frac{\partial}{\partial u}p(x,u) = -v\frac{\partial}{\partial x}p(x,u) + Dp\frac{\partial^\alpha}{\partial x^\alpha}p(x,u) + Dq\frac{\partial^\alpha}{\partial(-x)^\alpha}p(x,u) \tag{4.50}$$

and the probability densities $m(x,t)$ of the CTRW limit process $A'(E(t))$ solve the space-time fractional equation

$$\partial_t^\beta m(x,t) = -v\frac{\partial}{\partial x}m(x,t) + Dp\frac{\partial^\alpha}{\partial x^\alpha}m(x,t) + Dq\frac{\partial^\alpha}{\partial(-x)^\alpha}m(x,t). \tag{4.51}$$

If $\alpha = 2$ then the same equations apply, and in particular, the probability densities of the process $A'(E(t))$, a Brownian motion with drift where the time variable is replaced by an independent inverse stable subordinator, solve the time-fractional diffusion equation with drift

$$\partial_t^\beta m(x,t) = -v\frac{\partial}{\partial x}m(x,t) + D\frac{\partial^2}{\partial x^2}m(x,t). \tag{4.52}$$

Remark 4.24. There is an interesting connection between the CTRW scaling limit process $A(E(t))$ in the normal case $\alpha = 2$, and *iterated Brownian motion*. Given a Brownian motion $A(t)$, take another independent Brownian motion $B(t)$ and consider the subordinated process $A(|B(t)|)$. Allouba and Zheng [4] and Burdzy [43] develop governing equations and other properties of this process. Baeumer, Meerschaert and Nane [20] show that the process $A(E(t))$ with $\beta = 1/2$ has the same governing equation and the same one dimensional distributions. This is related to the fact that the first passage times of a Brownian motion are stable with index $\beta = 1/2$. Some related results for subordinated Brownian motion in a bounded domain are included in Meerschaert, Nane and Vellaisamy [140].

Details

Define the signum function $\operatorname{sgn}(k) = +1$ for $k \geq 0$ and $\operatorname{sgn}(k) = -1$ for $k < 0$. Write $(ik)^\alpha = (i\operatorname{sgn}(k)|k|)^\alpha = |k|^\alpha e^{i\operatorname{sgn}(k)\pi\alpha/2} = |k|^\alpha[\cos\theta + i\operatorname{sgn}(k)\sin\theta]$ where $\theta = \pi\alpha/2$. Then $(ik)^\alpha = a + ib$ where $a = |k|^\alpha\cos\theta$. A similar argument shows that $(-ik)^\alpha = |k|^\alpha e^{-i\operatorname{sgn}(k)\theta} = |k|^\alpha[\cos\theta - i\operatorname{sgn}(k)\sin\theta]$. Then $p(ik)^\alpha + q(-ik)^\alpha$ is a complex number with real part equal to $(p+q)|k|^\alpha\cos\theta = |k|^\alpha\cos(\pi\alpha/2)$. Hence

$$|\hat{p}(k,u)| = |\exp\left(Du[p(ik)^\alpha + q(-ik)^\alpha]\right)| = \exp(Du|k|^\alpha\cos(\pi\alpha/2)) = e^{-D_0u|k|^\alpha} \tag{4.53}$$

where $D_0 = -D\cos(\pi\alpha/2) > 0$: $D < 0$ and $\cos(\pi\alpha/2) > 0$ when $0 < \alpha < 1$; and $D > 0$ and $\cos(\pi\alpha/2) < 0$ when $1 < \alpha \leq 2$.

Differentiation inside the integral in (4.39) is justified as follows. Consider $y > 0$ (the case $y < 0$ is treated similarly). Since

$$p(x,u) = \frac{d}{dx}\mathbb{P}[A(u) \leq x]$$

we have

$$m(x,t) = \frac{d}{dx}\int_0^\infty \mathbb{P}[A(u) \leq x]h(u,t)\,du = \frac{d}{dx}\int_0^\infty\int_{-\infty}^x p(v,u)\,dv\,h(u,t)\,du.$$

Write the last expression as a difference quotient, and simplify to get

$$m(x,t) = \lim_{y\to 0}\int_0^\infty\left(y^{-1}\int_x^{x+y} p(v,u)\,dv\right)h(u,t)\,du.$$

It follows from (4.36) and (4.53) that

$$\begin{aligned}\left|y^{-1}\int_x^{x+y} p(v,u)\,dv\right| \leq \sup_{v\in[x,x+y]}|p(v,u)| &\leq \frac{1}{2\pi}\int_{-\infty}^\infty |\hat{p}(k,u)|\,dk \\ &\leq \frac{1}{2\pi}\int_{-\infty}^\infty e^{-D_0u|k|^\alpha}\,dk := C_0 < \infty\end{aligned}$$

since $D_0 > 0$. Then

$$\int_0^\infty\left|y^{-1}\int_x^{x+y} p(v,u)\,dv\right|h(u,t)\,du \leq \int_0^\infty C_0h(u,t)\,du = C_0,$$

and the dominated convergence theorem justifies differentiation under the integral.

The justification for the differentiation under the integral in the derivation of the LT of $E(t)$ in (4.42) is similar. Write

$$\lim_{y\to 0}\int_0^\infty e^{-st}\int_0^t \frac{g(r,u+y) - g(r,u)}{y}\,dr\,dt,$$

and

$$g(r,u) = \frac{1}{2\pi}\int_{-\infty}^{\infty} e^{ikr}\hat{g}(k,u)\,dk = \frac{1}{2\pi}\int_{-\infty}^{\infty} e^{ikr}e^{-u(ik)^\beta}\,dk$$

so that

$$\left|\frac{g(r,u+y)-g(r,u)}{y}\right| \le \frac{1}{2\pi}\int_{-\infty}^{\infty}\left|e^{-u(ik)^\beta}\right|\frac{\left|1-e^{-y(ik)^\beta}\right|}{|y|}\,dk.$$

Note that $\left|e^{-u(ik)^\beta}\right| = e^{-u|k|^\beta\cos(\pi\beta/2)}$ with $\cos(\pi\beta/2) > 0$ since $0 < \beta < 1$, and apply the mean value theorem to see that

$$\frac{\left|1-e^{-y(ik)^\beta}\right|}{|y|} \le \frac{e^{|y||k|^\beta\cos(\pi\beta/2)}-1}{|y|} \le |k|^\beta e^{(u/2)|k|^\beta\cos(\pi\beta/2)}$$

if $|y| < u/2$, which holds eventually since $y \to 0$, and $u > 0$ is fixed in this argument. It follows that

$$\left|\frac{g(r,u+y)-g(r,u)}{y}\right| \le \frac{1}{2\pi}\int_0^{\infty}|k|^\beta e^{(-u/2)|k|^\beta\cos(\pi\beta/2)}\,dk := C_1 < \infty$$

for any $r > 0$. Therefore

$$\int_0^\infty e^{-st}\int_0^t\left|\frac{g(r,u+y)-g(r,u)}{y}\right|\,dr\,dt \le \frac{1}{2\pi}\int_0^\infty e^{-st}tC_1\,dt < \infty,$$

and the argument can be completed using the dominated convergence theorem.

Suppose $c_nT_n \Rightarrow D$ where $T_n = J_1 + \cdots + J_n$ and J_n are iid with $J \in \mathrm{DOA}(D)$. If D has Lévy measure $\phi(r,\infty) = Br^{-\beta}$ concentrated on the positive real line (e.g., if $\mathbb{P}[J > t] = Bt^{-\beta}$ and $c_n = n^{-1/\beta}$), then Proposition 4.16 shows that $\mathbb{E}[e^{ikD}] = \exp(-B\Gamma(1-\beta)(-ik)^\beta)$. Define a new set of norming constants $\tilde{c}_n = [B\Gamma(1-\beta)]^{-1/\beta}c_n$ (this reduces to $[nB\Gamma(1-\beta)]^{-1/\beta}$ in the case of Pareto jumps) and note that $\tilde{c}_nT_n \Rightarrow [B\Gamma(1-\beta)]^{-1/\beta}D = \tilde{D}$. Write

$$\mathbb{E}[e^{ik\tilde{D}}] = \mathbb{E}[e^{ik[B\Gamma(1-\beta)]^{-1/\beta}D}] = \exp(-B\Gamma(1-\beta)(-ik[B\Gamma(1-\beta)]^{-1/\beta})^\beta) = e^{-(-ik)^\beta}$$

which shows that the limit is a standard stable subordinator.

For positive random variables, it is possible to develop an alternative theory of infinitely divisible laws based on Laplace transforms, see for example Sato [187]. The theory is similar to what was presented in Section 3.1, using Laplace transforms instead of characteristic functions. Since a positive random variable cannot have a normal distribution, the Lévy representation takes the simplified form $\mathbb{E}[e^{-sY}] = e^{\psi(s)}$, where $s > 0$ and

$$\psi(s) = -as + \int_0^\infty (e^{-sy}-1)\,\phi(dy) \tag{4.54}$$

for some $a \geq 0$, and some Lévy measure $\phi(dy)$. This Lévy representation is unique. The Lévy measure $\phi(dy)$ on $\{y : y > 0\}$ satisfies $\phi(R, \infty) < \infty$ and

$$\int_0^R y\phi(dy) < \infty \tag{4.55}$$

for all $R > 0$. A computation very similar to Proposition 3.10 shows that a centered one-sided stable law with Lévy measure (3.10) has Laplace symbol

$$\psi(s) = \int_0^\infty (e^{-sy} - 1)\, C\alpha y^{-\alpha-1} dy = -C\Gamma(1-\alpha)s^\alpha \tag{4.56}$$

for $0 < \alpha < 1$. If $C = 1/\Gamma(1-\alpha)$ we get a standard stable subordinator with Laplace transform $\mathbb{E}[e^{-sY}] = \exp(-s^\alpha)$.

One way to connect these two theories of infinitely divisible laws is to view the Laplace transform as a function of a complex variable. The Laplace transform

$$e^{-us^\beta} = \int_0^\infty e^{-st} g(t,u)\, dt \tag{4.57}$$

exists for any $s = ik + y$ with k real and $y > 0$, see Zolotarev [228, Lemma 2.2.1]. Hence we can substitute $s = ik$ into the formula (4.57) for the LT of the positive random variable $D(u)$, to get the corresponding FT formula (4.45).

To show that (4.47) holds, write $\mathbb{P}[E(t) \leq u] = \mathbb{P}[D \geq tu^{-1/\beta}] = 1 - G_\beta(tu^{-1/\beta})$ where $G_\beta(u)$ is the cdf of D, so that

$$g_\beta(u) = \frac{d}{du} G_\beta(u).$$

Then

$$\begin{aligned} h(t,u) &= \frac{d}{du}[1 - G_\beta(tu^{-1/\beta})] \\ &= -g_\beta(tu^{-1/\beta}) \frac{d}{du}[tu^{-1/\beta}] \end{aligned}$$

which reduces to (4.47).

5 Computations in R

In this chapter, we demonstrate computer codes that the reader can use to simulate random walks and their stochastic process limits, as well as the corresponding probability densities. These densities solve the fractional diffusion equations that are a main focus of this book.

5.1 R codes for fractional diffusion

The R programming language is a sophisticated and useful platform for probability and statistics [171]. This freely available open source code can be downloaded and installed on a wide variety of Unix, Windows, and Apple computer systems. See www.r-project.org for additional details. Once you have installed R on your computer, the easiest way to run a program is to type the code into a plain text file (or download), cut and paste the entire program into the R console window, and press the "Enter" key.

```
D=1.0
v=3.0
t=5.0
mu=v*t
sigma=sqrt(2*D*t)
x = seq(mu-4*sigma, mu+4*sigma, 0.1*sigma)
density=dnorm(x, mean = mu, sd = sigma)
plot(x,density,type="l",lwd=3)
```

Fig. 5.1: R code to plot solutions to the traditional diffusion equation with drift (5.1) at time $t = 5.0$ with velocity $v = 3.0$ and dispersion $D = 1.0$.

Example 5.1. The simple R code listed in Figure 5.1 plots the solution $p(x, t)$ to the traditional diffusion equation with drift

$$\frac{\partial}{\partial t}p(x,t) = -v\frac{\partial}{\partial x}p(x,t) + D\frac{\partial^2}{\partial x^2}p(x,t) \tag{5.1}$$

for any time $t > 0$, with drift velocity $v \in \mathbb{R}$ and dispersion $D > 0$. This code uses the fact that the solution to (5.1) is a normal pdf with mean $\mu = vt$ and standard deviation $\sigma = \sqrt{2Dt}$ for any $t > 0$. The R function `dnorm` produces a normal density with a specified mean and standard deviation. Efficient R code is based on vector mathematics. The vector `x` is a sequence of numbers from $\mu-4\sigma$ to $\mu+4\sigma$ in increments of 0.1σ. If you type `x` into the R console wind

"Enter" key, you will see this vector of $n = 81$ numbers. The command `dnorm` takes the vector `x` as input, and outputs a vector `density` consisting of the normal pdf at each value of the input vector. The command `plot` displays the points (`x[i]`, `density[i]`) for $i = 1, 2, \ldots, n$ and connects them with a curved line (graph `type="l"`). Figure 5.2 shows the output from running the R code in Figure 5.1. The same graph was also displayed as Figure 1.1 in Chapter 1. To obtain plots for other values of the input parameters D, v, and t, edit the file containing the source code, cut and paste this edited code back into the R console window, and press the "Enter" key.

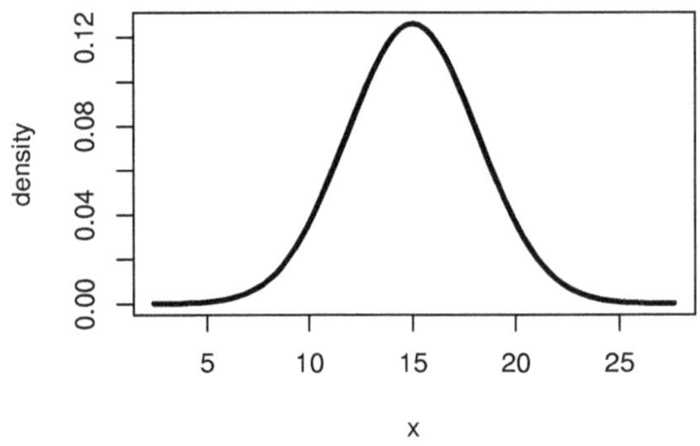

Fig. 5.2: Result of running the R code in Figure 5.1.

To save the output in the R graphics window, right-click and select a format for the graphics file (e.g., postscript). Production of this book used the freely available LaTeX package for mathematical typesetting, with encapsulated postscript (eps) graphics. See www.latex-project.org for more details, documentation, instructions on how to download and install LaTeX on your computer, and helpful examples.

Example 5.2. The next example compares the solution to the diffusion equation with drift (5.1) at different times $t_1, t_2, t_3 > 0$. The code in Figure 5.3 is very similar to Figure 5.1, repeated for each value of the time variable. The R command `lines` adds another curve to an existing graph. Figure 5.4 shows the graphical output. The same graph was also displayed as Figure 1.2. A good way to learn R is to start by running the same program listed here, and checking that the output is identical. Then modify the code slightly (e.g., change one of the input variables, or add a fourth curve) and check to see that the output is reasonable. This will also help build your intuition about the

```
D=1.0
v=3.0
t1=1.0
mu=v*t1
sigma=sqrt(2*D*t1)
x = seq(mu-4*sigma, mu+10*sigma, 0.1*sigma)
density=dnorm(x, mean = mu, sd = sigma)
plot(x,density,type="l",lwd=3)
t2=2.0
mu=v*t2
sigma=sqrt(2*D*t2)
x2 = seq(mu-4*sigma, mu+4*sigma, 0.1*sigma)
density=dnorm(x2, mean = mu, sd = sigma)
lines(x2,density,lty="dotted",lwd=3)
t3=3.0
mu=v*t3
sigma=sqrt(2*D*t3)
x3 = seq(mu-4*sigma, mu+4*sigma, 0.1*sigma)
density=dnorm(x3, mean = mu, sd = sigma)
lines(x3,density,lty="dashed",lwd=3)
```

Fig. 5.3: R code to compare solutions to the traditional diffusion equation with drift (5.1) at times $t_1 = 1.0$ (solid line), $t_2 = 2.0$ (dotted line), and $t_3 = 3.0$ (dashed line). The velocity $v = 3.0$ and dispersion $D = 1.0$.

underlying diffusion model. For example, you should be able to predict and check the result of changing the input parameter v.

Our next goal is to plot solutions to the fractional diffusion equation. This requires us to plot a stable density. There are existing R codes to plot stable densities, but they rely on an alternative parametrization, popularized by Samorodnitsky and Taqqu [185]. Recall that the signum function $\operatorname{sgn}(k) = +1$ for $k \geq 0$ and $\operatorname{sgn}(k) = -1$ for $k < 0$.

Proposition 5.3. *The characteristic function of a general stable random variable Y with Lévy representation $[a, 0, \phi]$ and Lévy measure (3.30) with index $0 < \alpha < 2$, $\alpha \neq 1$ can be written in the form*

$$\mathbb{E}[e^{ikY}] = \exp\left[ik\mu - \sigma^\alpha |k|^\alpha \left(1 - i\beta \operatorname{sgn}(k) \tan\left(\frac{\pi\alpha}{2}\right)\right)\right] \tag{5.2}$$

where $\mu = a$, $\beta = p - q$, and

$$\sigma^\alpha = C\frac{\Gamma(2-\alpha)}{1-\alpha}\cos\left(\frac{\pi\alpha}{2}\right). \tag{5.3}$$

In this case, we will write $Y \simeq S_\alpha(\beta, \sigma, \mu)$.

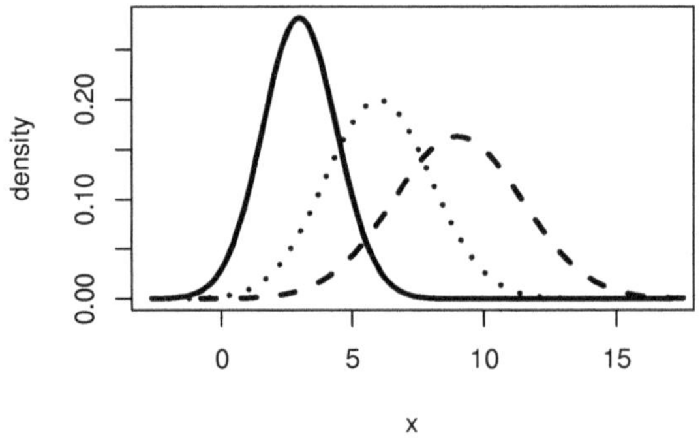

Fig. 5.4: Result of running the R code in Figure 5.3, displaying solutions to equation (5.1) at times $t_1 = 1.0$ (solid line), $t_2 = 2.0$ (dotted line), and $t_3 = 3.0$ (dashed line). The velocity $v = 3.0$ and dispersion $D = 1.0$.

Proof. If $0 < \alpha < 1$, then it follows from Example 3.27 that

$$\mathbb{E}[e^{ikY}] = \exp\left[ika + pA(-ik)^\alpha + qA(ik)^\alpha\right] \tag{5.4}$$

where $A = -C\Gamma(1-\alpha) < 0$. If $1 < \alpha < 2$, then it follows from Example 3.29 that (5.4) holds with $A = C\Gamma(2-\alpha)/(\alpha-1) > 0$. Since $\Gamma(2-\alpha) = (1-\alpha)\Gamma(1-\alpha)$, we can also write $A = C\Gamma(2-\alpha)/(\alpha-1)$ in the case $0 < \alpha < 1$. Use $e^{i\theta} = \cos\theta + i\sin\theta$ for $\theta \in \mathbb{R}$ to write

$$\begin{aligned}(ik)^\alpha &= (e^{i\pi/2}k)^\alpha \\ &= |k|^\alpha e^{i\,\mathrm{sgn}(k)\pi\alpha/2} \\ &= |k|^\alpha\left[\cos(\pi\alpha/2) + i\,\mathrm{sgn}(k)\sin(\pi\alpha/2)\right] \\ &= |k|^\alpha\cos(\pi\alpha/2)\left[1 + i\,\mathrm{sgn}(k)\tan(\pi\alpha/2)\right].\end{aligned} \tag{5.5}$$

Then $(-ik)^\alpha = |k|^\alpha\cos(\pi\alpha/2)\left[1 - i\,\mathrm{sgn}(k)\tan(\pi\alpha/2)\right]$ and so

$$\begin{aligned}pA(-ik)^\alpha + qA(ik)^\alpha &= pA|k|^\alpha\cos(\pi\alpha/2)\left[1 - i\,\mathrm{sgn}(k)\tan(\pi\alpha/2)\right] \\ &\quad + qA|k|^\alpha\cos(\pi\alpha/2)\left[1 + i\,\mathrm{sgn}(k)\tan(\pi\alpha/2)\right] \\ &= A\cos(\pi\alpha/2)|k|^\alpha\left[1 - i(p-q)\,\mathrm{sgn}(k)\tan(\pi\alpha/2)\right]\end{aligned} \tag{5.6}$$

and (5.2) follows. Note that the parameter $\sigma > 0$ for $0 < \alpha < 1$ and for $1 < \alpha < 2$, since $1 - \alpha$ and $\cos(\pi\alpha/2)$ both change sign at $\alpha = 1$. □

Remark 5.4. It is not hard to check, using characteristic functions, that if $Y \simeq S_\alpha(\beta, 1, 0)$ then $\sigma Y + \mu \simeq S_\alpha(\beta, \sigma, \mu)$. Hence σ is a scale parameter, and μ is a centering parameter. Some authors call $Y \simeq S_\alpha(\beta, 1, 0)$ a *standard stable law*. There are several additional parameterizations for stable laws. The seminal book of Zolotarev [228] lays out several useful parameterizations. The parametrization in Nolan [163] makes the density $f(y)$ a smooth function of all four parameters. The problem is that $e^{(ik)^\alpha} \to e^{ik}$ as $\alpha \to 1$, and this limit is the characteristic function of a point mass.

Remark 5.5. If $\alpha = 2$ then (5.2) also holds. Then $Y \simeq \mathcal{N}(\mu, 2\sigma^2)$ and the skewness β is irrelevant, since $\tan(\pi\alpha/2) = 0$ in this case. If $\alpha = 1$ then a formula somewhat different than (5.2) holds, since $\tan(\pi\alpha/2)$ is undefined. The characteristic function of a general stable random variable Y with Lévy representation $[a, 0, \phi]$ and Lévy measure (3.30) with index $\alpha = 1$ can be written in the form

$$\mathbb{E}[e^{ikY}] = \exp\left[ik\mu - \sigma^\alpha|k|\left(1 + i\beta\left(\frac{2}{\pi}\right)\operatorname{sgn}(k)\ln|t|\right)\right] \tag{5.7}$$

where $\mu = a$, $\beta = p - q$, and

$$\sigma^\alpha = C \cdot \frac{\pi}{2}, \tag{5.8}$$

see Meerschaert and Scheffler [146, Theorem 7.3.5] for complete details.

Remark 5.6. In Section 4.5 we defined the standard stable subordinator as the stable law with characteristic function $\hat{f}(k) = \exp(-(-ik)^\alpha)$ when $0 < \alpha < 1$. In Proposition 5.3 we can take $\mu = 0$, $\beta = 1$, and $\sigma^\alpha = \cos(\pi\alpha/2)$.

```
library(stabledist)
x = seq(-5, 10, 0.1)
density = dstable(x, alpha=1.5, beta=1.0, gamma=1.0, delta=0.0, pm=1)
plot(x,density,type="l")
grid()
```

Fig. 5.5: R code to plot a standard centered stable density with characteristic function (5.2), where $\mu = 0.0$, $\sigma = 1.0$, $\alpha = 1.5$, and $\beta = 1.0$.

Example 5.7. The R code in Figure 5.5 plots a stable density $f(y)$ for any values of the tail index $\alpha \in (0, 2]$, skewness $\beta \in [-1, 1]$, scale $\sigma > 0$, and center $\mu \in (-\infty, \infty)$. It relies on the `dstable` command from the R package `stabledist`, a freely available package of R codes for financial engineering and computational finance. See [221] for more details. You need to install the `stabledist` package on your R platform before you run the code in Figure 5.5. First try `Packages > Load package` to see if `stabledist` is available. If not, then use `Packages > Install package(s)` and select a convenient site for download to your computer. The calculation of the stable density behind the dstable command uses the sophisticated method of Nolan [163] to numerically compute the inverse Fourier transform. The option pm=1 specifies the Samorodnitsky and Taqqu parameterization (5.2). The parameters alpha and beta are as in equation (5.2). The scale parameter gamma is σ, and delta is the center μ, for this parametrization. Figure 5.6 shows the output from running the R code in Figure 5.5. Here we have set $\mu = 0.0$, $\sigma = 1.0$, $\alpha = 1.5$, and $\beta = 1.0$ to get a standard stable pdf that is totally positively skewed. This pdf represents the limit distribution of sums of iid positive jumps with power law tails $V_0(x) = \mathbb{P}[W > x] = Cx^{-\alpha}$ or, more generally, when $V_0(x)$ is RV$(-\alpha)$ and the right tail dominates.

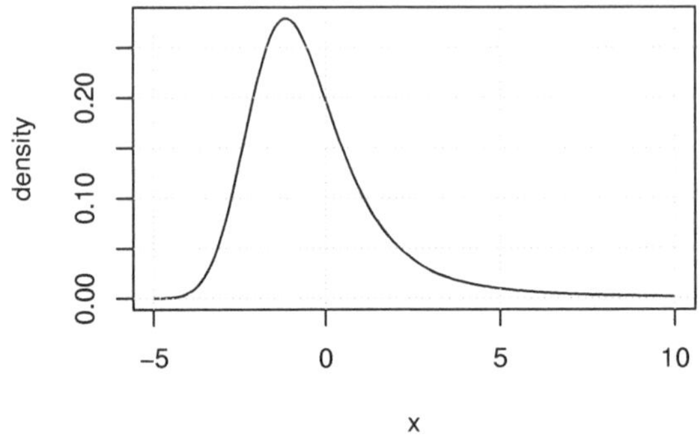

Fig. 5.6: Result of running the R code in Figure 5.5, a standard stable pdf with characteristic function (5.2), where $\mu = 0.0$, $\sigma = 1.0$, $\alpha = 1.5$, and $\beta = 1.0$.

In order to plot solutions to the fractional diffusion equation

$$\frac{\partial}{\partial t}p(x,t) = -v\frac{\partial}{\partial x}p(x,t) + Dp\frac{\partial^\alpha}{\partial x^\alpha}p(x,t) + Dq\frac{\partial^\alpha}{\partial(-x)^\alpha}p(x,t) \tag{5.9}$$

for $1 < \alpha < 2$, we need to convert to the parametrization of Proposition 5.3.

Proposition 5.8. *The solution $p(x,t)$ to the space-fractional diffusion equation* (5.9) *with index $1 < \alpha < 2$ is $S_\alpha(\beta, \sigma, \mu)$ with $\mu = vt$, $\beta = p - q$, and $\sigma^\alpha = Dt|\cos(\pi\alpha/2)|$.*

Proof. It follows from Example 3.29 that the point source solution $p(x,t)$ to (5.9) has characteristic function $\hat{p}(-k,t) = \exp\left[ikvt + pDt(-ik)^\alpha + qDt(ik)^\alpha\right]$. Write

$$pDt(-ik)^\alpha + qDt(ik)^\alpha = Dt\cos(\pi\alpha/2)|k|^\alpha\left[1 - i(p-q)\operatorname{sgn}(k)\tan(\pi\alpha/2)\right]$$

by an argument similar to Proposition 5.8. Now compare (5.2). □

Example 5.9. The R code in Figure 5.7 plots the solution to the space-fractional diffusion equation (5.9) for any time $t > 0$, with drift velocity $v \in \mathbb{R}$, dispersion $D > 0$, index $1 < \alpha \le 2$, and $0 \le q \le 1$. In this case, we have set $t = 5.0$ with velocity $v = 2.0$ and dispersion $D = 1.0$, for $\alpha = 1.5$ and $q = 0$ (totally positively skewed). The output of this code was displayed in Figure 1.3.

```
library(stabledist)
D=1.0
v=2.0
a=1.5
q=0.0
t=5.0
mu=v*t
pi=3.1415927
g=(D*t*abs(cos(pi*a/2)))^(1/a)
b=1-2*q
x = seq(mu-5*g, mu+5*g, 0.1*g)
p=dstable(x, alpha=a, beta=b, gamma = g, delta = mu, pm=1)
plot(x,p,type="l",lwd=3)
```

Fig. 5.7: R code to plot the solution $p(x, t)$ to the space-fractional diffusion equation (5.9) at time $t = 5.0$ with velocity $v = 2.0$ and dispersion $D = 1.0$, for $\alpha = 1.5$ and $q = 0$.

Remark 5.10. It follows from Example 3.27 that the solution to the fractional diffusion equation

$$\frac{\partial}{\partial t}p(x,t) = -v\frac{\partial}{\partial x}p(x,t) - Dp\frac{\partial^\alpha}{\partial x^\alpha}p(x,t) - Dq\frac{\partial^\alpha}{\partial(-x)^\alpha}p(x,t) \tag{5.10}$$

for $0 < \alpha < 1$ has characteristic function $\hat{p}(-k,t) = \exp\left[ikvt - pDt(-ik)^\alpha - qDt(ik)^\alpha\right]$. The only difference is a change of sign from D to $-D$ (we assume that $D > 0$). Then an argument similar to Proposition 5.8 shows that $p(x,t)$ is $S_\alpha(\beta,\sigma,\mu)$ with the same parameters as for the case $1 < \alpha < 2$, i.e., $\mu = vt$, $\beta = p - q$, and $\sigma^\alpha = Dt|\cos(\pi\alpha/2)|$. Hence the R code in Figure 5.7 can also be used to solve the fractional diffusion equation (5.10) in the case $0 < \alpha < 1$.

Example 5.11. The R code in Figure 5.8 compares the solution to the space-fractional diffusion equation (5.9) at times $t_1, t_2, t_3 > 0$, with drift velocity $v \in \mathbb{R}$, dispersion $D > 0$, index $1 < \alpha \le 2$, and $0 \le q \le 1$. The output of this code was displayed in Figure 1.4. It compares the solution at times $t_1 = 3.0$, $t_2 = 5.0$, and $t_3 = 8.0$ with velocity $v = 3.0$ and dispersion $D = 1.0$, for $\alpha = 1.5$ and $q = 0$ (positive skew). This is an

```
library(stabledist)
D=1.0
v=3.0
a=1.5
q=0.0
t1=3.0
t2=5.0
t3=8.0
pi=3.1415927
b=1-2*q
mu1=v*t1
g1=(D*t1*abs(cos(pi*a/2)))^(1/a)
x = seq(mu1-5*g1, mu1+10*g1, 0.1*g1)
p=dstable(x, alpha=a, beta=b, gamma = g1, delta = mu1, pm=1)
plot(x,p,type="l",lwd=3)
mu2=v*t2
g2=(D*t2*abs(cos(pi*a/2)))^(1/a)
p2=dstable(x, alpha=a, beta=b, gamma = g2, delta = mu2, pm=1)
lines(x,p2,lty="dotted",lwd=3)
mu3=v*t3
g3=(D*t3*abs(cos(pi*a/2)))^(1/a)
p3=dstable(x, alpha=a, beta=b, gamma = g3, delta = mu3, pm=1)
lines(x,p3,lty="dashed",lwd=3)
```

Fig. 5.8: R code to compare solutions $p(x, t)$ to the space-fractional diffusion equation (5.9) at times $t_1 = 3.0$, $t_2 = 5.0$, and $t_3 = 8.0$ with velocity $v = 3.0$ and dispersion $D = 1.0$, for $\alpha = 1.5$ and $q = 0$.

illustration of anomalous super-diffusion. The pdf spreads from its center of mass like $t^{1/1.5}$ which is faster than the $t^{1/2}$ spreading for a traditional diffusion.

Example 5.12. The R code in Figure 5.9 plots the density of a stable subordinator Y with characteristic function $E[\exp(ikY)] = \exp(-(-ik)^\alpha)$ for $0 < \alpha < 1$, using Remark 5.6 and the parametrization (5.2). Note that the Laplace transform of the density of Y is $E[\exp(-sY)] = \exp(-s^\alpha)$. Figure 5.10 plots the density of the stable subordinator with index $\alpha = 0.75$. Note that this density is always supported on the positive real line.

Example 5.13. The R code in Figure 5.11 plots the solution to the time-fractional diffusion equation

$$\partial_t^\beta p(x, t) = D \frac{\partial^2}{\partial x^2} p(x, t) \tag{5.11}$$

```
library(stabledist)
x = seq(0, 5, 0.01)
a=0.75
pi=3.1415927
g=(cos(pi*a/2))^(1/a)
density = dstable(x, alpha=a, beta=1.0, gamma=g, delta=0, pm=1)
plot(x,density,type="l")
grid()
```

Fig. 5.9: R code to plot the pdf of a standard stable subordinator with index $\alpha \in (0, 1)$.

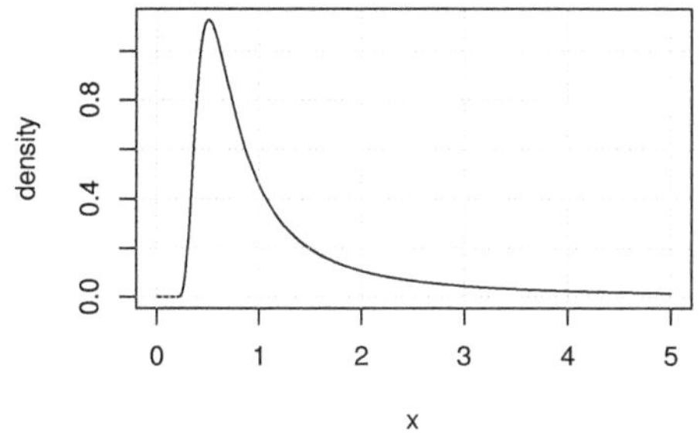

Fig. 5.10: Result of running the R code in Figure 5.9, pdf of a standard stable subordinator with index $\alpha = 0.75$.

for any time $t > 0$, with a Caputo derivative of order $0 < \beta < 1$, and dispersion $D > 0$. This is a special case of (4.44) with $\alpha = 2$. It represents the scaling limit of a CTRW with mean zero jumps in the domain of attraction of a normal law (e.g., mean zero finite variance jumps), separated by power law waiting times with index β. The solution to (5.11) is the pdf of $A(E(t))$, where $A(t)$ is a Brownian motion, and $E(t)$ is an inverse stable subordinator. The R code in Figure 5.11 is based on the formula (4.39) where $p(x, u)$ is the pdf of $A(u)$ and $h(u, t)$ is the pdf of $E(t)$. In the code, we use the fact that $p(x, u)$ is normal with mean zero and variance $2Du$ along with the alternative form

$$m(x, t) = \int_0^\infty p(x, (t/r)^\beta) g(r)\, dr \tag{5.12}$$

where $g(r)$ is a the standard stable subordinator pdf, see (4.48). This form is convenient for computations, because the pdf $g(r)$ can be calculated once, and used over and over for every value of the time $t > 0$. Since we have an analytical formula for the normal density, computing $p(x, (t/r)^\beta)$ is a simple matter. The integral in (5.12) is evaluated

```
library(stabledist)
dr=0.5
r=seq(dr,5000.0,dr)
b=0.75
pi=3.1415927
g=(cos(pi*b/2))^(1/b)
h=dstable(r, alpha = b, beta = 1.0, gamma = g, delta = 0, pm=1)
D=1.0
mcall <- function(y,t) {
          sum(dnorm(y, mean = 0.0, sd =sqrt(D*(t/r)^b) )*h*dr)
}
x=seq(-5.0,5.0,0.1); m=x; t=0.1
for (i in 1:length(x)){
m[i]=mcall(x[i],t)}
plot(x,m,type="l")
```

Fig. 5.11: R code to plot the solution to the time-fractional diffusion equation (5.11) for any time $t > 0$. Here $\beta = 0.75$ and $D = 1.0$.

numerically by a simple Euler (rectangle) approximation. Figure 5.12 shows the output for time $t = 0.1$ with $\beta = 0.75$ and dispersion $D = 1.0$. Note the sharp peak at $x = 0$, which is typical of the time-fractional diffusion profile. This same plot was shown previously as Figure 2.3.

Example 5.14. The R code in Figure 5.14 compares the solution to the time-fractional diffusion equation (5.11) at times $t_1, t_2, t_3 > 0$, with fractional derivative of order $0 < \beta < 1$ and dispersion $D > 0$. Figure 5.13 compares the solution at times $t_1 = 0.1$, $t_2 = 0.3$, and $t_3 = 0.8$ with $\beta = 0.75$ and dispersion $D = 1.0$. This plot illustrates anomalous sub-diffusion. The limit process $A(E(t))$ is self-similar with Hurst index $\beta/2 < 1/2$, so the solution spreads at a slower rate than a traditional diffusion.

5.2 Sample path simulations

This section introduces R codes to simulate the sample paths of stochastic processes, including random walks, Brownian motion, stable Lévy motion, CTRW, and CTRW limits. First we will simulate one dimensional processes, then we will explore the properties of two dimensional sample paths. The limit theory for two or more dimensions will be presented in Chapter 6.

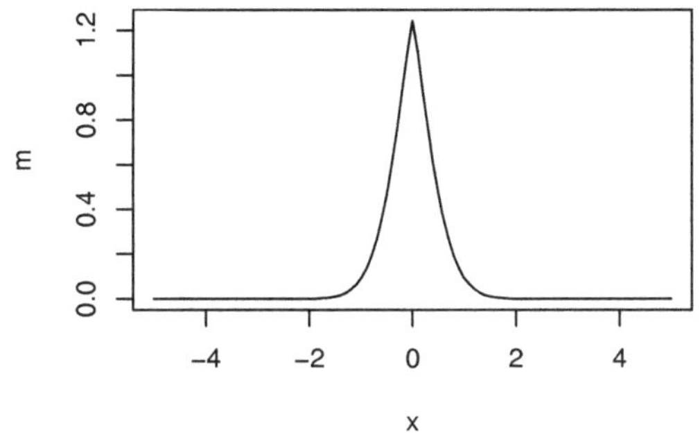

Fig. 5.12: Result of running the R code in Figure 5.11, the solution to time-fractional diffusion equation (5.11) at time $t = 0.1$ with $\beta = 0.75$ and dispersion $D = 1.0$

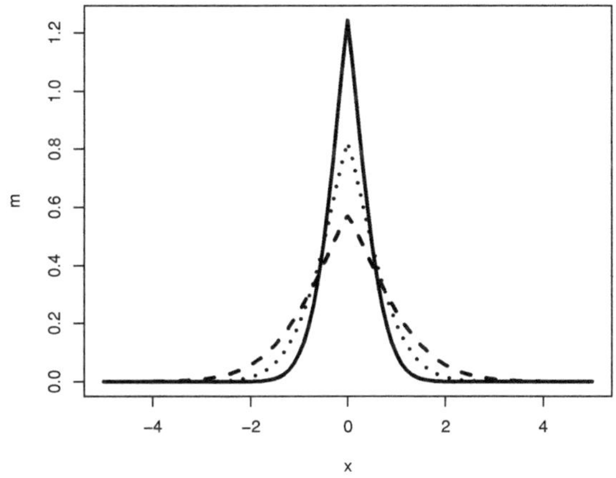

Fig. 5.13: Solution to time-fractional diffusion equation (5.11) at times $t_1 = 0.1$ (solid line), $t_2 = 0.3$ (dotted line), and $t_3 = 0.8$ (dashed line) with $\beta = 0.75$ and dispersion $D = 1.0$.

```
library(stabledist)
dr=0.5; b=0.75; D=1.0; pi=3.1415927; g=(cos(pi*b/2))^(1/b)
r=seq(dr,5000.0,dr)
h=dstable(r, alpha = b, beta = 1.0, gamma = g, delta = 0, pm=1)
mcall <- function(y,t) {
          sum(dnorm(y, mean = 0.0, sd =sqrt(D*(t/r)^b) )*h*dr)
}
x=seq(-5.0,5.0,0.1)
m=x; t1=0.1
for (i in 1:length(x)){
m[i]=mcall(x[i],t1)}
plot(x,m,type="l")
t2=0.3
m2=x
for (i in 1:length(x)){
m2[i]=mcall(x[i],t2)}
lines(x,m2,lty="dotted")
t3=0.8
m3=x
for (i in 1:length(x)){
m3[i]=mcall(x[i],t3)}
lines(x,m3,lty="dashed")
```

Fig. 5.14: R code to compare solutions to the time-fractional diffusion equation (5.11) at times at times $t_1 = 0.1$, $t_2 = 0.3$, and $t_3 = 0.8$. Here $\beta = 0.75$ and $D = 1.0$.

Example 5.15. We showed in Example 3.31 that a random walk $S(n) = W_1 + \cdots + W_n$ with iid mean zero finite variance jumps converges to a Brownian motion $A(t)$. In fact we have $c^{-1/2}S([ct]) \Rightarrow A(t)$ in $\mathbb{D}[0, \infty)$ with the Skorokhod J_1 topology (e.g., see Billingsley [37]). To illustrate this sample path convergence, we will use R to simulate a random walk. Figure 5.15 lists the R code to simulate a random walk whose iid jumps are uniform on the interval $[-1, 1]$. Since these jumps have mean zero and finite variance, the simulated random walk converges to a Brownian motion in the scaling limit. The `runif` command in R produces a vector of (iid) uniform random variates. The `cumsum` command returns the cumulative sum of a vector, i.e., given a vector $[W_i : i = 1, \ldots, n]$ it returns the vector with ith entry $S(i) = W_1 + \cdots + W_i$. Then the plot shows the points $[(i, S(i)) : i = 1, \ldots, n]$ connected by straight line segments. Figure 5.16 shows a typical output from running the R code in Figure 5.15. Since this *Monte Carlo simulation* involves random numbers, every run produces a different picture. However, these pictures all have similar features. Each plot can be considered

```
t=seq(1:100)
W=runif(t, min=-1, max=1)
S=cumsum(W)
plot(t,S,type="l")
```

Fig. 5.15: R code to simulate a random walk with iid uniform $[-1, 1]$ jumps.

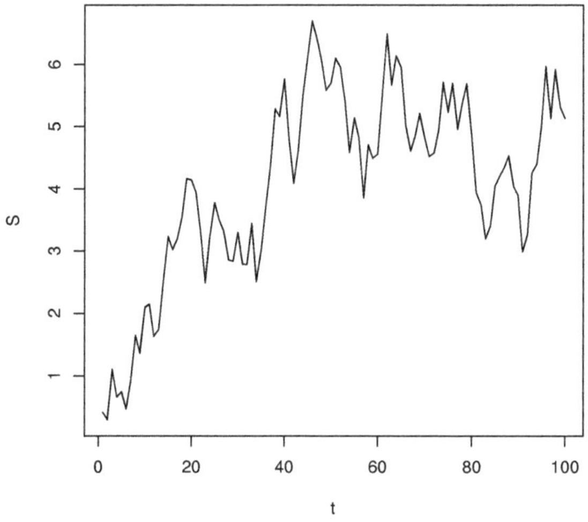

Fig. 5.16: Simulated random walk, the result of running the R code in Figure 5.15.

as a graphical representation of the path followed by a randomly selected particle. Running the same R simulation over and over shows paths of different particles.

One way to illustrate convergence to a Brownian motion is to vary the length of the random walk in the R code from Figure 5.15. Just change `t=seq(1:100)` to, say, `t=seq(1:10)` and then `t=seq(1:50)` and so forth. Once the sequence length is large enough, increasing it further does not significantly effect the general appearance of the graphical output. Of course the axis lengths will change. In fact, you can check that the scale on the vertical axis is roughly the square root of the horizontal scale.

Example 5.16. Figure 5.17 lists the R code to simulate a Brownian motion. In fact, we approximate the Brownian motion by a random walk with iid $\mathcal{N}(0, 1)$ jumps. Then $S(n) \simeq \mathcal{N}(0, n)$ approximates a standard Brownian motion $A(t)$. The approximation $A(t) \approx S([t])$ is exact when t is an integer, and the graph interpolates between these points. Since our simulation uses 1000 points, the difference between the exact and simulated sample path is indistinguishable to the human eye. Figure 5.18 shows a typ-

```
t=seq(1:1000)
W=rnorm(t, mean=0, sd=1.0)
A=cumsum(W)
plot(t,A,type="l")
```

Fig. 5.17: R code to simulate a standard Brownian motion.

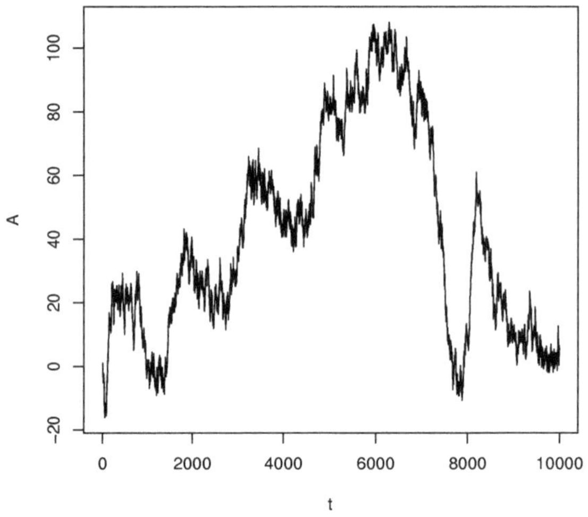

Fig. 5.18: Simulated Brownian motion, the result of running the R code in Figure 5.17.

ical output from running the R code in Figure 5.17. Running the same code over and over will generate statistically identical but individually distinct sample paths of a diffusing particle following a Brownian motion.

The sample paths of a Brownian motion have many interesting properties. The sample paths are (with probability one in the space $\mathbb{D}[0, \infty)$) everywhere continuous, but nowhere differentiable. They do not have bounded variation over finite intervals, i.e., the length of the path $\{(t, A(t)) : a \leq t \leq b\}$ is infinite. More specifically, if we subdivide the path into smaller increments and join these points by straight lines, the total length of these lines tends to infinity as the mesh of the partition tends to zero.

In fact, the graph is a (random) fractal of dimension $d = 3/2$. Fractals are sets whose dimension is not an integer. There are several notions of dimension (Hausdorff dimension, packing dimension, etc.) but the easiest to explain is the box dimension. Suppose that it requires $C(n)$ boxes of size $1/n$ to cover a set. If there is a number

$$d = \lim_{n\to\infty} \frac{\log C(n)}{\log n}$$

then we call d the box dimension. For example, it takes $C(n) = Ln$ boxes to cover a line of length L, so that $d = 1$. It takes $C(n) = Vn^3$ to cover a cube of volume V, so the cube has dimension $d = 3$. The proof that the graph of a Brownian motion has (almost surely) dimension $d = 3/2$ requires some deep analysis, e.g., see Falconer [65, Theorem 16.4].

One interesting property of fractals is self-similarity (or self-affinity) which means, essentially, that zooming in or out on the graph produces a similar shape. For our sample path simulations, we can illustrate self-similarity by increasing the length of the simulated Brownian motion (i.e., change `t=seq(1:1000)` to `t=seq(1:10000)`) or longer). The resulting graphs are more or less indistinguishable.

It is often overlooked that Brownian motion and the diffusion equation provide an *approximate* model for diffusing particles. The theory of relativity puts an upper bound on the distance a particle can travel in a finite time, but the normal pdf is positive on the entire real line. A real physical particle in the physical world cannot trace a path of unbounded variation (infinite length) in finite time. A real particle has a velocity, but the sample path of a Brownian motion does not, since the derivative is undefined. (It has, in some sense, an infinite speed.) From the point of view of probability, we understand that Brownian motion and the resulting diffusion equation are merely an approximation, valid at late time (after many particle jumps have accumulated). In the real world, the random walk is the fundamental physical model, and the limit process is a very useful approximation.

If you go back now to the simulation in Figure 5.15 and extend the length of the sequence of jumps simulated (i.e., change `t=seq(1:100)` to `t=seq(1:1000)`) or longer) you can see that the random walk becomes indistinguishable from a Brownian motion. If you change the distribution of the random walk jumps (e.g., change `runif(t, min=-1, max=1)` to `runif(t, min=-5, max=5)` or even to a different mean zero finite variance distribution) then the same effect persists. This illustrates the random walk convergence to a Brownian motion in a very concrete way.

Example 5.17. Figure 5.19 provides the R code to simulate a random walk with iid Pareto jumps. The simulation uses the fact that if W has cdf $F(y) = \mathbb{P}[W \le y]$ then $W \simeq F^{-1}(U)$ where U is uniform on $[0, 1]$. This is easy to check:

$$\mathbb{P}[F^{-1}(U) \le y] = \mathbb{P}[U \le F(y)] = F(y).$$

See Press et al. [170, Chapter 7] for more details. Applying this idea to a Pareto with $\mathbb{P}[W > x] = Cx^{-\alpha}$ we have

$$\mathbb{P}[(U/C)^{-1/\alpha} > x] = \mathbb{P}[U < Cx^{-\alpha}] = Cx^{-\alpha}$$

for $x > C^{1/\alpha}$ which shows that $(U/C)^{-1/\alpha}$ has a Pareto distribution, when U is uniform on $[0, 1]$. Using (3.52) we see that the Pareto has mean $\mu_1 = C^{1/\alpha}\alpha/(\alpha - 1)$ when $\alpha > 1$. The code simulates a random walk with iid Pareto jumps, corrected to mean zero. Then

```
C=1.0
alpha=1.5
t=seq(1:100)
U=runif(t)
Y=(U/C)^(-1/alpha)-(alpha/(alpha-1))*C^(1/alpha)
S=cumsum(Y)
plot(t,S,type="l")
```

Fig. 5.19: R code to simulate a random walk with iid Pareto jumps.

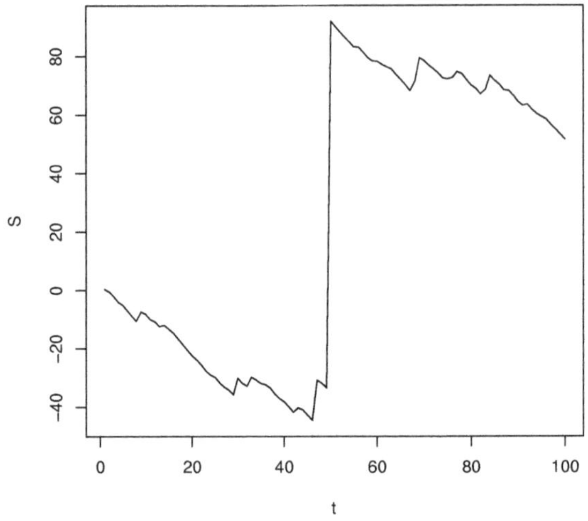

Fig. 5.20: Simulated random walk with Pareto jumps, the result of running the R code in Figure 5.19.

$c^{-1/\alpha}S([ct]) \Rightarrow A(t)$, a mean zero α-stable Lévy motion (see Section 4.4). Figure 5.20 shows a typical output from running the R code in Figure 5.19. The overall negative drift compensates for the occasional large positive jumps. For a Pareto with $0 < \alpha < 2$, these jumps persist in the long-time scaling limit. To check this, change the length of the simulated sequence and note that, unlike the finite variance random walk simulated previously, the large positive jumps remain prominent at any length scale.

To understand why this happens, consider the compound Poisson approximation. We have $S_n = X_{n1} + \cdots + X_{nn} \Rightarrow A = A(1)$ stable where $X_{nj} = a_n W_j$ are the rescaled random walk jumps. We can take $a_n = n^{-1/\alpha}$ in the Pareto case. Since $n\mathbb{P}[a_n W_j > R] \to \phi(R, \infty) = CR^{-\alpha}$ is the mean number of jumps of size greater than R, the probability of any one jump exceeding this threshold is approximately $n^{-1}CR^{-\alpha}$, but since there are n independent jumps, the probability of *at least one* of those jumps exceeding the threshold is approximately

$$1 - \left(1 - \frac{CR^{-\alpha}}{n}\right)^n \approx 1 - e^{-CR^{-\alpha}} \approx CR^{-\alpha}$$

for $R > 0$ sufficiently large. Furthermore, this is the probability that at least one jump W_j exceeds $n^{1/\alpha}R$, which is comparable to the total sum since $S(n) \approx n^{1/\alpha}A$. Hence the largest jump is comparable to the entire sum. Since $n\mathbb{P}[a_n W_j > R] \to 0$ for finite variance jumps, the largest jump there is a negligible part of the sum. This is one main distinguishing property of heavy tailed random walks.

```
C=1.0
alpha=1.5
p=0.3
t=seq(1:100)
U=runif(t)
Y=(U/C)^(-1/alpha)-(alpha/(alpha-1))*C^(1/alpha)
V=runif(t)
for (i in 1:length(t)){
if (V[i]>p) Y[i]=-Y[i]}
S=cumsum(Y)
plot(t,S,type="l")
```

Fig. 5.21: R code to simulate a random walk with iid power law jumps.

Example 5.18. Figure 5.21 provides the R code to simulate a random walk with iid power law jumps. The code is similar to Figure 5.19. First we simulate iid Pareto jumps (W_n) and correct to mean zero. Then we adjust by drawing a random number U uniform on $[0, 1]$ and changing the sign of this jump, to give a negative jump, with probability $q = 1 - p$. The resulting code simulates a random walk with iid power law jumps, corrected to mean zero, as in Theorem 3.41. Then $c^{-1/\alpha}S([ct]) \Rightarrow A(t)$, a mean zero α-stable Lévy motion with both positive and negative jumps. Figure 5.22 shows a typical output from running the R code in Figure 5.21. The sample path contains occasional large jumps, which can be either positive or negative. Again, if we lengthen the random walk sequence, we eventually get to the point where the resulting graphs are insensitive to the overall length of the simulation. This illustrates the convergence to a self-similar limit process.

Example 5.19. The R code in Figure 5.23 simulates a symmetric stable Lévy motion. The simulated process is actually a random walk with iid stable jumps, using the R

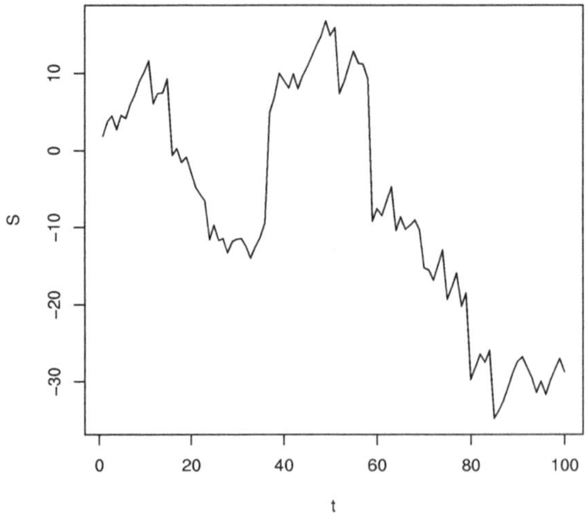

Fig. 5.22: Simulated random walk with power law jumps, the result of running the R code in Figure 5.21.

command `rstable` to generate a vector of iid stable random variates. This command is also part of the `stabledist` package introduced in Example 5.7. Figure 5.24 shows a typical sample path, obtained by running the R code in Figure 5.23. Note the occasional large jumps. Since we simulate a stable Lévy motion with $\beta = p - q = 0$ we have $p = 1/2$ and $q = 1/2$, i.e., the large jumps are equally likely to be positive or negative. Since we set μ = `delta` = 0, the process is compensated to mean zero.

The graph of a stable Lévy motion with index $1 < \alpha < 2$ is also a random fractal, with dimension $d = 2 - 1/\alpha$, see Falconer [65, Section 16.3]. This extends the result mentioned in Example 5.16 for Brownian motion, where $d = 2 - 1/2$. The fractal dimension describes the "roughness" of the particle traces. As α decreases from 2 to 1, the sample paths become smoother.

Example 5.20. The R code in Figure 5.25 simulates a continuous time random walk (CTRW) with iid Pareto waiting times and iid power law jumps. The method for simulating the jumps is the same as in Example 5.18. The method for simulating the waiting times is the same as Example 5.17. The CTRW is actually a random walk in space-time, i.e., a two-dimensional random walk in which the horizontal axis represents elapsed time, and the vertical axis represents the spatial location. Hence the R code is quite similar to what we have seen before. The only difference is that we plot the cumulative sum $S(i)$ of the jumps against the cumulative sum $T(i)$ of the waiting times, rather than

```
library(stabledist)
t=seq(1:1000)
Y=rstable(t, alpha = 1.5, beta = 0.0, gamma=1.0, delta=0.0, pm=1)
A=cumsum(Y)
plot(t,A,type="l")
```

Fig. 5.23: R code to simulate a stable Lévy motion.

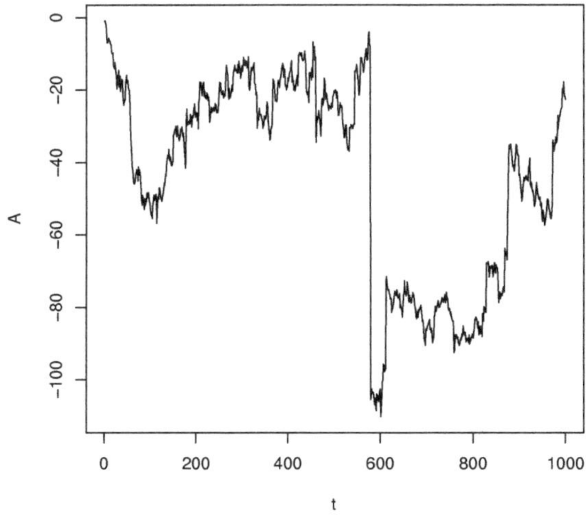

Fig. 5.24: Simulated stable Lévy motion, the result of running the R code in Figure 5.23.

plotting $S(i)$ versus i. Figure 5.26 shows a typical sample path, obtained by running the R code in Figure 5.25. Note the long jumps in space, and also the long jumps in time. Both will persist in the scaling limit, as the simulated sequence gets longer. Eventually, the character of the simulated sample paths becomes insensitive to the length of the sequence, an illustration of the CTRW limit (4.32). If we replace the simulated jumps by iid mean zero finite variance jumps as in Example 5.15, the outer process $A(t)$ in the scaling limit is a Brownian motion. In this case, the jumps in space disappear in the limit. If we replace the Pareto waiting times by some positive iid random variables with finite mean (e.g., use `runif(t, min=0, max=1)`) then the resulting sample paths very closely resemble those from Example 5.18, as we discussed in Section 4.3. The CTRW sample paths represents particle traces, in which a particle can stick at some point for a random period before the next jump. If the waiting time pdf has a sufficiently heavy tail, this significantly affects the movement of particles over the long term.

```
C=1.0; alpha=1.5; p=0.3; B=1.0; beta=0.8
t=seq(1:1000)
U=runif(t)
Y=(U/C)^(-1/alpha)-(alpha/(alpha-1))*C^(1/alpha)
V=runif(t)
for (i in 1:length(t)){
  if (V[i]>p) Y[i]=-Y[i]}
S=cumsum(Y)
U=runif(t)
J=(U/B)^(-1/beta)
T=cumsum(J)
plot(T,S,type="l")
```

Fig. 5.25: R code to simulate a continuous time random walk (CTRW).

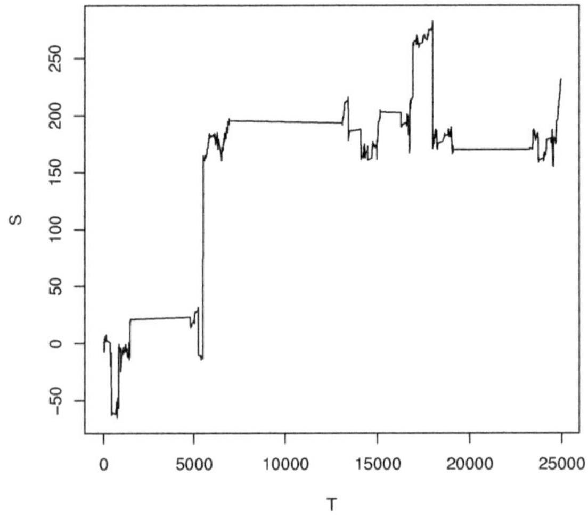

Fig. 5.26: Simulated CTRW, the result of running the R code in Figure 5.25.

```
library(stabledist)
a=1.5
skew=0.0
b=0.8
t=seq(1:1000)
Y=rstable(t,alpha=a, beta=skew)
A=cumsum(Y)
pi=3.1415927
g=(cos(pi*b/2))^(-1/b)
J=rstable(t, alpha=b, beta=1.0, gamma=g, delta=0, pm=1)
T=cumsum(J)
plot(T,A,type="l")
```

Fig. 5.27: R code to simulate the CTRW scaling limit process.

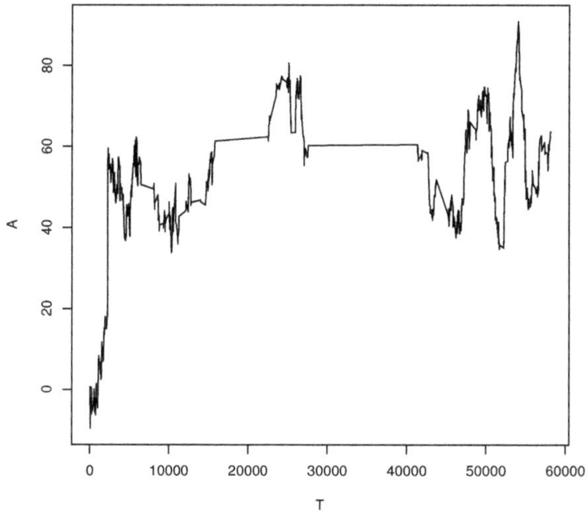

Fig. 5.28: Simulated CTRW limit, the result of running the R code in Figure 5.27.

Example 5.21. The R code in Figure 5.27 simulates the CTRW scaling limit process $A(E(t))$ from (4.32). Figure 5.28 shows a typical sample path, obtained by running the R code in Figure 5.27. The outer process $x = A(u)$ is symmetric stable with index $\alpha = 1.5$ and the inner process $u = E(t)$ is the inverse of a standard stable subordinator $t = D(u)$ with index $\beta = 0.8$. Actually the simulation approximates this process by a CTRW with stable particle jumps, and stable waiting times. Note that the graph of $(t, A(E(t))$ is essentially the same as the graph of $(D(u), A(u))$, since $E(D(u)) = u$ for all $u \geq 0$. The only difference is that the horizontal jumps in the graph of $(D(u), A(u))$ are connected by a continuous line in the graph of $(t, A(E(t))$, see Meerschaert, Nane and Xiao [143] for additional details. Since R code connects the plotted points with a continuous line, the resulting graph is approximate only in terms of the discretization of the Lévy processes: The code simulates the two independent Lévy processes $A(u)$ and $D(u)$ using random walks with iid stable jumps, as in Example 5.19. Note that the limit process retains the long jumps in both space and time. Some results on the fractal dimension of the CTRW limit process are contained in [143].

We conclude this section with two examples that illustrate the sample paths of vector-valued stochastic processes. From a physics point of view, it is quite natural to consider particle traces in two or three dimensions, since the real world is not one dimensional. Furthermore, we have already seen that the CTRW is fundamentally a random walk in two dimensions (one space and one time). Vector random walks, their limit processes, and their governing equations will be developed in Chapter 6.

```
t=seq(1:5000)
X=rnorm(t, mean=0, sd=1.0)
A1=cumsum(X)
Y=rnorm(t, mean=0, sd=1.0)
A2=cumsum(Y)
plot(A1,A2,type="l")
```

Fig. 5.29: R code to simulate a two dimensional Brownian motion.

Example 5.22. Figure 5.29 shows the R code to simulate a Brownian motion in two dimensions. The code is a simple modification of Example 5.16. The two dimensional Brownian motion is $A(t) = (A_1(t), A_2(t))$ where $A_1(t)$ and $A_2(t)$ are two independent one dimensional Brownian motions. A vector Brownian motion is the scaling limit of a random walk with vector jumps, when the iid jumps have mean zero and finite second moments. Figure 5.30 shows a typical sample path. The sample path of a Brownian motion in $\mathbb{R}^d$ for $d \geq 2$ is a random fractal with dimension two.

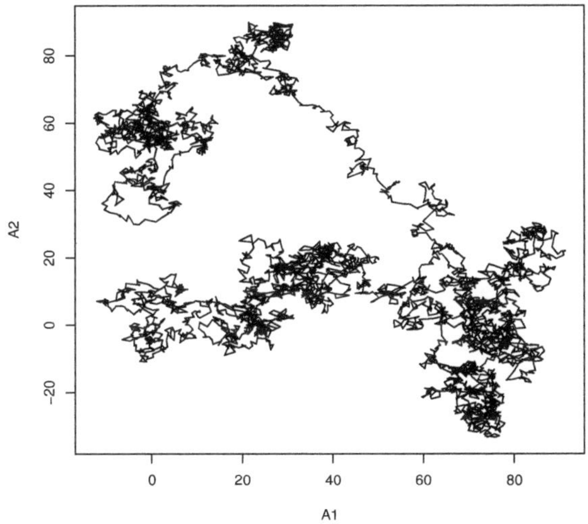

Fig. 5.30: Simulated Brownian motion in two dimensions, the result of running the R code in Figure 5.29.

Example 5.23. The R code in Figure 5.31 simulates a two dimensional stable Lévy motion with index $\alpha = 1.8$. The code is a simple modification of Example 5.19. This process is the scaling limit of a vector random walk with iid Pareto jumps in each coordinate. Figure 5.32 shows a typical sample path. In contrast to Brownian motion, the sample path of a vector stable Lévy motion shows occasional large jumps. The sample path of an α-stable Lévy motion in $\mathbb{R}^d$ for $d \geq 2$ is a random fractal with dimension α, extending the result for Brownian motion (see Blumenthal and Getoor [39] and Meerschaert and Xiao [156, Theorem 3.2]). Hence we can see that the power law index, the order of the fractional derivative, and the fractal dimension are all the same. The two dimensional stable Lévy motion is $A(t) = (A_1(t), A_2(t))$ where $A_1(t)$ and $A_2(t)$ are two

```
library(stabledist)
t=seq(1:5000)
X=rstable(t, alpha = 1.8, beta = 0.0, gamma=1.0, delta=0.0, pm=1)
A1=cumsum(X)
Y=rstable(t, alpha = 1.8, beta = 0.0, gamma=1.0, delta=0.0, pm=1)
A2=cumsum(Y)
plot(A1,A2,type="l")
```

Fig. 5.31: R code to simulate a two dimensional stable Lévy motion.

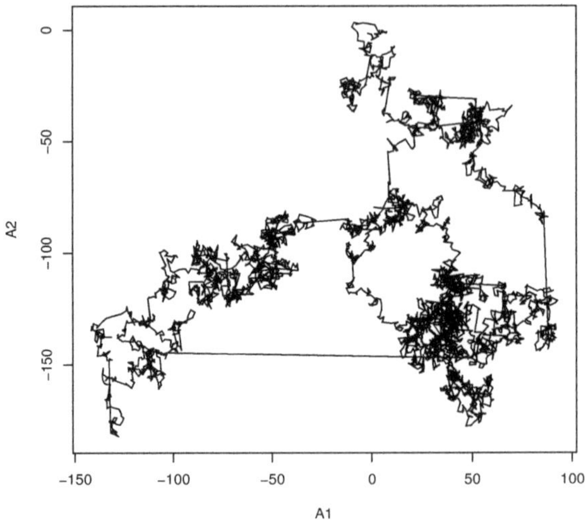

Fig. 5.32: Simulated stable Lévy motion in two dimensions, the result of running the R code in Figure 5.31.

independent one dimensional stable Lévy motions. If we take the index α_1 of $A_1(t)$ to be different than the index α_2 of the second component $A_2(t)$, the resulting process is called an *operator stable Lévy motion.* Operator stable Lévy motions are scaling limits of a vector random walk when the power law index of the Pareto jump pdf depends on the coordinate. It is a simple matter to simulate an operator stable Lévy motion, by editing the index `alpha` in the code. For more information on simulating operator stable sample paths, and additional examples, see Cohen et al. [51].

6 Vector Fractional Diffusion

Since many applied problems require a more realistic model in a 2-dimensional or 3-dimensional physical space, this chapter extends the fractional diffusion model developed in previous chapters to a vector setting.

6.1 Vector random walks

Suppose that (X_n) and (Y_n) are two independent sequences of iid random variables. The two dimensional random walk with coordinates $S_n = X_1 + \cdots + X_n$ and $R_n = Y_1 + \cdots + Y_n$ represents the position of a particle in the (x, y) plane after n jumps. Suppose that $\mathbb{E}[X_n] = \mathbb{E}[Y_n] = 0$ and $\mathbb{E}[X_n^2] = \mathbb{E}[Y_n^2] = 2D$ for some constant $D > 0$. Then it follows from Example 3.31 that

$$n^{-1/2} S_{[nt]} \Rightarrow Z_t \quad \text{and} \quad n^{-1/2} R_{[nt]} \Rightarrow W_t$$

where Z_t and W_t are two independent Brownian motions. In vector notation, we have

$$n^{-1/2} \begin{pmatrix} S_{[nt]} \\ R_{[nt]} \end{pmatrix} \Rightarrow \begin{pmatrix} Z_t \\ W_t \end{pmatrix} \tag{6.1}$$

as $n \to \infty$. The limit process in (6.1) is a two dimensional Brownian motion with independent components. A typical sample path was shown in Figure 5.30. If $p_1(x, t)$ is the pdf of Z_t and $p_2(y, t)$ is the pdf of W_t, then the vector limit has a pdf

$$\begin{aligned} p(x, y, t) = p_1(x, t) p_2(y, t) &= \frac{1}{\sqrt{4\pi Dt}} e^{-x^2/4Dt} \frac{1}{\sqrt{4\pi Dt}} e^{-y^2/4Dt} \\ &= \frac{1}{4\pi Dt} e^{-(x^2+y^2)/4Dt} \end{aligned} \tag{6.2}$$

whose Fourier transform

$$\begin{aligned} \hat{p}(k, \ell, t) &= \int\int e^{-ikx} e^{-i\ell y} p(x, y, t)\, dy\, dx \\ &= \int e^{-ikx} p_1(x, t)\, dx \int e^{-i\ell y} p_2(y, t)\, dy = e^{-Dtk^2} e^{-Dt\ell^2} \end{aligned}$$

solves the differential equation

$$\frac{d}{dt} \hat{p}(k, \ell, t) = [-Dk^2 - D\ell^2] \hat{p}(k, \ell, t) = [D(ik)^2 + D(i\ell)^2] \hat{p}(k, \ell, t).$$

Inverting the FT shows that $p(x, y, t)$ solves the two dimensional diffusion equation

$$\frac{\partial}{\partial t} p(x, y, t) = D \frac{\partial^2}{\partial x^2} p(x, y, t) + D \frac{\partial^2}{\partial y^2} p(x, y, t). \tag{6.3}$$

This is an *isotropic* diffusion equation. Figure 6.1 shows level sets of the isotropic pdf (6.2). It was produced using the R code in Figure 6.11, listed at the end of this chapter. Since the density $p(x, y, t)$ only depends on $x^2 + y^2$, the level sets are circles, and the pdf is rotationally symmetric. This means that the diffusion looks the same in any orthogonal coordinate system centered at the origin. Because Z_t is isotropic, any rotation and/or reflection in Figure 5.30 produces an equally likely sample path.

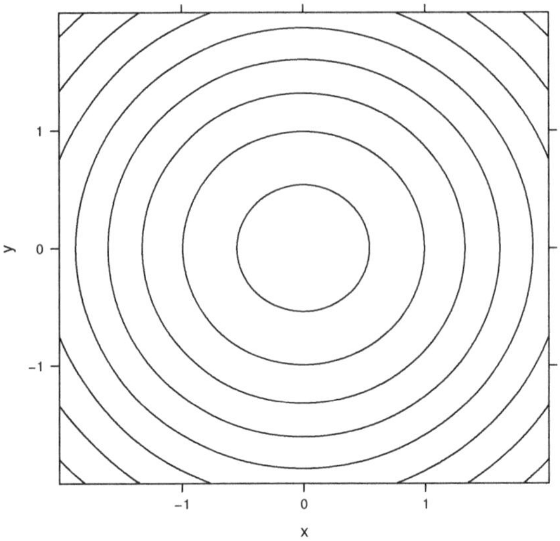

Fig. 6.1: Level sets of the solution (6.2) to the isotropic diffusion equation (6.3) at time $t = 1$ with dispersivity parameter $D = 2$.

To develop a more general, *anisotropic* diffusion equation, suppose that the two independent random walks have $\mathbb{E}[X_n] = \mathbb{E}[Y_n] = 0$ but $\mathbb{E}[X_n^2] = 2D_1 > 0$ and $\mathbb{E}[Y_n^2] = 2D_2 > 0$. Then a very similar argument shows that (6.1) holds and the limit has pdf $p(x, y, t)$ that solves

$$\frac{\partial}{\partial t}p(x,y,t) = D_1\frac{\partial^2}{\partial x^2}p(x,y,t) + D_2\frac{\partial^2}{\partial y^2}p(x,y,t). \tag{6.4}$$

Here we have

$$\begin{aligned} p(x,y,t) &= \frac{1}{\sqrt{4\pi D_1 t}}e^{-x^2/4D_1 t}\frac{1}{\sqrt{4\pi D_2 t}}e^{-y^2/4D_2 t} \\ &= \frac{1}{4\pi t\sqrt{D_1 D_2}}\exp\left[-\frac{1}{4t}\left(\frac{x^2}{D_1}+\frac{y^2}{D_2}\right)\right]. \end{aligned} \tag{6.5}$$

Figure 6.2 shows level sets of the anisotropic pdf (6.5). Now the level sets are ellipses, whose principal axes are the x and y coordinates, so there is a preferred coordinate system. Figure 6.2 was produced using the R code in Figure 6.12 at the end of this chapter.

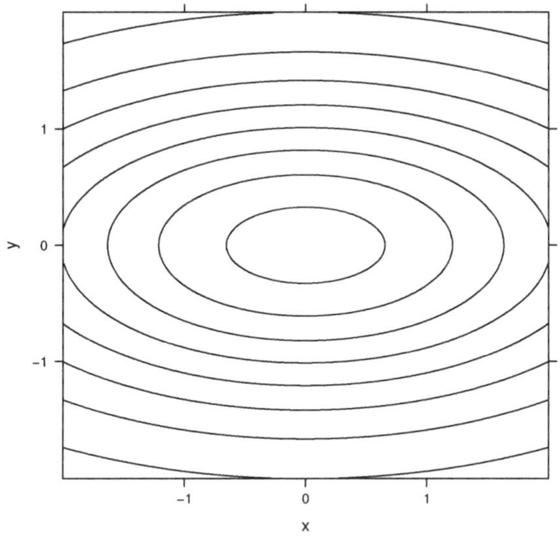

Fig. 6.2: Level sets of the solution (6.5) to the anisotropic diffusion equation (6.4) at time $t = 1$ with dispersivity parameters $D_1 = 2$ and $D_2 = 1/2$.

For vector random walks, it is natural to adopt a vector coordinate system. Given an $m \times n$ matrix

$$A = \begin{pmatrix} a_{11} & \cdots & a_{1n} \\ \vdots & & \vdots \\ a_{m1} & \cdots & a_{mn} \end{pmatrix} = [a_{ij}]$$

we define the *transpose*

$$A' = [a_{ji}] = \begin{pmatrix} a_{11} & \cdots & a_{m1} \\ \vdots & & \vdots \\ a_{1n} & \cdots & a_{mn} \end{pmatrix}.$$

The transpose of the column vector (a $d \times 1$ matrix)

$$x = \begin{pmatrix} x_1 \\ \vdots \\ x_d \end{pmatrix}$$

is the row vector $x' = (x_1, \ldots, x_d)$. The *inner product*

$$x \cdot y = x'y = (x_1, \ldots, x_d)\begin{pmatrix} y_1 \\ \vdots \\ y_d \end{pmatrix} = \sum_{j=1}^{d} x_j y_j$$

for two column vectors of the same dimension is defined by matrix multiplication. Then $x \cdot y = y \cdot x$. The *outer product*

$$xx' = \begin{pmatrix} x_1 \\ \vdots \\ x_d \end{pmatrix}(x_1, \ldots, x_d) = \begin{pmatrix} x_1 x_1 & \cdots & x_1 x_d \\ \vdots & & \vdots \\ x_d x_1 & \cdots & x_d x_d \end{pmatrix}$$

is a matrix, while the inner product is a scalar.

Given a d-dimensional random vector

$$X = \begin{pmatrix} X_1 \\ \vdots \\ X_d \end{pmatrix}$$

we define the mean vector

$$E[X] = \begin{pmatrix} \mathbb{E}[X_1] \\ \vdots \\ \mathbb{E}[X_d] \end{pmatrix} = \begin{pmatrix} \mu_1 \\ \vdots \\ \mu_d \end{pmatrix} = \mu \in \mathbb{R}^d,$$

and the covariance matrix (using the outer product)

$$\begin{aligned} Q = \operatorname{Cov}(X) &= \mathbb{E}[(X-\mu)(X-\mu)'] \\ &= \begin{pmatrix} \mathbb{E}[(X_1-\mu_1)(X_1-\mu_1)] & \cdots & \mathbb{E}[(X_1-\mu_1)(X_d-\mu_d)] \\ \vdots & & \vdots \\ \mathbb{E}[(X_d-\mu_d)(X_1-\mu_1)] & \cdots & \mathbb{E}[(X_d-\mu_d)(X_d-\mu_d)] \end{pmatrix}, \end{aligned}$$

a $d \times d$ matrix whose jj entry is the variance of X_j, and whose ij entry for $i \neq j$ is the covariance of X_i and X_j.

Now we can extend the simple arguments of Chapter 1 to the vector case. Later in this chapter, we will provide a more general treatment based on the theory of infinitely divisible random vectors and triangular arrays. Let $X = (X_1, \ldots, X_d)'$ be a random vector in $\mathbb{R}^d$ with cumulative distribution function (cdf)

$$F(x) = F(x_1, \ldots, x_d) = \mathbb{P}[X_1 \leq x_1, \ldots, X_d \leq x_d] = \mathbb{P}[X \leq x].$$

Then $F(x) = \mu\{y \in \mathbb{R}^d : y \leq x\}$ where $y \leq x$ means that $y_i \leq x_i$ for all $i = 1, 2, \ldots, d$. If the cdf $F(x)$ is differentiable, then the probability density function (pdf)

$$f(x) = f(x_1, \ldots, x_d) = \frac{\partial}{\partial x_1} \cdots \frac{\partial}{\partial x_d} F(x_1, \ldots, x_d)$$

and the probability measure

$$\mu(B) = \mathbb{P}[X \in B] = \int_{x \in B} F(dx) = \int_{x \in B} f(x)\, dx.$$

The characteristic function

$$\hat{\mu}(k) = \mathbb{E}[e^{ik \cdot X}] = \int e^{ik \cdot x} \mu(dx) = \int e^{ik \cdot x} F(dx)$$

so that, if the pdf $f(x)$ exists, then its Fourier transform (FT) is given by

$$\hat{f}(k) = \int e^{-ik \cdot x} f(x)\, dx = \hat{\mu}(-k).$$

Suppose that the d-dimensional random vector X has a pdf $f(x) = f(x_1, \ldots, x_d)$ and write the FT of X in the form

$$\begin{aligned}\hat{f}(k) = \mathbb{E}[e^{-ik \cdot X}] &= \int e^{-ik \cdot x} f(x)\, dx \\ &= \int \left(1 - ik \cdot x + \tfrac{1}{2}(-ik \cdot x)^2 + \cdots\right) f(x)\, dx \\ &= 1 - ik \cdot \mu - \tfrac{1}{2} \int k' x\, x' k f(x)\, dx + \cdots\end{aligned}$$

where k is a column vector with $k' = (k_1, \ldots, k_d)$. If the random vector X has mean $\mathbb{E}[X] = 0$ and covariance $Q = \mathbb{E}[XX'] = \int xx' f(x)\, dx$ then

$$\hat{f}(k) = 1 - \tfrac{1}{2} k' \mathbb{E}[XX'] k + \cdots = 1 - \tfrac{1}{2} k' Q k + \cdots$$

is the FT of X. If (X_n) are iid with X, then the vector sum $S_n = X_1 + \cdots + X_n$ has FT $\hat{f}(k)^n$ and the rescaled sum $n^{-1/2} S_n$ has FT

$$\hat{f}(k/\sqrt{n})^n = \left(1 - \frac{\tfrac{1}{2} k' Q k}{n} + \cdots\right)^n \to \exp(-\tfrac{1}{2} k' Q k) \tag{6.6}$$

which shows that

$$n^{-1/2} S_n \Rightarrow Y \tag{6.7}$$

where the limit has FT $\exp(-\tfrac{1}{2} k' Q k)$. The limit Y is a multidimensional Gaussian pdf with mean zero and covariance matrix Q. Its probability density function is

$$g(x) = (2\pi)^{-d/2}\, |\det(Q)|^{-1/2} \exp\left[-\tfrac{1}{2} x' Q^{-1} x\right]$$

where $\det(Q)$ is the determinant of the matrix Q, see details at the end of this section for more information.

Next, consider a vector random walk $S_{[nt]} = X_1 + \cdots + X_{[nt]}$ where (X_n) are iid with $\mu = \mathbb{E}[X_n] = 0$ and $\mathrm{Cov}(X_n) = \mathbb{E}[X_n X_n'] = 2D$ is invertible. Then the rescaled random walk $n^{-1/2} S_{[nt]}$ has FT

$$\hat{f}(k/\sqrt{n})^{[nt]} = \left(1 - \frac{k' D k}{n} + \cdots\right)^{[nt]} \to \exp(-k' D t k)$$

and the Lévy Continuity Theorem (see details) yields

$$n^{-1/2} S_{[nt]} \Rightarrow Z_t. \tag{6.8}$$

If $p(x, t)$ is the pdf of Z_t then

$$\hat{p}(k, t) = \exp(-k'Dtk) = \exp[(ik)'Dt(ik)], \tag{6.9}$$

which solves

$$\frac{d}{dt}\hat{p}(k, t) = (ik)'D(ik)\hat{p}(k, t).$$

Invert the FT to see that $p(x, t)$ solves

$$\frac{\partial}{\partial t}p(x, t) = \nabla \cdot D\nabla p(x, t), \tag{6.10}$$

the vector diffusion equation in natural vector notation. Here we use $x'y = x \cdot y$, the fact that $(ik)\hat{f}(k)$ is the FT of

$$\nabla f(x) = \begin{pmatrix} \frac{\partial}{\partial x_1} f(x_1, \dots, x_d) \\ \vdots \\ \frac{\partial}{\partial x_d} f(x_1, \dots, x_d) \end{pmatrix},$$

and the fact that $(ik) \cdot \hat{F}(k)$ is the FT of $\nabla \cdot F(x)$ when $F(x) = (f_1(x), \dots, f_d(x))'$ is a vector-valued function of the vector $x = (x_1, \dots, x_d)'$ (see details). Inverting the FT in (6.9) shows that

$$p(x, t) = (4\pi t)^{-d/2}\,|\det(D)|^{-1/2} \exp\left[-\frac{1}{4t}x'D^{-1}x\right],$$

see details at the end of this section.

We can also add a drift: The process $Z_t + vt$ has FT

$$\hat{p}(k, t) = \mathbb{E}[e^{-ik\cdot(vt+Z_t)}] = \exp(-ik \cdot vt - k'Dtk), \tag{6.11}$$

which solves

$$\frac{d}{dt}\hat{p}(k, t) = \left[-ik \cdot v + (ik)'D(ik)\right]\hat{p}(k, t).$$

Invert the FT to get the vector diffusion equation with drift

$$\frac{\partial}{\partial t}p(x, t) = -v \cdot \nabla p(x, t) + \nabla \cdot D\nabla p(x, t). \tag{6.12}$$

Inverting (6.11) shows that

$$p(x, t) = (4\pi t)^{-d/2}\,|\det(D)|^{-1/2} \exp\left[-\frac{1}{4t}(x - vt)'D^{-1}(x - vt)\right], \tag{6.13}$$

see details at the end of this section.

Remark 6.1. The geometry of the solution (6.13) is determined by the structure of the *dispersion tensor* D. For simplicity, suppose that the drift velocity $v = 0$. Since D is symmetric and positive definite, there is an orthonormal basis of eigenvectors $b_1 \ldots, b_d$ with corresponding eigenvalues a_i such that $Db_i = a_i b_i$ for $1 \le i \le d$. For any $x \in \mathbb{R}^d$ we can write $x = \sum_{j=1}^d x_j b_j$ where $x_j = x \cdot b_j$. Note that

$$b_i' D^{-1} b_j = b_i' a_j^{-1} b_j = a_j^{-1}(b_i \cdot b_j) = \begin{cases} 0 & \text{if } i \neq j; \\ a_j^{-1} & \text{if } i = j. \end{cases}$$

Then

$$\begin{aligned} x' D^{-1} x &= \Big(\sum_{i=1}^d x_i b_i \Big)' D^{-1} \Big(\sum_{j=1}^d x_j b_j \Big) \\ &= \sum_{i=1}^d \sum_{j=1}^d x_i x_j b_i' D^{-1} b_j \\ &= \sum_{i=1}^d a_i^{-1} x_i^2 \end{aligned}$$

and then (6.13) reduces to

$$p(x, t) = (4\pi t)^{-d/2} \left[\prod_{i=1}^d a_i^{-1/2} \right] \exp \left[-\frac{1}{4t} \sum_{i=1}^d \frac{x_i^2}{a_i} \right].$$

The level sets of this pdf are ellipsoids

$$\frac{x_1^2}{a_1} + \cdots + \frac{x_d^2}{a_d} = C$$

whose principal axes are the eigenvector coordinates $b_1, \ldots, b_d$. The level sets are widest in the direction of the eigenvector with the largest eigenvalue. Recall that $2D$ is also the covariance matrix of the random walk jumps X_n. You can check (e.g., using Lagrange multipliers) that this eigenvector maximizes the variance $\mathbb{E}[(X_n \cdot \theta)^2]$ over all unit vectors $\|\theta\| = 1$.

Remark 6.2. The Gaussian limit in (6.7) depends on the choice of norming. Assume as before that (X_n) are iid with mean $\mathbb{E}[X_n] = 0$ and covariance matrix $Q = \mathbb{E}[X_n X_n']$. If A is any matrix, then

$$(n^{-1/2} A) \sum_{j=1}^n X_j = n^{-1/2} \sum_{j=1}^n A X_j \Rightarrow AY \simeq \mathcal{N}(0, AQA')$$

since the iid random vectors AX_n have covariance matrix AQA'. Hence $n^{-1/2}A$ is another suitable sequence of norming operators. (We could also apply the Continuous Mapping Theorem 4.19: If $n^{-1/2} S_n \Rightarrow Y$, then $A(n^{-1/2} S_n) \Rightarrow AY$.) If we choose A so that $A'QA = I$ (see details), then $Z = AY \simeq \mathcal{N}(0, I)$. Since S_n is a vector, matrix norming is quite natural.

Remark 6.3. We say that the $d \times d$ matrix U is *orthogonal* if $U^{-1} = U'$. It is easy to check that $UZ \simeq Z$ for U orthogonal and $Z \simeq \mathcal{N}(0, I)$: Just note that UZ has FT

$$\mathbb{E}[e^{-ik\cdot UZ}] = \mathbb{E}[e^{-iU'k\cdot Z}] = \exp[-\tfrac{1}{2}(U'k)'(U'k)] = \exp[-\tfrac{1}{2}k'UU'k] = \exp[-\tfrac{1}{2}\|k\|^2]$$

using the fact that $U'U = U^{-1}U = I$. We say that the orthogonal matrix U is a *symmetry* of Z. Geometrically, orthogonal matrices U are the norm-preserving coordinate changes, i.e., rotations and reflections. Suppose that $S_n = X_1 + \cdots + X_n$ is a random walk whose iid jumps satisfy $\mathbb{E}[X_n] = 0$ and $\mathbb{E}[X_n^2] = I$, so that $n^{-1/2}S_n \Rightarrow Z \simeq \mathcal{N}(0, I)$. If U_n are orthogonal, then we also have $n^{-1/2}U_nS_n \Rightarrow Z$, so that $n^{-1/2}U_n$ is another suitable sequence of norming matrices. To check this, use FT and the fact that the orthogonal matrices form a compact set. For any subsequence, there is a further subsequence $U_n \to U$ along which the FT of $n^{-1/2}U_nS_n$ converges:

$$\hat{f}(n^{-1/2}U_n'k)^n = \left(1 - \frac{\frac{1}{2}(U_n'k)'I(U_n'k)}{n} + \cdots\right)^n \to \exp(-\tfrac{1}{2}k'UU'k) = \exp(-\tfrac{1}{2}\|k\|^2).$$

Since the every subsequence has a further subsequence that converges to the same limit, the Lévy Continuity Theorem implies that $n^{-1/2}U_nS_n \Rightarrow Y$. If $I_n \to I$, a similar argument shows that $n^{-1/2}U_nI_nS_n \Rightarrow Z$. For more information on symmetry, and the permissible sequences of norming operators, see Meerschaert and Scheffler [146, Section 2.3].

Details

In (6.3) we used the fact that the FT of $\partial f(x, y)/\partial x$ is $(ik)\hat{f}(k, \ell)$, and the FT of $\partial f(x, y)/\partial y$ is $(i\ell)\hat{f}(k, \ell)$. The proof is a direct application of the corresponding one dimensional formula (1.14). For example, suppose that $f(x, y)$ is integrable, and that $\partial f(x, y)/\partial y$ exists and is integrable. Then (1.14) implies

$$\int_{-\infty}^{\infty} e^{-i\ell y}\frac{\partial}{\partial y}f(x, y)\, dy = (i\ell)\int_{-\infty}^{\infty} e^{-i\ell y}f(x, y)\, dy$$

for each x, and then

$$\begin{aligned}\int_{-\infty}^{\infty}\int_{-\infty}^{\infty} e^{-ikx}e^{-i\ell y}\frac{\partial}{\partial y}f(x, y)\, dy\, dx &= \int_{-\infty}^{\infty} e^{-ikx}(i\ell)\int_{-\infty}^{\infty} e^{-i\ell y}f(x, y)\, dy\, dx\\ &= (i\ell)\hat{f}(k, \ell).\end{aligned}$$

If $\partial f(x, y)/\partial x$ also exists and is integrable, then it follows that the vector-valued function $\nabla f(x, y) = (\partial f(x, y)/\partial x, \partial f(x, y)/\partial y)'$ has FT $((ik)\hat{f}(k, \ell), (i\ell)\hat{f}(k, \ell))'$. Note that, for a vector-valued function $F(x, y) = (f(x, y), g(x, y))'$, we define the FT

$$\hat{F}(k, \ell) = \int_{-\infty}^{\infty}\int_{-\infty}^{\infty} e^{-ikx}e^{-i\ell y}F(x, y)\, dy\, dx = \begin{pmatrix}\hat{f}(k, \ell)\\ \hat{g}(k, \ell)\end{pmatrix}$$

where

$$\hat{f}(k,\ell) = \int_{-\infty}^{\infty}\int_{-\infty}^{\infty} e^{-ikx}e^{-i\ell y} f(x,y)\,dy\,dx$$

$$\hat{g}(k,\ell) = \int_{-\infty}^{\infty}\int_{-\infty}^{\infty} e^{-ikx}e^{-i\ell y} g(x,y)\,dy\,dx.$$

Now extending to $\mathbb{R}^d$ in vector notation shows that $(ik)\hat{f}(k)$ is the FT of the gradient vector $\nabla f(x)$.

In (6.10) we use the fact that $(ik)\cdot\hat{F}(k)$ is the FT of $\nabla\cdot F(x)$, when $F(x) = (f_1(x),\dots,f_d(x))'$ is a vector-valued function of the vector $x = (x_1,\dots,x_d)'$. Write

$$(ik)\cdot\hat{F}(k) = \begin{pmatrix} ik_1 \\ \vdots \\ ik_d \end{pmatrix} \cdot \begin{pmatrix} \hat{f}_1(k) \\ \vdots \\ \hat{f}_d(k) \end{pmatrix}$$

and note that $ik_j\hat{f}_j(k)$ is the FT of $\partial f_j(x)/\partial x_j$ for all $j = 1,2,\dots,d$. Then $(ik)\cdot\hat{F}(k)$ is the FT of

$$\sum_{j=1}^{d} \frac{\partial f_j(x)}{\partial x_j} = \begin{pmatrix} \partial/\partial x_1 \\ \vdots \\ \partial/\partial x_d \end{pmatrix} \cdot \begin{pmatrix} f_1(x) \\ \vdots \\ f_d(x) \end{pmatrix} = \nabla\cdot F(x).$$

In (6.7) we use the Lévy continuity theorem for the vector Fourier transform. The statement of this theorem is exactly the same as for random variables. Suppose that X_n, X are random vectors on $\mathbb{R}^d$. Let $\hat{f}_n(k) = \mathbb{E}[e^{-ik\cdot X_n}]$ and $\hat{f}(k) = \mathbb{E}[e^{-ik\cdot X}]$. The Lévy Continuity Theorem [146, Theorem 1.3.6] states that $X_n \Rightarrow X$ if and only if $\hat{f}_n(k)\to\hat{f}(k)$. More precisely, $X_n \Rightarrow X$ implies that $\hat{f}_n(k)\to\hat{f}(k)$ for each $k\in\mathbb{R}^d$, uniformly on compact subsets. Conversely, if X_n is a sequence of random vectors such that $\hat{f}_n(k)\to\hat{f}(k)$ for each $k\in\mathbb{R}^d$, and the limit $\hat{f}(k)$ is continuous at $k=0$, then $\hat{f}(k)$ is the FT of some X, and $X_n\Rightarrow X$.

The general solution to the diffusion equation (6.12) comes from inverting the FT to obtain a normal density. Since the limit in (6.6) is continuous at $k = 0$, the Lévy continuity theorem implies that it is the FT of some random vector Y, i.e., we have $\mathbb{E}[e^{-ik\cdot Y}] = \exp(-\frac{1}{2}k'Qk)$. Using the general fact that $(AB)' = B'A'$ for vectors and matrices, it is easy to see that the covariance matrix is *symmetric*: $Q' = E[(XX')'] = E[(X')'(X)'] = \mathbb{E}[XX'] = Q$. The covariance matrix is also *non-negative definite*: For any vector $a\in\mathbb{R}^d$ we have $a'Qa = \mathbb{E}[a'XX'a] = \mathbb{E}[(a\cdot X)^2] \ge 0$. Of course it is possible that X is supported on some lower dimensional subspace of $\mathbb{R}^d$ and, to avoid this, we will assume that Q is *positive definite*, meaning that $a'Qa > 0$ when $a \ne 0$. This is equivalent to assuming that the distribution of X is not concentrated on some lower dimensional affine subspace, i.e., there is no $a \ne 0$ such that $X\cdot a$ is almost surely constant. In this case, we say that the distribution of X is *full*. Then a deep result from linear algebra (the Principal Axis Theorem, e.g., see Curtis [53, Theorem 31.9]) implies that Q has an *orthonormal basis* of *eigenvectors* $v_1, \dots, v_d$ such that $\|v_i\|^2 = v_i \cdot v_i = 1$ and $v_i \cdot v_j = 0$ for $i \neq j$, with $Qv_i = \lambda_i v_i$ for some *eigenvalues* $\lambda_j > 0$.

For any vectors $x, y \in \mathbb{R}^d$ and any $d \times d$ matrix A we have $x \cdot Ay = x'Ay = (A'x)'y = A'x \cdot y$. Define A to be the unique matrix (linear operator) such that $Av_i = \lambda_i^{-1/2} v_i$ for every $i = 1, 2, \dots, d$. Note that $v_i \cdot Av_i = \lambda_i^{-1/2}$ and $v_i \cdot Av_j = 0$ for $i \neq j$. Then $A'v_i \cdot v_i = v_i \cdot Av_i = \lambda_i^{-1/2}$ for all $i = 1, 2, \dots, d$, and $A'v_i \cdot v_j = 0$ for $j \neq i$. It follows that $A'v_i = \lambda_i^{-1/2} v_i$ for all $i = 1, 2, \dots, d$. Then $AQA'v_i = v_i$ for all $i = 1, 2, \dots, d$. Since $v_1, \dots, v_d$ forms a basis for $\mathbb{R}^d$, it follows that $AQA' = I$, the $d \times d$ identity matrix. Then the FT of $Z = AY$ is

$$\begin{aligned}
\hat{f}(k) = \mathbb{E}\left[e^{-ik\cdot Z}\right] &= \mathbb{E}\left[e^{-ik\cdot AY}\right] \\
&= \mathbb{E}\left[e^{-iA'k\cdot Y}\right] \\
&= \exp\left[-\tfrac{1}{2}(A'k)'QA'k\right] \\
&= \exp\left[-\tfrac{1}{2}k'AQA'k\right] \\
&= \exp\left[-\tfrac{1}{2}k'Ik\right] = \exp\left[-\tfrac{1}{2}(k_1^2 + \cdots + k_d^2)\right] = \prod_{j=1}^{d} e^{-k_j^2/2}
\end{aligned}$$

which inverts to

$$f(z) = \prod_{j=1}^{d} \frac{1}{\sqrt{2\pi}} e^{-z_j^2/2} = (2\pi)^{-d/2} e^{-\|z\|^2/2}$$

the density of random vector in $\mathbb{R}^d$ with iid $\mathcal{N}(0, 1)$ components. This pdf is isotropic, since it only depends on z through its norm $\|z\|$. The pdf of Y comes from a change of variables $z = Ay$ with $dz = \det(A)\, dy$, so that for any Borel set $B \subseteq \mathbb{R}^d$ we have

$$\begin{aligned}
\mathbb{P}[Y \in B] = \mathbb{P}[A^{-1}Z \in B] &= \mathbb{P}[Z \in AB] \\
&= \int_{z\in AB} f(z)\, dz = \int_{Ay\in AB} f(Ay) \det(A)\, dy
\end{aligned}$$

where $\det(A) = \lambda_1^{-1/2} \cdots \lambda_d^{-1/2}$ is the determinant (product of the eigenvalues) of the matrix A. This shows that the random vector limit Y in (6.6) has pdf

$$\begin{aligned}
f(Ay)\det(A) &= (2\pi)^{-d/2} \det(A) e^{-(Ay)'(Ay)/2} \\
&= (2\pi)^{-d/2}\, |\det(Q)|^{-1/2} \exp\left[-\tfrac{1}{2}y'Q^{-1}y\right]
\end{aligned} \tag{6.14}$$

since $A'A = Q^{-1}$, which is easy to check, and two basic facts about determinants: $\det(A) = \det(A')$ and $\det(AB) = \det(A)\det(B)$ (e.g., see Curtis [53]). Since the limit Z_t in (6.8) has FT $\exp(-k'Dtk)$, we can set $Q = 2Dt$ in (6.14) to see that Z_t has pdf

$$p(x, t) = (4\pi t)^{-d/2}\, |\det(D)|^{-1/2} \exp\left[-\frac{1}{4t}x'D^{-1}x\right]$$

using the fact that $\det(2tD) = (2t)^d \det(D)$. Another change of variables shows that $Z_t + vt$ has pdf (6.13) with FT $\hat{p}(k, t) = \exp(-ikvt - k'Dtk)$. This shows that the pdf (6.13) solves the vector diffusion equation with drift (6.12).

6.2 Vector random walks with heavy tails

Suppose that (X_n) and (Y_n) are two independent sequences of zero mean iid random variables with heavy tails, such that

$$n^{-1/\alpha}S_{[nt]} \Rightarrow Z_t \quad \text{and} \quad n^{-1/\beta}R_{[nt]} \Rightarrow W_t$$

where $S_n = X_1 + \cdots + X_n$, $R_n = Y_1 + \cdots + Y_n$, and Z_t, W_t are independent stable Lévy motions with index $\alpha, \beta \in (1, 2)$. In vector notation, we have

$$\begin{pmatrix} n^{-1/\alpha}S_{[nt]} \\ n^{-1/\beta}R_{[nt]} \end{pmatrix} \Rightarrow \begin{pmatrix} Z_t \\ W_t \end{pmatrix} \tag{6.15}$$

as $n \to \infty$. Figure 5.32 shows a typical sample path of the vector limit process in (6.15) in the case $\alpha = \beta = 1.8$. Since the limit has independent components, it follows immediately from Theorem 3.41 that this process has a pdf $p(x, y, t)$ with FT

$$\hat{p}(k, \ell, t) = \int\int e^{-ikx}e^{-i\ell y}p(x, y, t)\,dy\,dx = e^{t\psi_1(-k)}e^{t\psi_2(-\ell)}$$

where

$$\psi_1(k) = p_1D_1(-ik)^\alpha + q_1D_1(ik)^\alpha \quad \text{and} \quad \psi_2(\ell) = p_2D_2(-i\ell)^\beta + q_2D_2(i\ell)^\beta$$

for some $D_i > 0$ and some $p_i, q_i \geq 0$ with $p_i + q_i = 1$. Then

$$\frac{d}{dt}\hat{p}(k, \ell, t) = [\psi_1(-k) + \psi_2(-\ell)]\hat{p}(k, \ell, t)$$

and inverting the FT shows that $p(x, y, t)$ solves the two dimensional fractional diffusion equation

$$\begin{aligned}\frac{\partial}{\partial t}p(x, y, t) &= p_1D_1\frac{\partial^\alpha}{\partial x^\alpha}p(x, y, t) + q_1D_1\frac{\partial^\alpha}{\partial(-x)^\alpha}p(x, y, t)\\ &\quad + p_2D_2\frac{\partial^\beta}{\partial y^\beta}p(x, y, t) + q_2D_2\frac{\partial^\beta}{\partial(-y)^\beta}p(x, y, t).\end{aligned} \tag{6.16}$$

The fractional partial differential equation (6.16) governs the densities of a two dimensional *operator stable Lévy motion.* If $\alpha = \beta$, then this reduces to a two dimensional *stable Lévy motion.* For $\alpha = \beta = 2$, equation (6.16) reduces to the two dimensional diffusion equation (6.3), whose solutions are rotationally symmetric (isotropic). The geometry for two dimensional stable Lévy motions is more complicated.

The solution $p(x, y, t)$ to the two dimensional diffusion equation (6.3) has FT

$$\hat{p}(k, \ell, t) = \exp\left[-Dt(k^2 + \ell^2)\right].$$

The rotational symmetry of solutions comes from the fact that the FT only depends on (k, ℓ) through $k^2 + \ell^2$ which is rotationally invariant. Even if we assume $\alpha = \beta$,

$D_1 = D_2$, and $p_i = q_i$ in (6.16), we only get a rotationally symmetric solution in the special case $\alpha = 2$. It follows from Proposition 5.8 that

$$\hat{p}(k, \ell, t) = \exp\left[D_1 t \cos(\pi\alpha/2)(|k|^\alpha + |\ell|^\alpha)\right].$$

The term $|k|^\alpha + |\ell|^\alpha$ is not rotationally symmetric unless $\alpha = 2$, making Brownian motion a very special case of a stable Lévy motion. Figure 6.3 shows level sets of the solution $p(x, y, t)$ to the two dimensional fractional diffusion equation (6.16) with $\alpha = \beta = 1.2$ and $p_i = q_i$. There is a clear anisotropy here, and a preferred coordinate system. The R code for Figure 6.3 is listed Figure 6.13 at the end of this chapter.

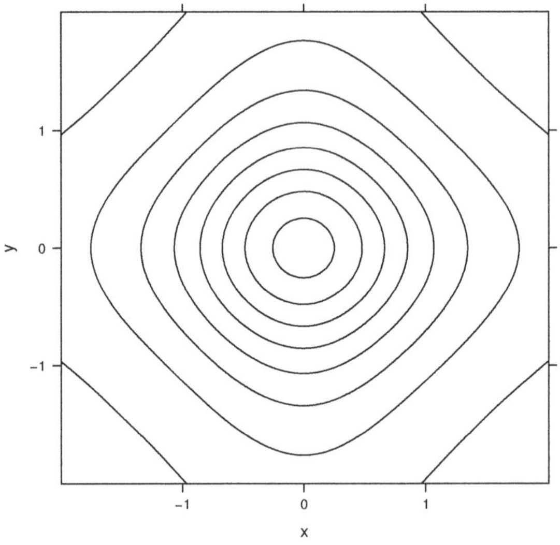

Fig. 6.3: Level sets of the solution $p(x, y, t)$ to the two dimensional fractional diffusion equation (6.16) at time $t = 5$ with parameters $\alpha = \beta = 1.2$, $D_1 = D_2 = 0.5$, and $p_1 = p_2 = q_1 = q_2 = 1/2$.

The general d-dimensional random walk $S_n = X_1 + \cdots + X_n$ is a sum of iid random vectors. Suppose that (X_n) are iid with X, and assume that X is full, i.e., there is no $a \neq 0$ such that $X \cdot a$ is almost surely constant. If there exist linear operators on $\mathbb{R}^d$ (i.e., $d \times d$ matrices) A_n and vectors $b_n \in \mathbb{R}^d$ such that $A_n S_n - b_n \Rightarrow Y$, we say that X belongs to the *generalized domain of attraction* of Y, and we write $X \in \text{GDOA}(Y)$. In the special case of scalar norming $A_n = a_n I$ for some real numbers $a_n > 0$, we say that X belongs to the *domain of attraction* of Y, and we write $X \in \text{DOA}(Y)$.

Example 6.4. Suppose that $X = (X_1, \ldots, X_d)'$ has independent components, where each $X_i \in \text{DOA}(Y_i)$ for some stable random variables Y_i with index $\alpha_i \in (0, 2]$. For ease of notation, suppose that the norming constants are of the form n^{-1/α_i} for each component. (In the general case, the norming sequence is $\mathrm{RV}(-1/\alpha_i)$.) Define the diagonal norming operators

$$A_n = \begin{pmatrix} n^{-1/\alpha_1} & 0 & 0 & \cdots & 0 \\ 0 & n^{-1/\alpha_2} & 0 & \cdots & 0 \\ 0 & 0 & n^{-1/\alpha_3} & & \vdots \\ \vdots & \vdots & & \ddots & \\ 0 & 0 & \cdots & 0 & n^{-1/\alpha_d} \end{pmatrix},$$

and note that, since $A_nS_n - b_n$ has independent components,

$$A_nS_n - b_n \Rightarrow Y = \begin{pmatrix} Y_1 \\ \vdots \\ Y_d \end{pmatrix}.$$

Remark 4.17 implies random walk convergence: $A_nS_{[nt]} - tb_n \Rightarrow Z_t$ where $Z_1 \simeq Y$. In view of Proposition 4.16, if all $1 < \alpha_i < 2$, we can take $b_n = nA_n\mathbb{E}[X]$, and if all $0 < \alpha_i < 1$, we can set $b_n = 0$. The ith component of the limit process Z_t is a stable Lévy motion with index α_i, and pdf $p_i(x_i, t)$. Since these components are independent, Z_t has pdf

$$p(x,t) = p(x_1,\ldots,x_d,t) = \prod_{i=1}^{d} p_i(x_i,t)$$

a product of one dimensional stable densities. Suppose all $1 < \alpha_i < 2$. Then $p(x, t)$ has FT

$$\hat{p}(k,t) = \mathbb{E}\left[e^{-ik\cdot Z_t}\right] = \exp\left(t\left[\psi_1(-k_1) + \cdots + \psi_d(-k_d)\right]\right)$$

where $\psi_j(k_j) = p_jD_j(-ik_j)^{\alpha_j} + q_jD_j(ik_j)^{\alpha_j}$ for each $1 \le j \le d$, for some $D_j > 0$ and some $p_j, q_j \ge 0$ with $p_j + q_j = 1$. Then

$$\frac{d}{dt}\hat{p}(k,t) = [\psi_1(-k_1) + \cdots + \psi_d(-k_d)]\hat{p}(k,t)$$

and inverting the FT shows that $p(x, t)$ solves the d-dimensional fractional diffusion equation

$$\frac{\partial}{\partial t}p(x,t) = \sum_{j=1}^{d}\left[p_jD_j\frac{\partial^{\alpha_j}}{\partial(x_j)^{\alpha_j}}p(x,t) + q_jD_j\frac{\partial^{\alpha_j}}{\partial(-x_j)^{\alpha_j}}p(x,t)\right]. \tag{6.17}$$

The fractional partial differential equation (6.17) governs the probability densities of a d-dimensional operator stable Lévy motion, whose components are independent stable Lévy motions with indices $\alpha_1, \ldots, \alpha_d$. If all $\alpha_j = \alpha$ and all $D_j = D$ and all $p_j = q_j = 1/2$, then it follows from Proposition 5.8 that

$$\hat{p}(k,t) = \exp\left[Dt\cos(\pi\alpha/2)\sum_{j=1}^{d}|k_j|^{\alpha}\right].$$

These solutions are not rotationally symmetric, since the sum $\sum_{j=1}^{d} |k_j|^\alpha$ is rotationally invariant only when $\alpha = 2$.

Example 6.5. Suppose that $B(t)$ is a Brownian motion in $\mathbb{R}^d$ such that $\mathbb{E}\left[e^{-ik\cdot B(t)}\right] = \exp\left[-k'tQk\right]$. Then $B(t) \simeq \mathcal{N}(0, 2tQ)$. Let D_t be a standard stable subordinator with pdf $g(u, t)$ such that

$$\tilde{g}(s, t) = \mathbb{E}\left[e^{-sD_t}\right] = \int_0^\infty e^{-su} g(u, t)\, du = e^{-ts^\beta}$$

for some $0 < \beta < 1$, as in (4.41). Define $Z_t = B(D_t)$ for $t \geq 0$. This subordinated process has FT

$$\begin{aligned} \hat{p}(k, t) = \mathbb{E}\left[e^{-ik\cdot Z_t}\right] &= \mathbb{E}\left[e^{-ik\cdot B(D_t)}\right] \\ &= \int_0^\infty \mathbb{E}[e^{-ik\cdot B(D_t)} | D_t = u] g(u, t)\, du \\ &= \int_0^\infty \mathbb{E}[e^{-ik\cdot B(u)}] g(u, t)\, du \\ &= \int_0^\infty e^{-(k'Qk)u} g(u, t)\, du = e^{-t(k'Qk)^\beta} \end{aligned} \tag{6.18}$$

for all $t \geq 0$. Suppose for example that $Q = c^{1/\beta} I$ for some $c > 0$. Then the subordinated process Z_t has characteristic function

$$\hat{p}(-k, t) = \mathbb{E}\left[e^{ik\cdot Z_t}\right] = e^{-tc\|k\|^{2\beta}} = e^{t\psi(k)}$$

where the Fourier symbol $\psi(k) = -c\|k\|^\alpha$ with $\alpha = 2\beta$. This is the *isotropic stable Lévy motion* in $\mathbb{R}^d$ with index $0 < \alpha < 2$, a natural extension of a standard, rotationally symmetric Brownian motion.

The *fractional Laplacian* operator Δ^β is defined by specifying that $\Delta^\beta f(x)$ has FT $-\|k\|^{2\beta}\hat{f}(k)$ for suitable functions $f(x)$. If $\beta = 1$, this reduces to the usual Laplacian $\Delta f(x) = \nabla \cdot \nabla f(x)$, whose FT is $(ik)\cdot(ik)\hat{f}(k) = -\|k\|^2\hat{f}(k)$. The subordinated process Z_t from Example 6.5, in the special case $Q = c^{1/\beta} I$, has a FT $\hat{p}(k, t) = e^{-tc\|k\|^{2\beta}}$ that solves

$$\frac{d}{dt}\hat{p}(k, t) = -c\|k\|^{2\beta}\hat{p}(k, t).$$

Invert to obtain the isotropic vector fractional diffusion equation

$$\frac{\partial}{\partial t} p(x, t) = c\Delta^\beta p(x, t) \tag{6.19}$$

for $0 < \beta < 1$. When $\beta = 1$, this reduces to the vector diffusion equation (6.10) with $D = cI$. Since the FT is rotationally symmetric, the solutions of (6.19) are invariant under rotations and reflections. The pdf in (6.18) has elliptical symmetry. Solutions to (6.17) in the case $p_i = q_i$ are symmetric with respect to reflection across the ith coordinate axis. For $0 < \alpha < 2$, there are three distinct Fourier symbols

$$-\|k\|^\alpha \neq (ik_1)^\alpha + \cdots + (ik_d)^\alpha \neq -|k_1|^\alpha - \cdots - |k_d|^\alpha$$

which are all equal in the case $\alpha = 2$. These symbols give rise to three different Lévy processes, corresponding to three different stable limits, when $0 < \alpha < 2$. See the details at the end of this section for more information.

The stable Lévy process with Fourier symbol $\psi_1(k) = (ik_1)^\alpha + \cdots + (ik_d)^\alpha$ is the limit of random walks whose jumps have iid components consisting of only positive power law jumps. The process with Fourier symbol $\psi_2(k) = -|k_1|^\alpha - \cdots - |k_d|^\alpha$ is the limit of random walks whose jumps have iid components consisting of symmetric power law jumps. The isotropic stable process constructed in Example 6.5 will be shown to arise as the limit of a random walk with iid spherically symmetric power law jumps. Take $X = R\Theta$ where $\mathbb{P}[R > r] = Cr^{-\alpha}$ and Θ is uniformly distributed over the unit sphere. We will show in Section 6.4 that $n^{-1/\alpha}(X_1 + \cdots + X_n) \Rightarrow Z_t$ when (X_n) are iid with X. In the case of finite second moments, a random walk with spherically symmetric jumps gives the same limit as a random walk whose jumps have iid (one-dimensional) symmetric components. In the case of heavy tails, these two limits are different. In the next section, we will build the necessary machinery of infinitely divisible laws and triangular arrays, to make these statements rigorous.

Details

Recall from Remark 6.3 that U is orthogonal if $U^{-1} = U'$. Then

$$\|Ux\|^2 = (Ux)'(Ux) = x'U'Ux = x'U^{-1}Ux = x'Ix = \|x\|^2$$

so that the linear transformation $x \mapsto Ux$ preserves the Euclidean norm. If X is a random vector on $\mathbb{R}^d$ with FT $\hat{f}(k) = \mathbb{E}[e^{-ik\cdot X}]$ and A is a $d \times d$ matrix then, since $k \cdot AX = A'k \cdot X$, the transformed random vector AX has FT

$$\mathbb{E}[e^{-ik\cdot AX}] = \mathbb{E}[e^{-iA'k\cdot X}] = \hat{f}(A'k).$$

The solution to (6.19) has FT $\hat{p}(k,t) = \mathbb{E}[e^{-ik\cdot Z_t}] = \exp\left[-tc\|k\|^{2\beta}\right]$. If U is orthogonal, then so is $U' = U^{-1}$, and it follows that UZ_t has the same FT $\hat{p}(U'k,t) = \exp\left[-tc\|k\|^{2\beta}\right]$ as Z_t. This proves that $UZ_t \simeq Z_t$, so that every orthogonal transformation (every rotation and reflection) is a symmetry of this process.

The process Z_t in (6.18) is elliptically symmetric. Apply the construction in Section 6.1 to obtain a matrix A such that $AQA' = I$. Then the process AZ_t has FT

$$\hat{p}(A'k,t) = e^{-t((A'k)'Q(A'k))^\beta} = e^{-t(k'AQA'k)^\beta} = e^{-t(k'Ik)^\beta} = e^{-t\|k\|^{2\beta}}$$

so that $UAZ_t \simeq AZ_t$ for every orthogonal U. It follows that $A^{-1}UAZ_t \simeq Z_t$ so that $A^{-1}UA$ is a symmetry of Z_t for every orthogonal U. The level sets of the pdf are ellipsoids whose principal axes are the eigenvectors of A (i.e., the eigenvectors of Q). More information on symmetries for stable and operator stable laws can be found in Cohen, Meerschaert and Rosiński [51], see also [146, Section 7.2] and references therein.

Solutions of (6.17) with $p_i = q_i$ are symmetric with respect to the linear transformation $(x_1, \ldots, x_i, \ldots, x_d) \mapsto (x_1, \ldots, -x_i, \ldots, x_d)$ since $\psi_i(k_i) = \psi_i(-k_i)$. However, they are not spherically or elliptically symmetric.

6.3 Triangular arrays of random vectors

In this section, we begin to develop the general theory of fractional diffusion in multiple dimensions, starting with the Lévy representation for infinitely divisible laws. We say that a random vector Y is infinitely divisible if $Y \simeq X_1 + \cdots + X_n$ for every positive integer n, where (X_n) are independent and identically distributed (iid) random vectors. If $X_n \simeq \mu_n$, then this is equivalent to $\hat{\mu}(k) = \hat{\mu}_n(k)^n$.

Example 6.6. If $Y \simeq \mathcal{N}(a, Q)$ (normal with mean a and covariance matrix Q), then $\hat{\mu}(k) = \exp(ik \cdot a - \frac{1}{2}k'Qk)$. Take $\hat{\mu}_n(k) = \exp(ik \cdot n^{-1}a - \frac{1}{2}k'(n^{-1}Q)k)$ to see that Y is infinitely divisible, the sum of n iid $\mathcal{N}(n^{-1}a, n^{-1}Q)$ random vectors.

Example 6.7. A compound Poisson random vector $Y = W_1 + \cdots + W_N = S_N$ is a random sum, where $S_n = W_1 + \cdots + W_n$, $(W_j) \simeq \omega(dy)$ are iid random vectors, and N is Poisson with $\mathbb{E}[N] = \lambda$, independent of (W_j). Then

$$\begin{aligned} F(y) = \mathbb{P}[Y \leq y] &= \mathbb{P}[S_N \leq y] \\ &= \sum_{j=0}^{\infty} \mathbb{P}[S_N \leq y | N = j]\mathbb{P}[N = j] \\ &= \sum_{j=0}^{\infty} \mathbb{P}[S_j \leq y] e^{-\lambda} \frac{\lambda^j}{j!}. \end{aligned}$$

Then Y has characteristic function

$$\begin{aligned} \hat{\mu}(k) &= \sum_{j=0}^{\infty} \hat{\omega}(k)^j e^{-\lambda} \frac{\lambda^j}{j!} \\ &= e^{-\lambda} \sum_{j=0}^{\infty} \frac{[\lambda \hat{\omega}(k)]^j}{j!} \\ &= e^{-\lambda} e^{\lambda \hat{\omega}(k)} = e^{\lambda[\hat{\omega}(k) - 1]}. \end{aligned}$$

Take $\hat{\mu}_n(k) = e^{(\lambda/n)[\hat{\omega}(k)-1]}$ to see that Y is infinitely divisible. This argument is identical to Example 3.3, using vector notation. Continuing as in Section 3.1, write

$$\begin{aligned}\hat{\mu}(k) = e^{\lambda[\hat{\omega}(k)-1]} &= \exp\left(\lambda\left[\int e^{ik\cdot x}\omega(dx) - 1\right]\right)\\ &= \exp\left(\lambda\left[\int \left(e^{ik\cdot x} - 1\right)\omega(dx)\right]\right)\\ &= \exp\left(\int \left(e^{ik\cdot x} - 1\right)\phi(dx)\right)\end{aligned}$$

where the *Lévy measure* $\phi(dx) = \lambda\,\omega(dx)$ (jump intensity) controls the number and size of jumps that make up the random sum. In particular, $\phi(B)$ is the expected number of jumps in B for any Borel set B bounded away from zero.

A Lévy measure $\phi(dy)$ on $\mathbb{R}^d$ is a σ-finite Borel measure such that

$$\int_{0<\|y\|\le R} \|y\|^2\phi(dy) < \infty \quad \text{and} \quad \phi\{y : \|y\| > R\} < \infty \tag{6.20}$$

for all $R > 0$. The next theorem extends the Lévy representation from Theorem 3.4 to random vectors.

Theorem 6.8 (Lévy representation for random vectors). *A random vector $Y \simeq \mu$ on $\mathbb{R}^d$ is infinitely divisible if and only if its characteristic function $\hat{\mu}(k) = \mathbb{E}[e^{ik\cdot Y}] = e^{\psi(k)}$ where*

$$\psi(k) = ik\cdot a - \tfrac{1}{2}k'Qk + \int\left(e^{ik\cdot y} - 1 - \frac{ik\cdot y}{1+\|y\|^2}\right)\phi(dy) \tag{6.21}$$

for some $a \in \mathbb{R}^d$, some symmetric nonnegative definite matrix Q, and some Lévy measure $\phi(dy)$. This Lévy representation $\mu \simeq [a, Q, \phi]$ is unique.

Proof. The proof is based on a compound Poisson approximation, see Meerschaert and Scheffler [146, Theorem 3.1.11]. □

Example 6.9. If $Y \simeq \mathcal{N}(a, Q)$ then Theorem 6.8 holds with $Y \simeq [a, Q, 0]$.

Example 6.10. If Y is compound Poisson, then Theorem 6.8 holds with $Y \simeq [a, 0, \phi]$, where $\phi(dy) = \lambda\,\omega(dy)$, and

$$a = \int \frac{y}{1+\|y\|^2}\,\phi(dy).$$

The Lévy representation (6.21) is a natural extension of the one dimensional formula (3.4). Note that the Lévy representation implies that any infinitely divisible law can be written as a sum of two independent components, one Gaussian, and one Poissonian.

In a *triangular array* of random vectors $\{X_{nj} : j = 1, \ldots, k_n; n = 1, 2, 3, \ldots\}$ the row sums $S_n = X_{n1} + \cdots + X_{nk_n}$ have independent summands for each $n \ge 1$, and

$k_n \to \infty$ as $n \to \infty$. A general result [146, Theorem 3.2.14] states that Y is infinitely divisible if and only if $S_n - a_n \Rightarrow Y$ for some $a_n \in \mathbb{R}^d$ and some triangular array that satisfies

$$\lim_{n\to\infty} \sup_{1\le j\le k_n} \mathbb{P}[\|X_{nj}\| > \varepsilon] = 0 \quad \text{for all } \varepsilon > 0. \tag{6.22}$$

Define the truncated random vectors $X_{nj}^R = X_{nj}I(\|X_{nj}\| \le R)$ and recall that a sequence of σ-finite Borel measures $\phi_n(dy) \to \phi(dy)$ on $\{y : y \ne 0\}$ if $\phi_n(B) \to \phi(B)$ for any Borel set B bounded away from zero such that $\phi(\partial B) = 0$ (vague convergence). The next result extends Theorem 3.33 to random vectors.

Theorem 6.11 (Triangular array convergence for random vectors). *Given a triangular array such that* (6.22) *holds, there exists a random vector Y such that $S_n - a_n \Rightarrow Y$ for some $a_n \in \mathbb{R}^d$ if and only if:*

(i) $\sum_{j=1}^{k_n} \mathbb{P}[X_{nj} \in dy] \to \phi(dy)$ *for some σ-finite Borel measure on $\{y : y \ne 0\}$; and*

(ii) $\lim_{\varepsilon\to 0}\limsup_{n\to\infty} \sum_{j=1}^{k_n} \mathrm{Cov}[X_{nj}^\varepsilon] = \lim_{\varepsilon\to 0}\liminf_{n\to\infty} \sum_{j=1}^{k_n} \mathrm{Cov}[X_{nj}^\varepsilon] = Q.$

In this case, Y is infinitely divisible with Lévy representation $[a, Q, \phi]$, where $a \in \mathbb{R}^d$ depends on the centering sequence (a_n). We can take

$$a_n = \sum_{j=1}^{k_n} \mathbb{E}[X_{nj}^R] \tag{6.23}$$

for any $R > 0$ such that $\phi\{y : \|y\| = R\} = 0$, and then $\mathbb{E}[e^{ik\cdot Y}] = e^{\psi_0(k)}$ where

$$\psi_0(k) = -\tfrac{1}{2}k'Qk + \int \left(e^{ik\cdot y} - 1 - ik\cdot yI(\|y\| \le R)\right)\phi(dy). \tag{6.24}$$

Proof. The proof follows the same ideas as the one dimensional case, using a Poisson approximation. The main ideas (see details) are similar to Remark 3.35. For the complete proof, see [146, Theorem 3.2.2]. □

Remark 6.12. To establish vague convergence (i), it suffices to show

$$\sum_{j=1}^{k_n} \mathbb{P}[X_{nj} \in A] \to \phi(A) \tag{6.25}$$

for sets of the form $A = \{t\theta : t > r, \theta \in B\}$ where $r > 0$ and B is a Borel subset of the unit sphere $S = \{y \in \mathbb{R}^d : \|y\| = 1\}$. Both (6.23) and (6.24) depend on the choice of $R > 0$. If the Lévy measure has a density, then any $R > 0$ may be used. To establish condition (ii), it is sufficient to show that

$$\lim_{\varepsilon\to 0}\limsup_{n\to\infty} k'Q_nk = \lim_{\varepsilon\to 0}\liminf_{n\to\infty} k'Q_nk = k'Qk \tag{6.26}$$

for all $k \in \mathbb{R}^d$, where

$$Q_n = \sum_{j=1}^{k_n} \operatorname{Cov}(X_{nj}^{\varepsilon}).$$

As an illustration, we prove the vector central limit theorem in the case of finite second moments.

Theorem 6.13 (Vector Central Limit Theorem). *Suppose that* (W_n) *are iid and that* $\mu = \mathbb{E}[W_n]$ *and* $Q = \mathbb{E}[(W_n - \mu)(W_n - \mu)']$ *exist. Then*

$$n^{-1/2} \sum_{j=1}^{n} (W_j - \mu) \Rightarrow Y \simeq \mathcal{N}(0, Q). \tag{6.27}$$

Proof. The proof is quite similar to Theorem 3.36, extending to vector notation. Define the triangular array row elements $X_{nj} = n^{-1/2} W_j$ for $j = 1, \dots, n$. Then condition (6.22) holds (see details), and so it suffices to check conditions (i) and (ii) in Theorem 6.11. For condition (i) we have for each $\varepsilon > 0$ that

$$\begin{aligned}
\sum_{j=1}^{k_n} \mathbb{P}[\|X_{nj}\| > \varepsilon] &= n\mathbb{P}[n^{-1/2}\|W_j\| > \varepsilon] \\
&= n\mathbb{P}[\|W_j\| > n^{1/2}\varepsilon] \\
&\le n\mathbb{E}\left[\left(\frac{\|W_j\|}{n^{1/2}\varepsilon}\right)^2 I(\|W_j\| > n^{1/2}\varepsilon)\right] \\
&= \varepsilon^{-2}\mathbb{E}\left[\|W_j\|^2 I(\|W_j\| > n^{1/2}\varepsilon)\right] \to 0
\end{aligned}$$

as $n \to \infty$, since $\mathbb{E}[\|W_n\|^2]$ exists (see details). Then (i) holds with $\phi = 0$.

As for condition (ii), use the general fact that $\operatorname{Cov}(X) = \mathbb{E}[XX'] - \mathbb{E}[X]\mathbb{E}[X]'$ (see details) to write

$$\begin{aligned}
\sum_{j=1}^{k_n} \operatorname{Cov}[X_{nj}^{\varepsilon}] &= nE\left[\left(X_{nj}^{\varepsilon}\right)\left(X_{nj}^{\varepsilon}\right)'\right] - n\mathbb{E}\left[X_{nj}^{\varepsilon}\right]\mathbb{E}\left[X_{nj}^{\varepsilon}\right]' \\
&= nE\left[(n^{-1/2}W_j)(n^{-1/2}W_j)' I(\|n^{-1/2}W_j\| \le \varepsilon)\right] \\
&\quad - nE\left[n^{-1/2}W_j I(\|n^{-1/2}W_j\| \le \varepsilon)\right] E\left[n^{-1/2}W_j I(\|n^{-1/2}W_j\| \le \varepsilon)\right]' \\
&= E\left[W_j W_j' I(\|W_j\| \le n^{1/2}\varepsilon)\right] \\
&\quad - E\left[W_j I(\|W_j\| \le n^{1/2}\varepsilon)\right] E\left[W_j I(\|W_j\| \le n^{1/2}\varepsilon)\right]' \\
&\to E\left[W_j W_j'\right] - E\left[W_j\right] E\left[W_j\right]' = \operatorname{Cov}(W_j) = Q
\end{aligned}$$

as $n \to \infty$. Then Theorem 6.11 implies that $S_n - a_n \Rightarrow Y \simeq [a, Q, 0] \simeq \mathcal{N}(a, Q)$ for some $a \in \mathbb{R}^d$. Since $\phi = 0$, for any $R > 0$ we can take

$$a_n = \sum_{j=1}^{k_n} \mathbb{E}[X_{nj}^R] = n\mathbb{E}\left[n^{-1/2} W_j I(\|W_j\| \le n^{1/2}R)\right]$$
$$= n^{1/2}\left\{\mu - \mathbb{E}\left[W_j I(\|W_j\| > n^{1/2}R)\right]\right\}$$

where $\mu = \mathbb{E}[W_j]$ and

$$\left\|n^{1/2}\mathbb{E}\left[W_j I(\|W_j\| > n^{1/2}R)\right]\right\| \le n^{1/2}\mathbb{E}\left[\|W_j\| I(\|W_j\| > n^{1/2}R)\right]$$
$$\le n^{1/2}\mathbb{E}\left[\|W_j\|\left(\frac{\|W_j\|}{n^{1/2}R}\right) I(\|W_j\| > n^{1/2}R)\right]$$
$$= R^{-1}\mathbb{E}\left[\|W_j\|^2 I(\|W_j\| > n^{1/2}R)\right] \to 0$$

since $\mathbb{E}[\|W_n\|^2]$ exists. This shows that $a_n - n^{1/2}\mu \to 0$, and then (6.27) follows. □

Corollary 6.14. *Suppose* (W_n) *are iid and* $\mu = \mathbb{E}[W_n]$ *and* $Q = \mathbb{E}[(W_n - \mu)(W_n - \mu)']$ *exist. Then*

$$n^{-1/2} \sum_{j=1}^{[nt]} (W_j - \mu) \Rightarrow Z_t \simeq \mathcal{N}(0, tQ). \tag{6.28}$$

for all $t > 0$.

Proof. The proof is essentially identical to Theorem 3.41. Theorem 6.13 shows that (6.28) holds for $t = 1$, with $Z_1 = Y$. Let $\hat{\mu}_n(k)$ be the characteristic function of $n^{-1/2}(W_j - \mu)$, so that $\hat{\mu}_n(k)^n \to \hat{\mu}(k) = \mathbb{E}[e^{ik\cdot Y}]$ for all $k \in \mathbb{R}^d$. Then we also have

$$\hat{\mu}_n(k)^{[nt]} = (\hat{\mu}_n(k)^n)^{[nt]/n} \to \hat{\mu}(k)^t$$

for any $t > 0$, which shows that (6.28) holds for any $t > 0$. □

Details

If X is any random vector, then the distribution of X is *tight*, meaning that

$$\mathbb{P}[\|X\| > r] \to 0 \quad \text{as } r \to \infty. \tag{6.29}$$

Equation (6.29) follows by a simple application of the dominated convergence theorem. It follows that

$$\mathbb{P}[\|X_{nj}\| > \varepsilon] = \mathbb{P}[\|W_j\| > n^{1/2}\varepsilon] \to 0$$

as $n \to \infty$, so that condition (6.22) holds.

If X is a random vector with $\mu = \mathbb{E}[X]$ then

$$\begin{aligned} Q = \mathrm{Cov}(X) &= \mathbb{E}\left[(X-\mu)(X-\mu)'\right] \\ &= \mathbb{E}\left[XX' - \mu X' - X\mu' + \mu\mu'\right] \\ &= \mathbb{E}\left[XX'\right] - \mu\mathbb{E}\left[X\right]' - \mathbb{E}\left[X\right]\mu' + \mu\mu' \\ &= \mathbb{E}\left[XX'\right] - \mu\mu' - \mu\mu' + \mu\mu' = \mathbb{E}\left[XX'\right] - \mu\mu' \end{aligned}$$

which we used in the proof of Theorem 6.13.

Let μ_i denote the ith coordinate of the mean vector $\mu = \mathbb{E}[X]$ and let $Q_{ij} = \mathbb{E}[(X_i - \mu_i)(X_j - \mu_j)']$ denote the ij entry of the covariance matrix $Q = \mathrm{Cov}(X)$. The proof of Theorem 6.13 also used the fact that, in this case,

$$\mathbb{E}\left[\|X\|^2\right] = \mathbb{E}\left[X_1^2 + \cdots + X_d^2\right] = \sum_{i=1}^{d}\left(Q_{ii} + \mu_i^2\right)$$

exists, since the mean and covariance matrix exist.

The proof of Theorem 6.11 uses a compound Poisson approximation $S_n \approx S_N$ where N is Poisson with $\mathbb{E}[N] = k_n$. We sketch the main ideas here. For the complete proof, see [146, Theorem 3.2.2]. Let $\hat{\mu}(k) = e^{\psi(k)} = \mathbb{E}[e^{ik\cdot Y}]$ and let $\hat{\mu}_n(k) = e^{\psi_n(k)}$ be the characteristic function of the appropriately shifted compound Poisson random vector $S_N \simeq [b_n, Q_n, \phi_n]$. Then $\mu_n \Rightarrow \mu$ if and only if $\psi_n(k) \to \psi(k)$ [146, Lemma 3.1.10]. Write

$$f(y,k) = e^{ik\cdot y} - 1 - \frac{ik\cdot y}{1+\|y\|^2}$$

and note that $y \mapsto f(y,k)$ is a bounded continuous function such that

$$f(y,k) = -\tfrac{1}{2}(k\cdot y)^2 + O((k\cdot y)^2) \quad \text{as } y \to 0$$

for any fixed k. If condition (i) holds, then it is not hard to show that

$$\int_{\|y\|>\varepsilon} f(y,k)\phi_n(dy) \to \int_{\|y\|>\varepsilon} f(y,k)\phi(dy)$$

whenever $\phi\{\|y\| = \varepsilon\} = 0$, which must be true for almost every $\varepsilon > 0$. Then

$$\lim_{\varepsilon\to 0}\lim_{n\to\infty}\int_{\|y\|>\varepsilon} f(y,k)\phi_n(dy) = \lim_{\varepsilon\to 0}\int_{\|y\|>\varepsilon} f(y,k)\phi(dy) = \int f(y,k)\phi(dy)$$

since $\int \|y\|^2 I(0 < \|y\| \le \varepsilon)\,\phi(dy)$ exists for a Lévy measure. Now observe that

$$\begin{aligned} &\lim_{\varepsilon\to 0}\lim_{n\to\infty}\left[-\tfrac{1}{2}k'Q_nk + \int_{0<\|y\|\le\varepsilon} f(y,k)\phi_n(dy)\right] \\ &= \lim_{\varepsilon\to 0}\lim_{n\to\infty}\left[-\tfrac{1}{2}k'Q_nk - \tfrac{1}{2}\int_{0<\|y\|\le\varepsilon} k'yy'k\,\phi_n(dy)\right] = -\tfrac{1}{2}k'Qk \end{aligned}$$

whenever

$$\lim_{\varepsilon\to 0}\lim_{n\to\infty}\left[Q_n + \int_{0<\|y\|\le\varepsilon} yy'\phi_n(dy)\right] = Q. \tag{6.30}$$

Then it can be shown that $S_N - a_n \Rightarrow Y$ for suitable $a_n \in \mathbb{R}^d$ if this condition holds. Note that

$$\int_{0<\|y\|\le\varepsilon} yy'\phi_n(dy) = k_n\mathbb{E}[X_{nj}X'_{nj}I(\|X_{nj}\| \le \varepsilon)]$$

is the un-centered covariance matrix of the truncated row element. This leads to condition (ii). Finally, argue that convergence of the random sum implies convergence without the Poisson randomization [146, Theorem 3.2.12].

As in the one variable case, some alternative forms of the Lévy representation (6.21) are also useful.

Theorem 6.15. *Suppose $Y \simeq \mu$ is infinitely divisible with characteristic function $\hat{\mu}(k) = e^{\psi(k)}$ and (6.21) holds. Then we can also write $\hat{\mu}(k) = e^{\psi_0(k)}$ where*

$$\psi_0(k) = ik\cdot a_0 - \tfrac{1}{2}k'Qk + \int\left(e^{ik\cdot y} - 1 - ik\cdot yI(\|y\|\le R)\right)\phi(dy) \tag{6.31}$$

for any $R > 0$, for some unique a_0 depending on R and a. Furthermore:

(a) If

$$\int_{0<|y|\le R} \|y\|\phi(dy) < \infty \tag{6.32}$$

then we can also write $\hat{\mu}(k) = e^{\psi_1(k)}$ where

$$\psi_1(k) = ik\cdot a_1 - \tfrac{1}{2}k'Qk + \int\left(e^{ik\cdot y} - 1\right)\phi(dy) \tag{6.33}$$

for some unique a_1 depending on a_0; and

(b) If

$$\int_{|y|>R} \|y\|\phi(dy) < \infty \tag{6.34}$$

then we can also write $\hat{\mu}(k) = e^{\psi_2(k)}$ where

$$\psi_2(k) = ik\cdot a_2 - \tfrac{1}{2}k'Qk + \int\left(e^{ik\cdot y} - 1 - ik\cdot y\right)\phi(dy) \tag{6.35}$$

for some unique a_2 depending on a_0.

Proof. The proof is similar to Theorem 3.8. The integral

$$\delta_0 = \int\left(\frac{y}{1+\|y\|^2} - yI(\|y\|\le R)\right)\phi(dy)$$

exists, since the integrand is bounded and $O(\|y\|^3)$ as $y \to 0$. If we take $a_0 = a - \delta_0$, then $\psi(k) = \psi_0(k)$. If (6.32) holds, then $\psi_0(k) = \psi_1(k)$, where

$$a_1 = a_0 - \int_{0<\|y\|\le R} y\phi(dy).$$

If (6.34) holds, then $\psi_0(k) = \psi_2(k)$, where

$$a_2 = a_0 + \int_{\|y\|>R} y\phi(dy).$$

Uniqueness follows from Theorem 6.8. □

Remark 6.16. It can be shown by differentiating the characteristic function that $\mathbb{E}[Y] = a_2$ for any infinitely divisible law that satisfies condition (6.34) in Theorem 6.15, see [146, Remark 3.1.15] for details.

6.4 Stable random vectors

Stable random vectors are the weak limits of random walks with power law jumps. Each jump is of the form $X = W\Theta$, where $\mathbb{P}[W > r] = Cr^{-\alpha}$ for some $C > 0$ and some $0 < \alpha < 2$, and Θ is a random unit vector. The distribution of the stable limit is determined, up to centering, by C, α, and the distribution of Θ.

Theorem 6.17. *Suppose $X_n = W_n\Theta_n$ are iid random vectors in $\mathbb{R}^d$ with $\mathbb{P}[W_n > r] = Cr^{-\alpha}$ iid Pareto for some $0 < \alpha < 2$, and Θ_n are iid random unit vectors with probability measure $M(d\theta)$, independent of (W_n). Then*

$$n^{-1/\alpha}(X_1 + \cdots + X_n) - a_n \Rightarrow Y \tag{6.36}$$

for some $a_n \in \mathbb{R}^d$, where Y is infinitely divisible with Lévy representation $[a, 0, \phi]$ and

$$\phi\{t\theta : t > r, \theta \in B\} = Cr^{-\alpha}M(B) \tag{6.37}$$

for any $r > 0$ and any Borel subset B of the unit sphere. If $0 < \alpha < 1$, we can choose $a_n = 0$, and then the limit Y is centered stable with characteristic function

$$\mathbb{E}[e^{ik\cdot Y}] = \exp\left[-C\Gamma(1-\alpha)\int_{\|\theta\|=1}(-ik\cdot\theta)^\alpha M(d\theta)\right]. \tag{6.38}$$

If $1 < \alpha < 2$, we can choose $a_n = n^{1-1/\alpha}\mathbb{E}[X_n]$, and then the limit Y is centered stable with mean zero and characteristic function

$$\mathbb{E}[e^{ik\cdot Y}] = \exp\left[C\frac{\Gamma(2-\alpha)}{\alpha-1}\int_{\|\theta\|=1}(-ik\cdot\theta)^\alpha M(d\theta)\right]. \tag{6.39}$$

Proof. Consider a triangular array $X_{nj} = n^{-1/\alpha} X_j$ for $1 \le j \le n = k_n$. Condition (6.22) holds (see details), and so we only need to check conditions (i) and (ii) from Theorem 6.11, see also Remark 6.12. For condition (i) it suffices to prove that (6.25) holds for $A = \{t\theta : t > r, \theta \in B\}$, where B is a Borel subset of the unit sphere. For n sufficiently large we have

$$\begin{aligned}\sum_{j=1}^{k_n} \mathbb{P}[X_{nj} \in A] &= n\mathbb{P}[n^{-1/\alpha} X_j \in A]\\ &= n\mathbb{P}[n^{-1/\alpha} W_j \Theta_j \in A]\\ &= n\mathbb{P}[n^{-1/\alpha} W_j > r, \Theta_j \in B]\\ &= n\mathbb{P}[W_j > n^{1/\alpha} r]\mathbb{P}[\Theta_j \in B]\\ &= nC(n^{1/\alpha} r)^{-\alpha} M(B) = Cr^{-\alpha} M(B)\end{aligned}$$

which shows that (i) holds with the Lévy measure (6.37).

To prove condition (ii), write

$$\begin{aligned}k' Q_n k = nk' \operatorname{Cov}(X_{nj}^{\varepsilon})k &= n \operatorname{Var}(k \cdot X_{nj}^{\varepsilon})\\ &\le n\mathbb{E}\left[(k \cdot X_{nj}^{\varepsilon})^2\right]\\ &= n\mathbb{E}\left[(n^{-1/\alpha} W_j)^2 I(|W_j| \le n^{1/\alpha}\varepsilon)(\Theta_j \cdot k)^2\right]\\ &= n\mathbb{E}\left[(n^{-1/\alpha} W_j)^2 I(|W_j| \le n^{1/\alpha}\varepsilon)\right] \mathbb{E}\left[(\Theta_j \cdot k)^2\right]\\ &\le n^{1-2/\alpha}\, \mathbb{E}\left[W_j^2 I(|W_j| \le n^{1/\alpha}\varepsilon)\right] \|k\|^2\\ &= \left(\varepsilon^{2-\alpha} \frac{C\alpha}{2-\alpha} - n^{1-2/\alpha} \frac{\alpha}{2-\alpha} C^{2/\alpha}\right) \|k\|^2\end{aligned} \tag{6.40}$$

by (3.45) and the fact that $(k \cdot \theta)^2 \le \|k\|^2$ for any unit vector $\|\theta\| = 1$. It follows that

$$\begin{aligned}\lim_{\varepsilon\to 0} \limsup_{n\to\infty} k' Q_n k &\le \lim_{\varepsilon\to 0} \lim_{n\to\infty} \left(\varepsilon^{2-\alpha} \frac{C\alpha}{2-\alpha} - n^{1-2/\alpha} \frac{\alpha}{2-\alpha} C^{2/\alpha}\right) \|k\|^2\\ &= \lim_{\varepsilon\to 0} \varepsilon^{2-\alpha} \frac{C\alpha}{2-\alpha} \|k\|^2 = 0\end{aligned}$$

since $0 < \alpha < 2$, so that $1 - 2/\alpha < 0$ and $2 - \alpha > 0$. Then Theorem 6.11 implies that $S_n - a_n \Rightarrow Y_0$ holds for some sequence (a_n), where $S_n = n^{-1/\alpha}(X_1 + \cdots + X_n)$, $Y_0 \simeq [a, 0, \phi]$, and (6.37) holds.

Suppose $0 < \alpha < 1$. It follows from (6.37) that the Lévy measure

$$\phi(dy) = \alpha C r^{-\alpha-1} dr M(d\theta) \tag{6.41}$$

in polar coordinates $y = r\theta$ with $r > 0$ and $\|\theta\| = 1$ (see details), so we can choose (a_n) according to (6.23) for any $R > 0$. Then $\mathbb{E}[e^{ik\cdot Y_0}] = e^{\psi_0(k)}$ where

$$\begin{aligned}\psi_0(k) &= \int_{\|\theta\|=1} \int_0^\infty \left(e^{ik\cdot r\theta} - 1 - ik \cdot r\theta I(\|r\theta\| \le R)\right) \alpha C r^{-\alpha-1} dr M(d\theta)\\ &= \int_{\|\theta\|=1} \psi(k, \theta) M(d\theta)\end{aligned} \tag{6.42}$$

using (3.46), where

$$\psi(k,\theta) = \int_0^\infty \left(e^{i(k\cdot\theta)r} - 1 - i(k\cdot\theta)rI(r \le R)\right)\alpha C r^{-\alpha-1}\,dr$$
$$= -C\Gamma(1-\alpha)(-ik\cdot\theta)^\alpha - (ik\cdot\theta)a \tag{6.43}$$

and a is given by (3.47). It follows that

$$\psi_0(k) = -C\Gamma(1-\alpha)\int_{\|\theta\|=1}(-ik\cdot\theta)^\alpha M(d\theta) - ik\cdot b$$

where

$$b = a\int_{\|\theta\|=1}\theta M(d\theta) = \frac{C\alpha}{1-\alpha}R^{1-\alpha}\,\mathbb{E}[\Theta_j].$$

Then $Y = Y_0 + b$ is a centered stable random vector with characteristic function (6.38).

Use (6.23) along with (3.48) to write

$$a_n = \sum_{j=1}^{k_n}\mathbb{E}\left[X_{nj}^R\right] = n\mathbb{E}\left[n^{-1/\alpha}W_jI(|W_j| \le n^{1/\alpha}R)\right]\mathbb{E}[\Theta_j]$$
$$= \left[\frac{C\alpha}{1-\alpha}R^{1-\alpha} - n^{1-1/\alpha}\frac{\alpha}{1-\alpha}C^{1/\alpha}\right]\mathbb{E}[\Theta_j] \to \frac{C\alpha}{1-\alpha}R^{1-\alpha}\,\mathbb{E}[\Theta_j] = b \tag{6.44}$$

as $n \to \infty$, since $1 - 1/\alpha < 0$ in this case. Then $S_n - b = S_n - a_n + (a_n - b) \Rightarrow Y_0$, so $S_n = S_n - b + b \Rightarrow Y_0 + b = Y$. Hence we can take $a_n = 0$ in this case, and then the limit has characteristic function (6.38).

Now suppose that $1 < \alpha < 2$. Theorem 6.11 shows that, if we choose (a_n) according to (6.23), then $\mathbb{E}[e^{ik\cdot Y_0}] = e^{\psi_0(k)}$ where (6.42) holds with

$$\psi(k,\theta) = C\frac{\Gamma(2-\alpha)}{\alpha-1}(-ik\cdot\theta)^\alpha + (ik\cdot\theta)a \tag{6.45}$$

by (3.49), where a is given by (3.50). It follows that

$$\psi_0(k) = C\frac{\Gamma(2-\alpha)}{\alpha-1}\int_{\|\theta\|=1}(-ik\cdot\theta)^\alpha M(d\theta) + ik\cdot b$$

where

$$b = a\,\mathbb{E}[\Theta_j] = \frac{C\alpha}{\alpha-1}R^{1-\alpha}\,\mathbb{E}[\Theta_j].$$

Then $Y = Y_0 - b$ is a centered stable random vector with characteristic function (6.39). Using (6.44) we have

$$a_n = \left[\frac{C\alpha}{1-\alpha}R^{1-\alpha} - n^{1-1/\alpha}\frac{\alpha}{1-\alpha}C^{1/\alpha}\right]\mathbb{E}[\Theta_j] = -b + n^{1-1/\alpha}\mu \tag{6.46}$$

where

$$\mu = \mathbb{E}[X_j] = \mathbb{E}[W_j]\,E[\Theta_j] = \frac{\alpha}{\alpha - 1} C^{1/\alpha}\,\mathbb{E}[\Theta_j] \tag{6.47}$$

by (3.52). Since $S_n - a_n \Rightarrow Y_0$ and $a_n + b = n^{1-1/\alpha}\mu$, it follows that $S_n - n^{1-1/\alpha}\mu = S_n - a_n - b \Rightarrow Y_0 - b = Y$. Hence we can take $a_n = n^{1-1/\alpha}\mathbb{E}[X_j]$ in this case, and then the limit has characteristic function (6.39). Then it follows from Remark 6.16 that $\mathbb{E}[Y] = 0$. □

Proposition 6.18. *The characteristic function of a general stable random vector Y with Lévy measure (6.37) and index $0 < \alpha < 2$, $\alpha \neq 1$ can be written in the form*

$$\mathbb{E}[e^{ik\cdot Y}] = \exp\left[ik\cdot\mu - \int\limits_{\|\theta\|=1} |\theta\cdot k|^{\alpha}\left(1 - i\,\mathrm{sgn}(\theta\cdot k)\tan\left(\frac{\pi\alpha}{2}\right)\right)\Lambda(d\theta)\right] \tag{6.48}$$

with center μ and spectral measure

$$\Lambda(d\theta) = C\frac{\Gamma(2-\alpha)}{1-\alpha}\cos\left(\frac{\pi\alpha}{2}\right)M(d\theta). \tag{6.49}$$

In this case, we will write $Y \simeq S_\alpha(\Lambda, \mu)$.

Proof. If $1 < \alpha < 2$, then (6.39) implies

$$\mathbb{E}[e^{ik\cdot Y}] = \exp\left[A\int\limits_{\|\theta\|=1} (-ik\cdot\theta)^{\alpha} M(d\theta)\right] \tag{6.50}$$

with $A = C\Gamma(2-\alpha)/(\alpha-1)$. If $0 < \alpha < 1$, then (6.38) implies that (6.50) holds with $A = -C\Gamma(1-\alpha) = C\Gamma(2-\alpha)/(\alpha-1)$. Now use (5.5) to write

$$\mathbb{E}[e^{ik\cdot Y}] = \exp\left[A\int\limits_{\|\theta\|=1} |k\cdot\theta|^{\alpha}\cos(\pi\alpha/2)[1 - i\,\mathrm{sgn}(k\cdot\theta)\tan(\pi\alpha/2)]M(d\theta)\right] \tag{6.51}$$

so that $Y + \mu$ satisfies (6.48) and (6.49) holds. □

Remark 6.19. The *spectral measure* $\Lambda(d\theta)$ in (6.49) is an arbitrary positive finite Borel measure on the unit sphere, since both $1 - \alpha$ and $\cos(\pi\alpha/2)$ change sign at $\alpha = 1$. In one dimension, we have $\Lambda\{+1\} = p\sigma^{\alpha}$ and $\Lambda\{-1\} = q\sigma^{\alpha}$ where the skewness $\beta = p - q$ in the notation of Proposition 5.3. The spectral measure plays a role similar to the covariance matrix, i.e., it controls the dependence of the components of the stable random vector $Y = (Y_1, \ldots, Y_d)'$. If $\Lambda(d\theta)$ is a discrete measure that only assigns positive weight to the coordinate axes, then it follows easily from (6.50) that $Y_1, \ldots, Y_d$ are independent. In fact, $Y_1, \ldots, Y_d$ are independent if and only if Λ is concentrated on the coordinate axes, see Meerschaert and Scheffler [147, Lemma 2.3].

Remark 6.20. If $\alpha = 2$, then (6.48) reduces to the characteristic function of a normal random vector $Y \simeq \mathcal{N}(\mu, Q)$ where the covariance matrix

$$Q = 2 \int\limits_{\|\theta\|=1} \theta\theta' M(d\theta). \tag{6.52}$$

The characteristic function of a general stable random vector Y with Lévy measure (6.37) and index $\alpha = 1$ can be written in the form

$$\mathbb{E}[e^{ik\cdot Y}] = \exp\left[ik \cdot \mu - \int\limits_{\|\theta\|=1} |\theta \cdot k| \left(1 + i\left(\frac{2}{\pi}\right) \operatorname{sgn}(\theta \cdot k) \ln |\theta \cdot k|\right) \Lambda(d\theta)\right] \tag{6.53}$$

with center μ and spectral measure

$$\Lambda(d\theta) = C\left(\frac{\pi}{2}\right) M(d\theta). \tag{6.54}$$

These formulas (6.48) and (6.53) describe the entire class of limit distributions for sums of iid random vectors with scalar norming, see [146, Theorem 7.3.16].

Theorem 6.21. *Suppose $X_n = W_n\Theta_n$ are iid random vectors with $\mathbb{P}[W_n > r] = Cr^{-\alpha}$ for some $0 < \alpha < 2$, and Θ_n are iid with probability measure $M(d\theta)$ on the unit sphere, independent of W_n.*

(a) If $0 < \alpha < 1$, then

$$n^{-1/\alpha} \sum_{j=1}^{[nt]} X_j \Rightarrow Z_t \tag{6.55}$$

for all $t > 0$, where

$$\mathbb{E}\left[e^{ikZ_t}\right] = \exp\left[-tD \int\limits_{\|\theta\|=1} (-ik \cdot \theta)^\alpha M(d\theta)\right] \tag{6.56}$$

and $D = C\Gamma(1 - \alpha)$;

(b) If $1 < \alpha < 2$, then $\mu = \mathbb{E}[X_n]$ exists and

$$n^{-1/\alpha} \sum_{j=1}^{[nt]} (X_j - \mu) \Rightarrow Z_t \tag{6.57}$$

for all $t > 0$, where

$$\mathbb{E}\left[e^{ikZ_t}\right] = \exp\left[tD \int\limits_{\|\theta\|=1} (-ik \cdot \theta)^\alpha M(d\theta)\right] \tag{6.58}$$

and $D = C\Gamma(2 - \alpha)/(\alpha - 1)$.

Proof. The proof is essentially identical to Theorem 3.41. For example, in the case $0 < \alpha < 1$, Theorem 6.17 shows (6.55) and (6.56) hold for $t = 1$, with $Z_1 = Y$. Let $\hat{\mu}_n(k)$ be the characteristic function of $n^{-1/\alpha}X_j$, so that $\hat{\mu}_n(k)^n \to \hat{\mu}(k) = \mathbb{E}[e^{ik\cdot Y}]$ for all $k \in \mathbb{R}^d$. Then we also have

$$\hat{\mu}_n(k)^{[nt]} = (\hat{\mu}_n(k)^n)^{[nt]/n} \to \hat{\mu}(k)^t$$

for any $t > 0$, which shows that (6.55) and (6.56) hold for any $t > 0$. □

Details

Since $X_j = W_j\Theta_j$ is tight for any fixed j, so that (6.29) holds with $X = X_j$, it follows that

$$\mathbb{P}[\|X_{nj}\| > \varepsilon] = \mathbb{P}[\|X_j\| > n^{1/\alpha}\varepsilon] \to 0$$

as $n \to \infty$, so that condition (6.22) holds.

In (6.40) we used the fact that

$$X^{\varepsilon}_{nj} = n^{-1/\alpha}W_j\Theta_j I(\|n^{-1/\alpha}W_j\Theta_j\| \le \varepsilon) = n^{-1/\alpha}W_j I(|W_j| \le n^{1/\alpha}\varepsilon)\Theta_j$$

since $\|\Theta_j\| = 1$. We also used the general fact that, if $Q = \text{Cov}(X) = \mathbb{E}[(X-\mu)(X-\mu)']$ with $\mu = \mathbb{E}[X]$, then $\mathbb{E}[k \cdot X] = k \cdot \mu$ and

$$k'Qk = \mathbb{E}[k'(X-\mu)(X-\mu)'k] = \mathbb{E}[(k\cdot(X-\mu))^2] = \text{Var}[k \cdot X]$$

for any fixed $k \in \mathbb{R}^d$.

To establish (6.41), write $A = \{t\theta : t > r, \theta \in B\}$ and note that

$$\int_{t\theta\in A} \alpha C t^{-\alpha-1}\,dt M(d\theta) = \int_r^\infty \alpha C t^{-\alpha-1}\,dt \int_{\theta\in B} M(d\theta) = Cr^{-\alpha}M(B).$$

This is sufficient to prove (6.41) since sets of this form determine the measure ϕ.

6.5 Vector fractional diffusion equation

Theorem 6.21 shows that a vector random walk with power law jumps converges to a vector stable Lévy motion Z_t. Suppose $1 < \alpha < 2$. Then (6.58) shows that the pdf $p(x, t)$ of Z_t has FT

$$\hat{p}(k, t) = \exp\left[tD \int_{\|\theta\|=1} (ik\cdot\theta)^\alpha M(d\theta)\right]$$

and then

$$\frac{d}{dt}\hat{p}(k, t) = \psi(-k)\hat{p}(k, t) \tag{6.59}$$

where the Fourier symbol

$$\psi(-k) = D \int_{\|\theta\|=1} (ik \cdot \theta)^\alpha M(d\theta). \tag{6.60}$$

Equation (6.59) represents the FT of the equation

$$\frac{\partial}{\partial t} p(x, t) = Lp(x, t)$$

where the generator $Lf(x)$ has FT $\psi(-k)\hat{f}(k)$. We would like to understand the meaning of this generator in terms of fractional derivatives.

First we consider the FT $(ik \cdot \theta)^\alpha \hat{f}(k)$. If $\alpha = 1$, then $(ik \cdot \theta)\hat{f}(k) = (ik)\hat{f}(k) \cdot \theta$ is the FT of the *directional derivative* (use the chain rule)

$$\begin{aligned}
\mathbb{D}_\theta f(x) &= \frac{d}{dt} f(x + t\theta)\big|_{t=0} \\
&= \frac{d}{dt} f(x_1 + t\theta_1, \ldots, x_d + t\theta_d)\big|_{t=0} \\
&= \left[\frac{\partial}{\partial x_1} f(x + t\theta)\theta_1 + \cdots + \frac{\partial}{\partial x_d} f(x + t\theta)\theta_d\right]_{t=0} \\
&= \nabla f(x) \cdot \theta
\end{aligned} \tag{6.61}$$

defined for any unit vector $\theta \in \mathbb{R}^d$. We will define the *fractional directional derivative* $\mathbb{D}_\theta^\alpha f(x)$ to be the function with FT $(ik \cdot \theta)^\alpha \hat{f}(k)$. It is not hard to check (see details) that $\mathbb{D}_\theta^\alpha f(x)$ is the (positive Riemann-Liouville) fractional derivative of the function $t \mapsto f(x + t\theta)$ evaluated at $t = 0$.

Take $e_1 = (1, 0, \ldots, 0)'$, $e_2 = (0, 1, 0, \ldots, 0)'$, and so forth, the standard coordinate vectors. If $\theta = e_j$, then $k \cdot e_j = k_j$ and

$$\mathbb{D}_{e_j}^\alpha f(x) = \frac{\partial^\alpha}{\partial (x_j)^\alpha} f(x_1, \ldots, x_d)$$

is the fractional partial derivative in this coordinate. Now define

$$\nabla_M^\alpha f(x) = \int_{\|\theta\|=1} \mathbb{D}_\theta^\alpha f(x) M(d\theta). \tag{6.62}$$

Then $D\nabla_M^\alpha f(x)$ has FT $\psi(-k)\hat{f}(k)$, where the Fourier symbol $\psi(-k)$ is given by (6.60), see details at the end of this section. Inverting the FT in (6.59) shows that the density $p(x, t)$ of Z_t solves the *vector fractional diffusion equation*

$$\frac{\partial}{\partial t} p(x, t) = D\nabla_M^\alpha p(x, t) \tag{6.63}$$

for $1 < \alpha < 2$. Next we add a drift: For $v \in \mathbb{R}^d$ the FT of $vt + Z_t$ is

$$\hat{p}(k, t) = \mathbb{E}\left[e^{-ik\cdot(vt+Z_t)}\right] = \exp\left(-ik \cdot vt + Dt \int_{\|\theta\|=1} (ik \cdot \theta)^\alpha M(d\theta)\right).$$

Then

$$\frac{d}{dt}\hat{p}(k,t) = \left(-ik \cdot v + D \int_{\|\theta\|=1} (ik \cdot \theta)^\alpha M(d\theta) \right) \hat{p}(k,t). \tag{6.64}$$

Inverting the FT in (6.64) shows that the density $p(x,t)$ of $vt + Z_t$ solves the vector fractional diffusion equation with drift

$$\frac{\partial}{\partial t}p(x,t) = -v \cdot \nabla p(x,t) + D\nabla_M^\alpha p(x,t) \tag{6.65}$$

for $1 < \alpha < 2$. This equation was introduced in Meerschaert, Benson and Baeumer [137]. It was originally applied to describe the movement of contaminant particles in ground water in a heterogeneous aquifer by Schumer et al. [194]. It has also been applied by Cushman and Moroni [56] to model particle traces in a laboratory setting. If $0 < \alpha < 1$, then (6.65) governs $vt + Z_t$ with $D < 0$.

Example 6.22. Suppose that $M\{e_j\} = 1/d$ for $j = 1, \dots, d$ where $e_1, \dots e_d$ are the standard coordinate vectors. Then

$$\int_{\|\theta\|=1} (ik \cdot \theta)^\alpha M(d\theta) = \sum_{j=1}^{d} (ik \cdot e_j)^\alpha d^{-1} = d^{-1} \sum_{j=1}^{d} (ik_j)^\alpha$$

since $k_j = k \cdot e_j$. Then

$$\nabla_M^\alpha f(x) = d^{-1} \sum_{j=1}^{d} \frac{\partial^\alpha}{\partial (x_j)^\alpha} f(x)$$

and the vector fractional diffusion equation

$$\frac{\partial}{\partial t}p(x,t) = D_0 \sum_{j=1}^{d} \frac{\partial^\alpha}{\partial (x_j)^\alpha} p(x,t) \tag{6.66}$$

with $D_0 = D/d$ governs the scaling limit Z_t of a random walk with Pareto jumps evenly scattered over the positive coordinate axes. Here the components of Z_t are iid α-stable Lévy motions that are totally positively skewed ($p = 1$ and $q = 0$, so that the skewness $\beta = 1$). Figure 6.4 shows a typical solution on $\mathbb{R}^2$ in the case $\alpha = 1.3$, obtained using a small modification of the R code from Figure 6.13: Set `a1=1.3`, `a2=1.3`, `q1=0.0`, `q2=0.0`, and `t=2.0`. The mean of the pdf in Figure 6.4 is zero, but the mode is shifted into the negative, to balance the heavy positive tail.

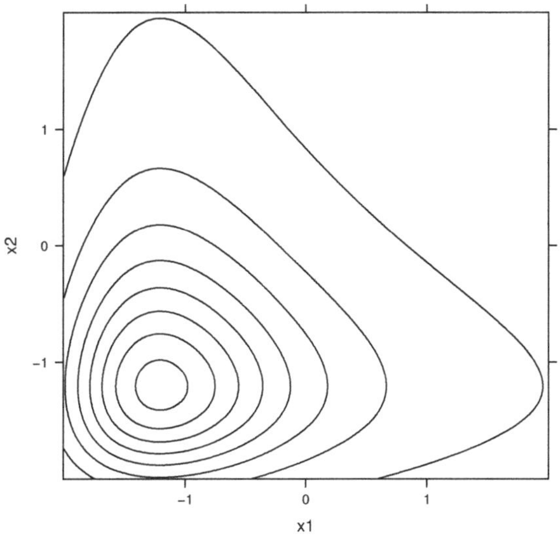

Fig. 6.4: Level sets of the solution $p(x, t)$ to the fractional diffusion equation (6.66) in dimension $d = 2$ at time $t = 2$ with $\alpha = 1.3$ and $D_0 = 0.5$.

Example 6.23. Suppose that $M(d\theta) = M(-d\theta)$ (origin symmetric) for all $\|\theta\| = 1$. Then it follows using (5.6) with $p = q = 1/2$ that

$$\begin{aligned}\int_{\|\theta\|=1} (ik\cdot\theta)^\alpha M(d\theta) &= \int_{\|\theta\|=1} (ik\cdot\theta)^\alpha M(-d\theta)\\ &= \int_{\|\theta\|=1} (-ik\cdot\theta)^\alpha M(d\theta)\\ &= \int_{\|\theta\|=1} [\tfrac{1}{2}(ik\cdot\theta)^\alpha + \tfrac{1}{2}(-ik\cdot\theta)^\alpha] M(d\theta)\\ &= \cos(\pi\alpha/2)\int_{\|\theta\|=1} |k\cdot\theta|^\alpha M(d\theta).\end{aligned}$$

For example, if $M\{e_j\} = M\{-e_j\} = 1/(2d)$ for for $j = 1, \ldots, d$ then

$$\int_{\|\theta\|=1} (ik\cdot\theta)^\alpha M(d\theta) = d^{-1}\sum_{j=1}^{d} [\tfrac{1}{2}(ik_j)^\alpha + \tfrac{1}{2}(-ik_j)^\alpha] = d^{-1}\cos(\pi\alpha/2)\sum_{j=1}^{d} |k_j|^\alpha$$

and (6.63) reduces to

$$\frac{\partial}{\partial t}p(x,t) = D_1\sum_{j=1}^{d}\left[\frac{\partial^\alpha}{\partial(x_j)^\alpha}p(x,t) + \frac{\partial^\alpha}{\partial(-x_j)^\alpha}p(x,t)\right] \tag{6.67}$$

where $D_1 = D/(2d)$. Some authors define the fractional Laplacian in one dimension, $df(x)/d|x|^\alpha$, as the inverse FT of $-|k|^\alpha \hat{f}(k)$. This is also called the *Riesz fractional derivative*. Then we can rewrite (6.67) in the form

$$\frac{\partial}{\partial t} p(x,t) = D_0 \sum_{j=1}^{d} \frac{\partial^\alpha}{\partial |x_j|^\alpha} p(x,t) \tag{6.68}$$

where $D_0 = -D\cos(\pi\alpha/2)/d$. Equation (6.67) governs the scaling limit Z_t of a random walk with Pareto jumps evenly scattered over the positive and negative coordinate axes. The components of Z_t are iid symmetric α-stable Lévy motions. A typical solution was graphed in Figure 6.3.

Example 6.24. Suppose that $M(d\theta)$ is uniform over the unit sphere $\|\theta\| = 1$. Write $k = \rho\omega$ in polar coordinates with $\rho > 0$ and $\|\omega\| = 1$. Then

$$\begin{aligned} \int_{\|\theta\|=1} (ik\cdot\theta)^\alpha M(d\theta) &= \cos(\pi\alpha/2) \int_{\|\theta\|=1} |k\cdot\theta|^\alpha M(d\theta) \\ &= \cos(\pi\alpha/2)\rho^\alpha \int_{\|\theta\|=1} |\omega\cdot\theta|^\alpha M(d\theta) \\ &= B\rho^\alpha = B\|k\|^\alpha \end{aligned}$$

where

$$\begin{aligned} B &= \cos(\pi\alpha/2) \int_{\|\theta\|=1} |\omega\cdot\theta|^\alpha M(d\theta) \\ &= \cos(\pi\alpha/2) \int_{\|\theta\|=1} |\theta_1|^\alpha M(d\theta) \end{aligned}$$

by symmetry, since the integral in the first line does not depend on choice of ω, so that we can set $\omega = e_1$. Note that B is a constant that only depends on α and the dimension d, with $B > 0$ for $0 < \alpha < 1$ and $B < 0$ for $1 < \alpha < 2$. Now (6.59) becomes

$$\frac{d}{dt}\hat{p}(k,t) = DB\|k\|^\alpha \hat{p}(k,t).$$

If $1 < \alpha < 2$ then this inverts to

$$\frac{\partial}{\partial t} p(x,t) = D_3 \Delta^{\alpha/2} p(x,t),$$

a version of (6.19) with $c = D_3 = -BD > 0$, involving the fractional Laplacian of order $\beta = \alpha/2$. The case $0 < \alpha < 1$ leads to the same differential equation, with $D_3 = BD > 0$. This isotropic vector fractional diffusion equation governs the scaling limit Z_t of a random walk with power law jumps, whose angle is evenly scattered over the entire unit sphere. The components of Z_t are symmetric α-stable Lévy motions, but they are not independent. This is clear because the FT

$$\mathbb{E}\left[e^{-ik\cdot Z_t}\right] = e^{-tD_3\|k\|^\alpha} \neq \prod_{j=1}^{d} e^{-tD_3|k_j|^\alpha}$$

and the quantity on the right-hand side is the product of the FT of the components.

It is instructive to contrast the normal case $\alpha = 2$ with the stable case $1 < \alpha < 2$. If $\alpha = 2$ then

$$D \int_{\|\theta\|=1} (ik\cdot\theta)^\alpha M(d\theta) = -D \int_{\|\theta\|=1} k'\theta\,\theta' k M(d\theta) = -k'Qk$$

where the dispersion tensor $Q = D\int \theta\,\theta' M(d\theta)$. Then (6.63) reduces to the vector diffusion equation

$$\frac{\partial}{\partial t} p(x,t) = \nabla\cdot Q\nabla p(x,t). \tag{6.69}$$

This equation governs the scaling limit Z_t of a random walk whose jumps have finite second moments, see Corollary 6.14. The dispersion tensor Q controls particle spreading, see Remark 6.1. This also reflects the jump distribution: The longest jumps tend to be in the direction of the eigenvector corresponding to the largest eigenvalue of the matrix Q. If $Q = cI$, then (6.63) reduces to the isotropic diffusion equation

$$\frac{\partial}{\partial t} p(x,t) = c\Delta p(x,t) \tag{6.70}$$

since $\nabla\cdot\nabla = \Delta$. Here Z_t is an isotropic Brownian motion. If the jump distribution is spherically symmetric, or if the jumps have iid symmetric components, or if the jumps have iid positive components (corrected to mean zero), then we get the same limit process, since all three types of jumps have the same covariance matrix. This stands in direct contrast to the stable case, where these three types of random walks lead to three different limits.

Remark 6.25. Please note that ∇_M^α is an extension of the common (abuse of) notation $\nabla^2 = \nabla\cdot\nabla = \Delta$, so that $\nabla_M^\alpha f(x)$ is scalar-valued. The operator ∇_M^α is an asymmetric version of the fractional Laplacian, not a fractional gradient vector.

Details

Fix $\|\theta\| = 1$ and define $g(t) = f(x + t\theta)$ for $t \in \mathbb{R}$. Then the positive Riemann-Liouville fractional derivative of order $0 < \alpha < 1$ is given by the generator form

$$\begin{aligned}\frac{d^\alpha g(t)}{dt^\alpha} &= \frac{1}{\Gamma(1-\alpha)}\int_0^\infty [g(t) - g(t-r)]\,\alpha r^{-\alpha-1}\,dr \\ &= \frac{1}{\Gamma(1-\alpha)}\int_0^\infty [f(x+t\theta) - f(x+(t-r)\theta)]\,\alpha r^{-\alpha-1}\,dr.\end{aligned} \tag{6.71}$$

A simple substitution $y = x - a$ shows that $f(x - a)$ has FT

$$\int e^{-ik\cdot x} f(x-a)\,dx = \int e^{-ik\cdot(y+a)} f(y)\,dy = e^{-ik\cdot a}\hat{f}(k).$$

Use this fact to show that the last expression in (6.71) for $t = 0$ has FT

$$\frac{1}{\Gamma(1-\alpha)}\int_0^\infty \hat{f}(k)\left[1 - e^{-ik\cdot r\theta}\right]\alpha r^{-\alpha-1}\,dr = (ik\cdot\theta)^\alpha \hat{f}(k)$$

using (3.14):

$$I(\alpha) = \int_0^\infty \left(e^{iky} - 1\right)\alpha y^{-\alpha-1}\,dy = -\Gamma(1-\alpha)(-ik)^\alpha.$$

The proof for $1 < \alpha < 2$ is similar.

A rigorous proof of the generator form (6.62) for the vector fractional derivative $\nabla_M^\alpha f(x)$ relies on the theory of semigroups and generators. The following result is the vector version of Theorem 3.17.

Theorem 6.26. *Suppose that Z_t is a Lévy process on $\mathbb{R}^d$, and that $\mathbb{E}[e^{ik\cdot Z_1}] = e^{\psi(k)}$ where $\psi(k)$ is given by (6.21). Then $T_t f(x) = \mathbb{E}[f(x - Z_t)]$ defines a C_0 semigroup on $C_0(\mathbb{R}^d)$ with generator*

$$Lf(x) = -a\cdot\nabla f(x) + \tfrac{1}{2}\nabla\cdot Q\nabla f(x) + \int\left(f(x-y) - f(x) + \frac{y\cdot\nabla f(x)}{1+\|y\|^2}\right)\phi(dy). \tag{6.72}$$

If f and all its partial derivatives up to order two are elements of $C_0(\mathbb{R}^d)$, then $f \in \mathrm{Dom}(L)$. If f and all its partial derivatives up to order two are also elements of $L^1(\mathbb{R}^d)$, then $\psi(-k)\hat{f}(k)$ is the FT of $Lf(x)$.

Proof. The proof is essentially identical to the one variable case presented in Theorem 3.17, see Sato [187, Theorem 31.5] and Hille and Phillips [90, Theorem 23.14.2]. □

As in the one variable case, there are some alternative forms of the generator. The next result extends Theorem 3.23.

Theorem 6.27. *Suppose that Z_t is a Lévy process on $\mathbb{R}^d$, and that $\mathbb{E}[e^{ik\cdot Z_1}] = e^{\psi(k)}$ where $\psi(k)$ is given by* (6.21). *Then we can also write the generator* (6.72) *in the form*

$$Lf(x) = -a_0 \cdot \nabla f(x) + \tfrac{1}{2}\nabla \cdot Q\nabla f(x) + \int \left(f(x-y) - f(x) + y \cdot \nabla f(x) I(\|y\| \le R)\right)\phi(dy) \tag{6.73}$$

for any $R > 0$, for some unique a_0 depending on R and a. Furthermore:

(a) If (6.32) *holds, then we can also write*

$$Lf(x) = -a_1 \cdot \nabla f(x) + \tfrac{1}{2}\nabla \cdot Q\nabla f(x) + \int \left(f(x-y) - f(x)\right)\phi(dy) \tag{6.74}$$

for some unique a_1 depending on a_0; and

(b) If (6.34) *holds, then we can also write*

$$Lf(x) = -a_2 \cdot \nabla f(x) + \tfrac{1}{2}\nabla \cdot Q\nabla f(x) + \int \left(f(x-y) - f(x) + y \cdot \nabla f(x)\right)\phi(dy) \tag{6.75}$$

for some unique a_2 depending on a_0.

Proof. The proof is very similar to Theorem 6.15. In view of Theorem 6.26, we know that the generator formula (6.72) holds. Since the integral

$$\delta_0 = \int \left(\frac{y}{1+\|y\|^2} - yI(\|y\| \le R)\right)\phi(dy)$$

exists, we can take $a_0 = a - \delta_0$, and then (6.73) follows. If (6.32) holds, the integral

$$a_1 = a_0 - \int_{0<|y|\le R} y\phi(dy)$$

exists, and then (6.74) follows from (6.73). If condition (6.34) holds, then

$$a_2 = a_0 + \int_{|y|>R} y\phi(dy)$$

exists, and (6.75) follows from (6.73). □

Example 6.28. Suppose that Z_1 is centered stable with index $0 < \alpha < 1$ and characteristic function (6.38). Use (6.74) to write the generator the corresponding stable semigroup in the form

$$Lf(x) = \int \left(f(x-y) - f(x)\right)\phi(dy)$$

where $\phi(dy)$ is given by (6.41). Make a change of variable $y = r\theta$ to see that

$$Lf(x) = \int_{\|\theta\|=1}\int_0^\infty \left(f(x-r\theta) - f(x)\right)\alpha C r^{-\alpha-1}\,dr M(d\theta).$$

If we take $C = 1/\Gamma(1-\alpha)$, then this shows that $Lf(x) = -\nabla_M^\alpha f(x)$.

Example 6.29. Suppose that Z_1 is centered stable with index $1 < \alpha < 2$ and characteristic function (6.39). Use Theorem 6.27 (b) to write the generator in the form

$$Lf(x) = \int (f(x-y) - f(x) + y \cdot \nabla f(x))\, \phi(dy)$$

where $\phi(dy)$ is given by (6.41). A change of variable $y = r\theta$ leads to

$$Lf(x) = \int_{\|\theta\|=1} \int_0^\infty (f(x - r\theta) - f(x) + r\theta \cdot \nabla f(x))\, \alpha C r^{-\alpha-1} dr M(d\theta).$$

If we take $C = (\alpha - 1)/\Gamma(2-\alpha)$, then $Lf(x) = \nabla_M^\alpha f(x)$.

6.6 Operator stable laws

Suppose that (X_n) are iid with some full random vector X on $\mathbb{R}^d$. Recall from Section 6.2 that $X \in \mathrm{GDOA}(Y)$ if

$$A_n S_n - b_n \Rightarrow Y \tag{6.76}$$

for some linear operators A_n and vectors b_n. In this case, we say that Y is *operator stable*. If $A_n = a_n I$ for some $a_n > 0$, then Y is stable with index $\alpha \in (0, 2]$.

Example 6.30. If the components of X are independent Pareto random variables with different indices $\alpha_i \in (0, 1)$, Example 6.4 shows that (6.76) holds with $b_n = 0$ and

$$A_n = \mathrm{diag}(n^{-1/\alpha_1}, \ldots, n^{-1/\alpha_d}) = \begin{pmatrix} n^{-1/\alpha_1} & & 0 \\ & \ddots & \\ 0 & & n^{-1/\alpha_d} \end{pmatrix}.$$

and furthermore,

$$A_n S_{[nt]} \Rightarrow Z(t) \tag{6.77}$$

where the limit $Z(t)$ is an operator stable Lévy motion with independent components, and $Z(1) \simeq Y$. The pdf $p(x, t)$ of $Z(t)$ has FT

$$\hat{p}(k, t) = \mathbb{E}\left[e^{-ik\cdot Z(t)}\right] = \exp\left[-t \sum_{j=1}^d D_j (ik_j)^{\alpha_j}\right].$$

This pdf $p(x, t)$ solves the vector fractional diffusion equation

$$\frac{\partial}{\partial t} p(x, t) = \sum_{j=1}^d \left[-D_j \frac{\partial^{\alpha_j}}{\partial x_j^{\alpha_j}} p(x, t)\right] \tag{6.78}$$

for some $D_i > 0$. Since the pdf

$$p(x, t) = \prod_{j=1}^d p_j(x_j, t)$$

is the product of stable densities with different indices α_i, the right tail $x_j \mapsto p(x,t)$ falls off at a different rate $\approx x_j^{-\alpha_j-1}$ in each coordinate. Figure 6.5 shows level sets of a typical solution $p(x,t)$ in $\mathbb{R}^2$ with $\alpha_1 = 0.8$ and $\alpha_2 = 0.6$, obtained using the R code from Figure 6.14 at the end of this chapter.

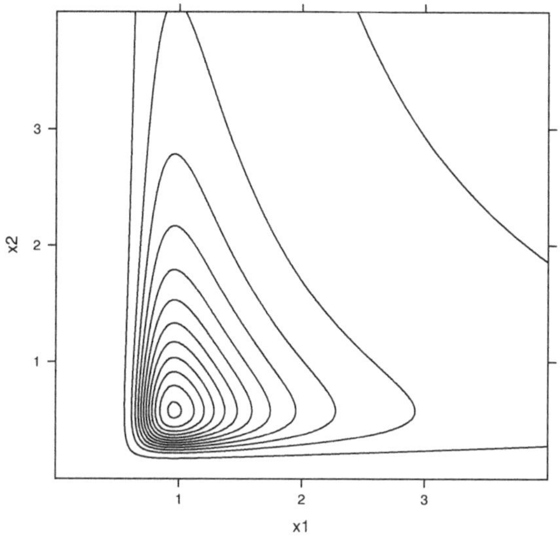

Fig. 6.5: Level sets of the solution $p(x,t)$ to the fractional diffusion equation (6.78) at time $t = 3$ in dimension $d = 2$, with $\alpha_1 = 0.8$, $\alpha_2 = 0.6$, and $D_1 = D_2 = 0.5$.

The scaling also varies with the coordinate. In fact, the operator stable Lévy motion $Z(t)$ has *operator scaling*

$$Z(ct) \simeq c^B Z(t) \tag{6.79}$$

where the scaling matrix

$$B = \text{diag}(1/\alpha_1, \ldots, 1/\alpha_d) = \begin{pmatrix} 1/\alpha_1 & & 0 \\ & \ddots & \\ 0 & & 1/\alpha_d \end{pmatrix}$$

and we define the matrix power

$$c^B = \text{diag}(c^{1/\alpha_1}, \ldots, c^{1/\alpha_d}) = \begin{pmatrix} c^{1/\alpha_1} & & 0 \\ & \ddots & \\ 0 & & c^{1/\alpha_d} \end{pmatrix}.$$

To check this, let

$$Z(t) = \begin{pmatrix} Z_1(t) \\ \vdots \\ Z_d(t) \end{pmatrix}$$

and recall that each component is self-similar with $Z_j(ct) \simeq c^{1/\alpha_j} Z_j(t)$ for all $c > 0$ and $t > 0$. Then

$$Z(ct) = \begin{pmatrix} Z_1(ct) \\ \vdots \\ Z_d(ct) \end{pmatrix} = \begin{pmatrix} c^{1/\alpha_1} Z_1(t) \\ \vdots \\ c^{1/\alpha_d} Z_d(t) \end{pmatrix} = \begin{pmatrix} c^{1/\alpha_1} & & 0 \\ & \ddots & \\ 0 & & c^{1/\alpha_d} \end{pmatrix} \begin{pmatrix} Z_1(t) \\ \vdots \\ Z_d(t) \end{pmatrix} = c^B Z(t).$$

Remark 6.31. The random walk convergence (6.77) extends easily to finite dimensional distributions. The argument is essentially identical to (4.28). The operator scaling (6.79) also holds in the sense of finite dimensional distributions, i.e., for any $0 < t_1 < t_2 < \cdots < t_n < \infty$ we have

$$(Z(ct_1), \ldots, Z(ct_n)) \simeq (c^B Z(t_1), \ldots, c^B Z(t_n)).$$

To see this, note that $Z(ct_k) - Z(ct_{k-1}) \simeq Z(c(t_k - t_{k-1})) \simeq c^B Z(t_k - t_{k-1})$ since $Z(t)$ has stationary increments. Since $Z(t)$ has independent increments, it follows that

$$(Z(ct_k) - Z(ct_{k-1}) : k = 1, \ldots, n) \simeq (c^B [Z(t_k) - Z(t_{k-1})] : k = 1, \ldots, n)$$

and then apply the Continuous Mapping Theorem 4.19. Then $Z(t)$ is *operator self-similar* with *exponent* B. For more on operator self-similar processes, see Embrechts and Maejima [64].

Remark 6.32. The random walk convergence (6.77) also extends to convergence in the Skorokhod space. Let $\mathbb{D}([0, \infty), \mathbb{R}^d)$ denote the set of real-valued functions $x : [0, \infty) \to \mathbb{R}^d$ which are continuous from the right with left-hand limits. Equip with the Skorokhod J_1 topology, defined exactly as in Section 4.4. Then we also have $A_n S_{[nt]} \Rightarrow Z(t)$ in $\mathbb{D}([0, \infty), \mathbb{R}^d)$ with this topology, see [146, Theorem 4.1] for complete details.

To proceed further, we need to introduce some additional notation. The *matrix exponential* is defined by

$$\exp(A) = \sum_{n=0}^{\infty} \frac{A^n}{n!} = I + A + \frac{A^2}{2!} + \cdots \tag{6.80}$$

for any $d \times d$ matrix A. The *matrix power* is defined by

$$t^A = \exp(A \log t) = I + A \log t + \frac{(\log t)^2}{2!} A^2 + \cdots \tag{6.81}$$

for any $t > 0$.

Example 6.33. If

$$A = \operatorname{diag}(a, b) = \begin{pmatrix} a & 0 \\ 0 & b \end{pmatrix}$$

then

$$A^n = \operatorname{diag}(a^n, b^n) = \begin{pmatrix} a^n & 0 \\ 0 & b^n \end{pmatrix}$$

and

$$\begin{aligned} \exp(A) &= \begin{pmatrix} 1 & 0 \\ 0 & 1 \end{pmatrix} + \begin{pmatrix} a & 0 \\ 0 & b \end{pmatrix} + \frac{1}{2!}\begin{pmatrix} a^2 & 0 \\ 0 & b^2 \end{pmatrix} + \cdots \\ &= \begin{pmatrix} 1 + a + a^2/2! + \cdots & 0 \\ 0 & 1 + b + b^2/2! + \cdots \end{pmatrix} = \begin{pmatrix} e^a & 0 \\ 0 & e^b \end{pmatrix}. \end{aligned}$$

Then

$$t^A = \exp(A \log t) = \exp\left[\begin{pmatrix} a \log t & 0 \\ 0 & b \log t \end{pmatrix}\right] = \begin{pmatrix} e^{a \log t} & 0 \\ 0 & e^{b \log t} \end{pmatrix} = \begin{pmatrix} t^a & 0 \\ 0 & t^b \end{pmatrix}.$$

More generally, if $A = \operatorname{diag}(a_1, \ldots, a_d)$, then $t^A = \operatorname{diag}(t^{a_1}, \ldots, t^{a_d})$. Some typical *orbits* $t \mapsto t^A x$ for different unit vectors x are shown as solid lines in Figure 6.6. Each orbit intersects the unit circle (dashed line) exactly once at the point $t = 1$. The R code for plotting these orbits is shown in Figure 6.15 at the end of this chapter.

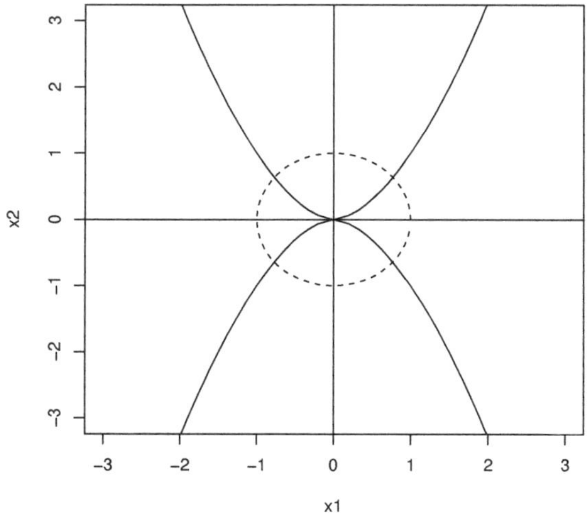

Fig. 6.6: Eight orbits $t \mapsto t^A x$ from Example 6.33 with $a = 0.7$ and $b = 1.2$ grow out from the origin as t increases. Each orbit intersects the unit circle (dashed line) at $t = 1$ when x is a unit vector.

Example 6.34. Suppose that

$$A = \begin{pmatrix} a & 1 \\ 0 & a \end{pmatrix} = \begin{pmatrix} a & 0 \\ 0 & a \end{pmatrix} + \begin{pmatrix} 0 & 1 \\ 0 & 0 \end{pmatrix} = D + N$$

where $D = aI$ is a diagonal matrix, and $DN = ND$. It is not hard to check, using the definition of the matrix exponential, that $DN = ND$ implies $\exp(N + D) = \exp(N)\exp(D)$. The matrix N is a *nilpotent matrix*, i.e., $N^k = 0$ for any sufficiently large integer $k > 0$. In fact we have

$$N^2 = \begin{pmatrix} 0 & 1 \\ 0 & 0 \end{pmatrix}\begin{pmatrix} 0 & 1 \\ 0 & 0 \end{pmatrix} = \begin{pmatrix} 0 & 0 \\ 0 & 0 \end{pmatrix}$$

so that $N^k = 0$ for all $k > 1$. Then

$$t^N = I + N \log t + 0 + \cdots = \begin{pmatrix} 1 & 0 \\ 0 & 1 \end{pmatrix} + \begin{pmatrix} 0 & \log t \\ 0 & 0 \end{pmatrix} = \begin{pmatrix} 1 & \log t \\ 0 & 1 \end{pmatrix}$$

and $t^D = \operatorname{diag}(t^a, t^a) = t^a I$ so that

$$t^A = \begin{pmatrix} t^a & 0 \\ 0 & t^a \end{pmatrix}\begin{pmatrix} 1 & \log t \\ 0 & 1 \end{pmatrix} = \begin{pmatrix} t^a & t^a \log t \\ 0 & t^a \end{pmatrix}.$$

Some typical orbits $t^A x$ are shown as solid lines in Figure 6.7. Each orbit $t^A x$ with $x \neq 0$ passes through the unit circle (dashed line) exactly once. The R code for plotting these orbits is shown in Figure 6.16 at the end of this chapter.

Example 6.35. Suppose that

$$A = \begin{pmatrix} a & -1 \\ 1 & a \end{pmatrix} = D + Q$$

where $D = aI$ is a diagonal matrix,

$$Q = \begin{pmatrix} 0 & -1 \\ 1 & 0 \end{pmatrix}$$

is skew-symmetric, and $DQ = QD$. Write

$$\begin{aligned} Q^2 &= \begin{pmatrix} 0 & -1 \\ 1 & 0 \end{pmatrix}\begin{pmatrix} 0 & -1 \\ 1 & 0 \end{pmatrix} = \begin{pmatrix} -1 & 0 \\ 0 & -1 \end{pmatrix} \\ Q^3 &= \begin{pmatrix} -1 & 0 \\ 0 & -1 \end{pmatrix}\begin{pmatrix} 0 & -1 \\ 1 & 0 \end{pmatrix} = \begin{pmatrix} 0 & 1 \\ -1 & 0 \end{pmatrix} \\ Q^4 &= \begin{pmatrix} 0 & 1 \\ -1 & 0 \end{pmatrix}\begin{pmatrix} 0 & -1 \\ 1 & 0 \end{pmatrix} = \begin{pmatrix} 1 & 0 \\ 0 & 1 \end{pmatrix} \\ Q^5 &= Q^4 Q = Q \\ Q^6 &= Q^4 Q^2 = Q^2 \end{aligned}$$

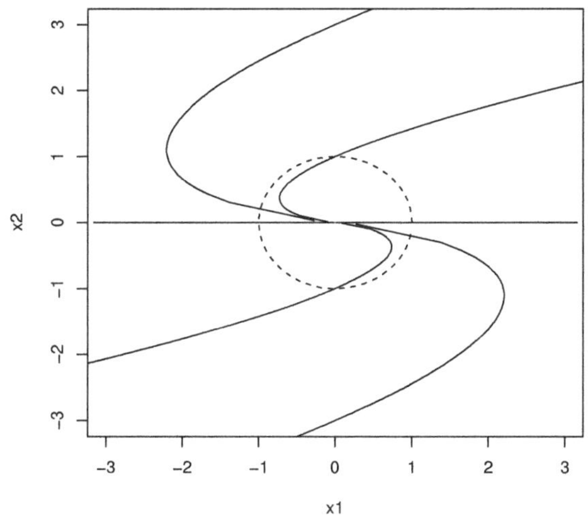

Fig. 6.7: Six orbits $t \mapsto t^A x$ from Example 6.34 with $a = 0.5$ grow out from the origin as t increases. Each orbit intersects the unit circle (dashed line) exactly once at the point $t = 1$ when x is a unit vector.

and so forth, so that

$$\begin{aligned}\exp(cQ) &= \begin{pmatrix}1 & 0\\ 0 & 1\end{pmatrix} + c\begin{pmatrix}0 & -1\\ 1 & 0\end{pmatrix} + \frac{c^2}{2!}\begin{pmatrix}-1 & 0\\ 0 & -1\end{pmatrix} + \frac{c^3}{3!}\begin{pmatrix}0 & 1\\ -1 & 0\end{pmatrix} + \cdots\\ &= \begin{pmatrix}1 - c^2/2! + c^4/4! + \cdots & -c + c^3/3! - c^5/5! + \cdots\\ c - c^3/3! + c^5/5! + \cdots & 1 - c^2/2! + c^4/4! + \cdots\end{pmatrix}\\ &= \begin{pmatrix}\cos c & -\sin c\\ \sin c & \cos c\end{pmatrix} = R_c\end{aligned}$$

the rotation matrix that rotates each vector $x = (r\cos\theta, r\sin\theta)'$ counterclockwise by an angle c: $R_c x = (r\cos(c+\theta), r\sin(c+\theta))'$. Then $t^Q = \exp(Q\log t) = R_{\log t}$ and

$$t^A = \begin{pmatrix}t^a & 0\\ 0 & t^a\end{pmatrix} R_{\log t} = \begin{pmatrix}t^a\cos(\log t) & -t^a\sin(\log t)\\ t^a\sin(\log t) & t^a\cos(\log t)\end{pmatrix}.$$

The orbits $t^A x$ are counter-clockwise spirals, see Figure 6.8. The R code for plotting these orbits is shown in Figure 6.17 at the end of this chapter.

Remark 6.36. The computations in Examples 6.33–6.35 can be extended to explicitly compute the matrix power t^A for any $d \times d$ matrix A, using the Jordan decomposition, see [146, Section 2.2]. The matrix exponential is also important in the theory of linear differential equations. The vector differential equation $x' = Ax$; $x(0) = x_0$ has a unique

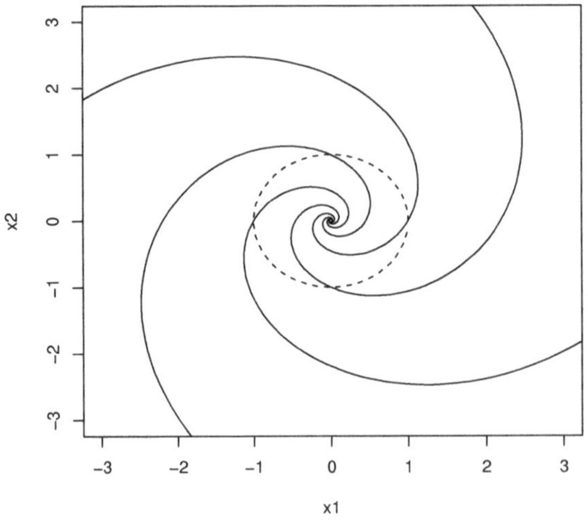

Fig. 6.8: Four orbits $t \mapsto t^A x$ from Example 6.35 with $a = 0.5$ grow out from the origin as t increases. Each orbit intersects the unit circle (dashed line) exactly once at $t = 1$ when x is a unit vector.

solution $x(t) = \exp(At)x_0$, so the orbits $t \mapsto \exp(At)x_0$ are the solution curves for this system of linear differential equations (e.g., see Hirsch and Smale [91]). The orbits $s \mapsto s^A x_0$ trace out the same curves with a different parametrization $t = \log s$.

Theorem 6.17 showed that α-stable random vectors are random walk limits with jumps of the form $X = W\Theta$, where W is a Pareto random variable with tail index α, and Θ is a random unit vector. Operator stable random vectors are limits of random walks with a more general jump distribution that allows the tail index α to vary with the coordinate. Let $B = \operatorname{diag}(1/\alpha_1, \dots, 1/\alpha_d)$ for some $\alpha_i \in (0, 2)$. That is, B is a diagonal matrix whose eigenvalues $\lambda_i = 1/\alpha_i > 1/2$. If all $\alpha_i \in (1, 2)$ then every eigenvalue $\lambda_i \in (1/2, 1)$.

Suppose $\mathbb{P}[W > r] = Cr^{-1}$ is a Pareto random variable with index $\alpha = 1$, and $\Theta = (\theta_1, \dots, \theta_d)'$ is a random unit vector with distribution $M(d\theta)$, independent of W. Write

$$X = W^B\Theta = \begin{pmatrix} W^{1/\alpha_1} & & 0 \\ & \ddots & \\ 0 & & W^{1/\alpha_d} \end{pmatrix} \begin{pmatrix} \theta_1 \\ \vdots \\ \theta_d \end{pmatrix} \tag{6.82}$$

and note that $\mathbb{P}[W^{1/\alpha} > r] = \mathbb{P}[W > r^\alpha] = Cr^{-\alpha}$ so that the ith diagonal entry in the matrix W^B is a Pareto random variable with index α_i. Note also that these entries are not independent!

Take (X_n) iid with $X = W^B\Theta$ in (6.82), and let $S_n = X_1 + \cdots + X_n$. In Section 6.7, we will prove that $X \in \text{GDOA}(Y)$ and (6.76) holds for some $b_n \in \mathbb{R}^d$, and in fact

$$n^{-B}S_n - b_n \Rightarrow Y. \tag{6.83}$$

This operator stable limit Y has Lévy representation $[a, 0, \phi]$, where a depends on the choice of centering b_n, and the Lévy measure ϕ reflects the operator scaling. Next we will compute this Lévy measure, using condition (i) of Theorem 6.11. The proof of condition (ii) is more complicated, and will be deferred to Section 6.7.

To establish the vague convergence condition (i) in Theorem 6.11, it suffices to show

$$\sum_{j=1}^{k_n} \mathbb{P}[X_{nj} \in U] \to \phi(U) \tag{6.84}$$

for sets of the form $U = \{t^B\theta : t > r, \theta \in V\}$ where $r > 0$ and V is a Borel subset of the unit sphere. A substitution $s = nt$ shows that

$$\begin{aligned} n^B U &= \{n^B t^B \theta : t > r, \theta \in V\} \\ &= \{(nt)^B\theta : t > r, \theta \in V\} \\ &= \{s^B\theta : s/n > r, \theta \in V\} \\ &= \{s^B\theta : s > nr, \theta \in V\}. \end{aligned}$$

Then for n sufficiently large we have

$$\begin{aligned} n\mathbb{P}[n^{-B}X \in U] &= n\mathbb{P}[X \in n^B U] \\ &= n\mathbb{P}[W^B\Theta \in n^B U] \\ &= n\mathbb{P}[W > nr, \Theta \in V] = n\,C(nr)^{-1}M(V) = Cr^{-1}M(V). \end{aligned}$$

This proves that condition (i) holds with

$$\phi\{t^B\theta : t > r, \theta \in V\} = Cr^{-1}M(V). \tag{6.85}$$

Example 6.37. If we take $\alpha_1 = \cdots = \alpha_d = \alpha \in (0, 2)$ in (6.82), then $B = (1/\alpha)I$ and $W^B\Theta = W^{1/\alpha}\Theta$, where $W^{1/\alpha}$ is a Pareto random variable with index α. Then Theorem 6.17 applies to show that (6.83) holds, where the α-stable random vector Y in the limit has Lévy measure (6.37). Substitute $s = t^\alpha$ in (6.85) to see that

$$\begin{aligned} \phi\{t\theta : t > r, \theta \in V\} &= \phi\{(t^\alpha)^B\theta : t > r, \theta \in V\} \\ &= \phi\{s^B\theta : s^{1/\alpha} > r, \theta \in V\} \\ &= \phi\{s^B\theta : s > r^\alpha, \theta \in V\} = C(r^\alpha)^{-1}M(V) = Cr^{-\alpha}M(V). \end{aligned}$$

Hence the operator stable Lévy measure (6.85) reduces to the stable Lévy measure (6.37) when the exponent B is a scalar multiple of the identity.

Example 6.38. Suppose that $\mathbb{P}[\Theta = e_j] = 1/d$ for $j = 1, \ldots, d$, where $e_1, \ldots, e_d$ are the standard coordinate vectors. Since $t^B e_j = t^{1/\alpha_j} e_j$ for all $t > 0$, it follows from (6.85) that ϕ is concentrated on the positive coordinate axes. A substitution $s = t^{\alpha_j}$ yields

$$\begin{aligned}\phi\{te_j : t > r\} &= \phi\{(t^{\alpha_j})^B \theta : t > r\}\\ &= \phi\{s^B\theta : s > r^{\alpha_j}, \theta \in V\} = Cr^{-\alpha_j} M(e_j).\end{aligned}$$

Then it follows from the Lévy representation (6.21) that $\mathbb{E}[e^{ik\cdot Y}] = e^{\psi(k)}$ where

$$\begin{aligned}\psi(k) &= ik\cdot a + \sum_{j=1}^{d} \int \left(e^{ik\cdot re_j} - 1 - \frac{ik\cdot re_j}{1+r^2}\right) C\alpha_j r^{-\alpha_j-1} dr M(e_j)\\ &= \sum_{j=1}^{d}\left[ik_j a_j + d^{-1}\int\left(e^{ik_j r} - 1 - \frac{ik_j r}{1+r^2}\right) C\alpha_j r^{-\alpha_j - 1} dr\right] = \sum_{j=1}^{d}\psi_j(k_j).\end{aligned}$$

Note that the jth component of Y has characteristic function $\mathbb{E}[e^{ik_jY_j}] = \mathbb{E}[e^{ik(e_j\cdot Y)}]$. Then Y has independent stable components $Y_j = e_j \cdot Y$ with index α_j and Fourier symbol $\psi_j(-k_j)$.

Remark 6.39. The formula (6.85) implies that ϕ has operator scaling:

$$c\phi(dy) = \phi(c^{-B}dy) \quad \text{for all } c > 0. \tag{6.86}$$

To see this, substitute $s = t/c$ to get

$$\begin{aligned}\phi(c^{-B}U) &= \phi\{c^{-B}t^B\theta : t > r, \theta \in V\}\\ &= \phi\{(t/c)^B\theta : t > r, \theta \in V\}\\ &= \phi\{s^B\theta : s > r/c, \theta \in V\} = C(r/c)^{-1}M(V) = c\,\phi(U).\end{aligned}$$

In fact, it is easy to check that the operator scaling relation (6.86) is equivalent to (6.85) with $CM(V) = \phi\{t^B\theta : t > 1, \theta \in V\}$.

Remark 6.40. We have noted previously in (6.41) that the stable Lévy measure $\phi(dy) = \alpha C r^{-\alpha-1} dr M(d\theta)$ in polar coordinates $y = r\theta$ with $r > 0$ and $\|\theta\| = 1$. The operator stable Lévy measure can be written in a similar manner

$$\phi(dy) = Cr^{-2} dr M(d\theta) \tag{6.87}$$

where $y = r^B\theta$ for some $r > 0$ and $\|\theta\| = 1$. These are called the *Jurek coordinates*, see Jurek and Mason [100]. For these coordinates to make sense, the function $r \mapsto \|r^B x\|$ must be strictly increasing for all $x \neq 0$. Then there is a unique unit vector θ such that $x = r^B\theta$ for some unique $r > 0$. This can be accomplished with a specific non-Euclidean norm [146, Lemma 6.1.5]. For the usual Euclidean norm in $\mathbb{R}^2$, $r \mapsto \|r^B x\|$ is always strictly increasing in the coordinate system that puts B in Jordan form, see [146,

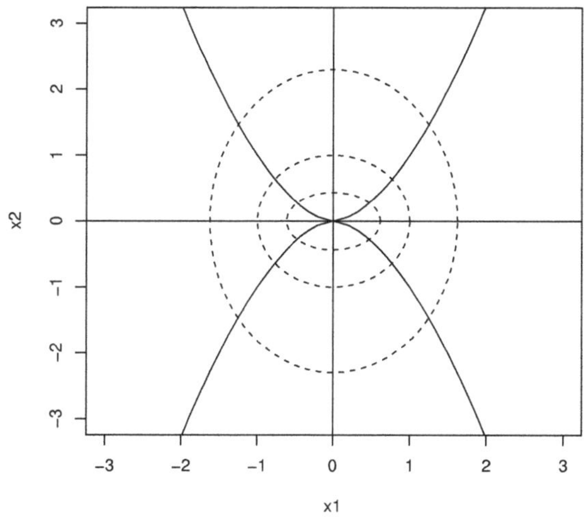

Fig. 6.9: Jurek coordinates $x = r^B\theta$ in the case $B = \mathrm{diag}(0.7, 1.2)$. Dashed lines show the sets $r = 1/2, 1, 2$.

Remark 6.1.6]. This was the case in Examples 6.33–6.35, since all those matrices are in canonical Jordan form. The Jurek coordinates are illustrated in Figure 6.9. The R code for plotting these orbits is shown in Figure 6.18 at the end of this chapter. Since ϕ is the jump intensity, the Jurek coordinates describe particle jumps in a curved coordinate system with operator scaling. They reduce to the usual polar coordinate system if $B = I$.

Remark 6.41. The operator scaling of the Lévy measure can also be visualized using Figure 6.9. Suppose $C = 1$ in (6.85), so that the exterior of the unit circle S in Figure 6.9 has ϕ-measure equal to 1. The exterior of the larger dashed curve is the set $\{t^B\theta : t > 2, \theta \in S\}$, so (6.85) implies that it has ϕ-measure 1/2. The exterior of the smaller dashed curve is the set $\{t^B\theta : t > 1/2, \theta \in S\}$, so it has ϕ-measure 2.

Remark 6.42. The name *operator stable* comes from a paper of Sharpe [198]. Exponents and symmetries of operator stable laws were characterized by Holmes, Hudson and Mason [92]. Hudson, Jurek and Veeh [93] showed that there is some exponent that commutes with every symmetry.

6.7 Operator regular variation

In this section, we show that operator stable laws with no normal component are the weak limits of random walks with operator scaling power law jumps of the form $X = W^B\Theta$. The following result extends Theorem 6.17 to operator stable limits.

Theorem 6.43. *Suppose that B is a $d \times d$ matrix whose eigenvalues $\lambda_j = a_j + ib_j$ all have real part $a_j > 1/2$. Suppose $X_n = W_n^B\Theta_n$ where (W_n) are iid with $\mathbb{P}[W_n > r] = Cr^{-1}$ for some $C > 0$, and Θ_n are iid random unit vectors with probability measure $M(d\theta)$, independent of (W_n). Then*

$$n^{-B}(X_1 + \cdots + X_n) - a_n \Rightarrow Y \tag{6.88}$$

for some $a_n \in \mathbb{R}^d$, where Y is infinitely divisible with Lévy representation $[a, 0, \phi]$ and Lévy measure (6.85).

The proof of Theorem 6.43 requires some regular variation tools. We say that a random vector X *varies regularly* if

$$n\mathbb{P}[A_nX \in dy] \to \phi(dy) \quad \text{as } n \to \infty \tag{6.89}$$

where A_n is invertible, $\|A_n\| \to 0$, and ϕ is a σ-finite Borel measure on $\{y \neq 0\}$ that is not concentrated on any lower dimensional subspace. The next result is the vector version of Proposition 4.15.

Proposition 6.44. *Suppose that X varies regularly, so that* (6.89) *holds. Then:*
(a) The limit measure $\phi(dy)$ satisfies (6.85) *for some B whose eigenvalues all have positive real part;*
(b) The sequence (A_n) can be chosen to be RV$(-B)$, *that is,*

$$A_{[\lambda n]}A_n^{-1} \to \lambda^{-B} \quad \text{as } n \to \infty \tag{6.90}$$

for all $\lambda > 0$.

Proof. This is [146, Theorem 6.1.24]. □

In the situation of Proposition 6.44, we say that (the probability distribution of) X varies regularly with *exponent* B, and we write $X \in \text{RV}(B)$. The matrix norming in (6.89) is critical, as it allows the tails of X to fall off at a different power law rate in different directions. For more information on regularly varying probability measures, see [146, Chapter 6].

Proof of Theorem 6.43. Condition (i) of Theorem 6.11 was already established in Section 6.6. The proof of condition (ii) uses a vector version of the Karamata Theorem 4.4. Define the truncated moments and tail moments

$$\begin{aligned} U_\zeta(r, \theta) &= \mathbb{E}\left[|X \cdot \theta|^\zeta I(|X \cdot \theta| \le r)\right] \\ V_\eta(r, \theta) &= \mathbb{E}\left[|X \cdot \theta|^\eta I(|X \cdot \theta| > r)\right] \end{aligned} \tag{6.91}$$

and note that these are just the truncated and tail moments of the one dimensional projection $X \cdot \theta$. Order the eigenvalues $\lambda_j = a_j + ib_j$ so that $a_1 \leq \cdots \leq a_d$. Then [146, Theorem 6.3.4] shows that

$$V_0(r,\theta) = \mathbb{P}[|X \cdot \theta| > r]$$

is *uniformly R-O varying*: For any $\delta > 0$ there exist $0 < m < M < \infty$ and $r_0 > 0$ such that

$$m\lambda^{-\delta-1/a_1} \leq \frac{V_0(r\lambda,\theta)}{V_0(r,\theta)} \leq M\lambda^{\delta-1/a_d} \quad \text{for all } \lambda \geq 1 \tag{6.92}$$

for any $r \geq r_0$ and any $\|\theta\| = 1$. Now define $\alpha_j = 1/a_j$ so that $\alpha_1 \geq \cdots \geq \alpha_d$. Then we also have

$$r^{-\delta-\alpha_1} < \mathbb{P}[|X \cdot \theta| > r] < r^{\delta-\alpha_d} \tag{6.93}$$

for all $r > 0$ sufficiently large. Since every $a_i > 1/2$, we also have $\alpha_i \in (0,2)$.

Suppose U_ζ and V_η exist. Then the vector Karamata theorem [146, Theorem 6.3.8] implies that, if V_η is uniformly R-O varying, then for some $C > 0$ and $r_0 > 0$ we have

$$\frac{r^{\zeta-\eta}V_\eta(r,\theta)}{U_\zeta(r,\theta)} \geq C \quad \text{for all } r \geq r_0 \text{ and all } \|\theta\| = 1. \tag{6.94}$$

In order to prove condition (*ii*), fix $k \in \mathbb{R}^d$ and write $k = \rho\vartheta$ for some $\rho > 0$ and $\|\vartheta\| = 1$. Then

$$\begin{aligned} n\operatorname{Var}\left[k \cdot (n^{-B}X)^\varepsilon)\right] &\leq n\mathbb{E}\left[\{k \cdot (n^{-B}X)^\varepsilon)\}^2\right] \\ &= n\mathbb{E}\left[(k \cdot n^{-B}X)^2 I(\|n^{-B}X\| \leq \varepsilon)\right] \\ &\leq n\mathbb{E}\left[(k \cdot n^{-B}X)^2 I(|n^{-B}X \cdot \vartheta| \leq \varepsilon)\right] \\ &= n\mathbb{E}\left[(k \cdot n^{-B}X)^2 I(|n^{-B}X \cdot k| \leq \varepsilon_1)\right] \end{aligned}$$

where $\varepsilon_1 = \rho\varepsilon$. It is not hard to check, using the definition of the matrix exponential, that $(t^B)' = t^{B'}$. Write $n^{-B'}k = r_n\theta_n$ where $r_n > 0$ and $\|\theta_n\| = 1$, and recall the general fact that $x \cdot Ay = A'x \cdot y$. Then

$$\begin{aligned} n\mathbb{E}\left[(k \cdot n^{-B}X)^2 I(|n^{-B}X \cdot k| \leq \varepsilon_1)\right] &= n\mathbb{E}\left[(n^{-B'}k \cdot X)^2 I(|X \cdot n^{-B'}k| \leq \varepsilon_1)\right] \\ &= n\mathbb{E}\left[(r_n\theta_n \cdot X)^2 I(|r_n\theta_n \cdot X| \leq \varepsilon_1)\right] \\ &= nr_n^2\mathbb{E}\left[|X \cdot \theta_n|^2 I(|X \cdot \theta_n| \leq r_n^{-1}\varepsilon_1)\right] \\ &= nr_n^2 U_2(r_n^{-1}\varepsilon_1, \theta_n). \end{aligned}$$

Now apply (6.94) to see that

$$\begin{aligned} nr_n^2 U_2(r_n^{-1}\varepsilon_1,\theta_n) &\leq nr_n^2 C^{-1}(r_n^{-1}\varepsilon_1)^2 V_0(r_n^{-1}\varepsilon_1,\theta_n) \\ &= C^{-1}\varepsilon_1^2 nV_0(r_n^{-1}\varepsilon_1,\theta_n) \\ &= C^{-1}\varepsilon_1^2\, nV_0(r_n^{-1},\theta_n)\,\frac{V_0(r_n^{-1}\varepsilon_1,\theta_n)}{V_0(r_n^{-1},\theta_n)} \end{aligned}$$

where

$$\begin{aligned}
nV_0(r_n^{-1}, \theta_n) &= n\mathbb{P}[|X \cdot \theta_n| > r_n^{-1}] \\
&= n\mathbb{P}[|X \cdot r_n\theta_n| > 1] \\
&= n\mathbb{P}[|X \cdot n^{-B'}k| > 1] \\
&= n\mathbb{P}[|n^{-B}X \cdot k| > 1] \\
&= n\mathbb{P}[n^{-B}X \in U] \to \phi(U)
\end{aligned}$$

with $U = \{y : |y \cdot k| > 1\}$. Since every eigenvalue of B has positive real part, it follows from [146, Theorem 2.2.4] that $r_n^{-1} \to \infty$ as $n \to \infty$. Then (6.92) implies that for any $\delta > 0$ we have

$$\frac{V_0(r_n^{-1}\varepsilon_1, \theta_n)}{V_0(r_n^{-1}, \theta_n)} = \frac{V_0(r_n^{-1}\varepsilon_1, \theta_n)}{V_0(\varepsilon_1^{-1}(\varepsilon_1 r_n^{-1}), \theta_n)} \le \frac{1}{m}(\varepsilon_1^{-1})^{\delta+\alpha_1}$$

for all n sufficiently large, where $\alpha_1 = 1/a_1 \in (0, 2)$. Then we have

$$\lim_{\varepsilon\to 0} \limsup_{n\to\infty} n \operatorname{Var}\left[k \cdot (n^{-B}X)^{\varepsilon})\right] \le \lim_{\varepsilon\to 0} C^{-1}\varepsilon_1^2\, \phi(U)\, \frac{1}{m}\, \varepsilon_1^{-\delta-\alpha_1} = 0$$

which proves condition (*ii*). Then Theorem 6.11 implies that (6.88) holds for some $a_n \in \mathbb{R}^d$, where Y is infinitely divisible with Lévy representation $[a, 0, \phi]$ and Lévy measure (6.85). □

Remark 6.45. If every eigenvalue of B has real part $a_i > 1$, then every $\alpha_i < 1$, and we can set $a_n = 0$ in (6.88). In this case, the limit has characteristic function $\mathbb{E}[e^{ik\cdot Y}] = e^{\psi(k)}$ with

$$\psi(k) = \int \left(e^{ik\cdot y} - 1\right)\phi(dy) = \int_{\|\theta\|=1} \int_0^\infty \left(e^{ik\cdot r^B\theta} - 1\right) Cr^{-2}dr\, M(d\theta). \tag{6.95}$$

If every eigenvalue of B has real part $a_i \in (1/2, 1)$, then every $\alpha_i \in (1, 2)$, and we can set $a_n = n\mathbb{E}[n^{-B}X]$ in (6.88) (if $\mathbb{E}[X] = 0$, we can set $a_n = 0$). In this case, the limit has characteristic function $\mathbb{E}[e^{ik\cdot Y}] = e^{\psi(k)}$ with

$$\begin{aligned}
\psi(k) &= \int \left(e^{ik\cdot y} - 1 - ik\cdot y\right)\phi(dy) \\
&= \int_{\|\theta\|=1} \int_0^\infty \left(e^{ik\cdot r^B\theta} - 1 - ik\cdot r^B\theta\right) Cr^{-2}dr\, M(d\theta)
\end{aligned}$$

and $\mathbb{E}[Y] = 0$. The proof is similar to Theorem 6.17, using vector regular variation, see [146, Theorem 8.2.7].

Remark 6.46. Suppose that (6.88) holds with $a_n = 0$. Then an argument very similar to Theorem 6.21 shows that we also get random walk convergence

$$n^{-B}S_{[nt]} \Rightarrow Z_t$$

where Z_t is an operator stable Lévy motion with $Z_1 \simeq Y$. Suppose that every $a_i > 1$. Then

$$\hat{p}(k,t) = \mathbb{E}[e^{-ik\cdot Z_t}] = e^{t\psi(-k)}$$

where $\psi(k)$ is given by (6.95). It follows that

$$\frac{d}{dt}\hat{p}(k,t) = \int \left(e^{ik\cdot y} - 1\right)\hat{p}(k,t)\,\phi(dy)$$

which inverts to

$$\begin{aligned}\frac{\partial}{\partial t}p(x,t) &= \int [p(x-y,t) - p(x,t)]\,\phi(dy)\\ &= \int_{\|\theta\|=1}\int_0^\infty \left[p(x - r^B\theta, t) - p(x,t)\right] Cr^{-2}\,dr\,M(d\theta). \qquad (6.96)\end{aligned}$$

If we define the *generalized fractional derivative*

$$\nabla_M^B f(x) = \int_{\|\theta\|=1}\int_0^\infty \left[f(x) - f(x - r^B\theta)\right] r^{-2}\,dr\,M(d\theta) \qquad (6.97)$$

using (6.74), then we can write (6.96) in the form

$$\frac{\partial}{\partial t}p(x,t) = -C\nabla_M^B p(x,t).$$

This *generalized fractional diffusion equation* governs the densities of operator stable Lévy motions with no normal component. If $B = (1/\alpha)I$, then (6.96) reduces to the vector fractional diffusion equation (6.63) that governs a vector stable Lévy motion: Substitute $s = r^{1/\alpha}$ to get

$$\begin{aligned}\nabla_M^B p(x,t) &= \int_{\|\theta\|=1}\int_0^\infty \left[p(x,t) - p(x - r^{1/\alpha}\theta, t)\right] r^{-2}\,dr\,M(d\theta)\\ &= \int_{\|\theta\|=1}\int_0^\infty [p(x,t) - p(x - s\theta, t)]\,(s^\alpha)^{-2}\alpha s^{\alpha-1}\,ds\,M(d\theta)\\ &= \int_{\|\theta\|=1}\int_0^\infty [p(x,t) - p(x - s\theta, t)]\,\alpha s^{-\alpha-1}\,ds\,M(d\theta)\\ &= \Gamma(1-\alpha)\nabla_M^\alpha p(x,t).\end{aligned}$$

When all $a_i \in (1/2, 1)$, the generalized fractional derivative is defined by

$$\nabla_M^B f(x) = \int_{\|\theta\|=1}\int_0^\infty \left[f(x - r^B\theta) - f(x) + r^B\theta\cdot\nabla f(x)\right] r^{-2}\,dr\,M(d\theta) \qquad (6.98)$$

using Theorem 6.27 (b), and the generalized fractional diffusion equation

$$\frac{\partial}{\partial t}p(x,t) = C\nabla_M^B p(x,t)$$

governs the densities of an operator stable Lévy motion with this exponent B.

6.8 Generalized domains of attraction

Recall from Section 6.2 that $X \in \text{GDOA}(Y)$ if

$$A_n S_n - b_n \Rightarrow Y \tag{6.99}$$

for some linear operators A_n and vectors b_n. Here $S_n = X_1 + \cdots + X_n$ and (X_n) are iid with some full random vector X on $\mathbb{R}^d$. In this case, we say that Y is *operator stable*. The necessary and sufficient conditions for $X \in \text{GDOA}(Y)$ are written in terms of regular variation. Recall from Section 6.7 that $X \in \text{RV}(B)$ if and only if

$$n\mathbb{P}[A_n X \in dy] \to \phi(dy) \quad \text{as } n \to \infty \tag{6.100}$$

where

$$A_{[\lambda n]}A_n^{-1} \to \lambda^{-B} \quad \text{for all } \lambda > 0 \tag{6.101}$$

for some linear operator B whose eigenvalues all have positive real part. Then we also have

$$c\,\phi(dy) = \phi(c^{-B}dy) \quad \text{for all } c > 0. \tag{6.102}$$

The next result extends Theorem 4.5 to random vectors. It also shows that the limits of power law random walks in Theorem 6.43 cover all possible limits in (6.99) when Y has no normal component.

Theorem 6.47 (Generalized CLT for Random Vectors). *If $X \in \text{GDOA}(Y)$, then Y is infinitely divisible with Lévy representation $[a, Q, \phi]$.*

(a) If Y is normal and $E[X] = 0$, then $X \in \text{GDOA}(Y)$ and (6.99) holds for some $b_n \in \mathbb{R}^d$ if and only if

$$nF(A_n' k_n) \to k'Qk \quad \text{for all } k_n \to k \neq 0 \tag{6.103}$$

where $F(k) = \mathbb{E}[|X \cdot k|^2 I(|X \cdot k| \le 1)]$;

(b) If Y has no normal component, then $X \in \text{GDOA}(Y)$ and (6.99) holds for some $b_n \in \mathbb{R}^d$ if and only if $X \in \text{RV}(B)$ for some B whose eigenvalues all have real part $a_i > 1/2$, and (6.100) holds.

Proof. Define the triangular array row elements $X_{nj} = A_n X_j$ for $j = 1, \ldots, n$. Then condition (6.22) holds (see details). If $X \in \text{RV}(B)$ for some B whose eigenvalues all have real part $a_i > 1/2$, then condition (*i*) from Theorem 6.11 holds, since this condition is identical to (6.100). The proof of condition (*ii*) is exactly the same as Theorem 6.43, using A_n in place of n^{-B}. Conversely, if $X \in \text{GDOA}(Y)$ and Y has no normal component, it follows from condition (*i*) in Theorem 6.11 that $X \in \text{RV}(B)$ and (6.102) holds. Since ϕ is a Lévy measure, (6.20) holds, and a simple estimate (a special case of [146, Lemma 7.1.7]) shows that every eigenvalue of B has real part $a_i > 1/2$.

The proof of part (a) is similar to Theorem 4.5, using the vector Karamata theorem. Condition (6.103) is equivalent to condition (*ii*) from Theorem 6.11 when condition (*i*) holds with $\phi = 0$, and the vector Karamata theorem is used to show that condition (*i*) holds with $\phi = 0$, see [146, Theorem 8.1.3]. □

Remark 6.48. The convergence criterion (6.103) in Theorem 6.47 (a) can also be stated in terms of regular variation. A real-valued (Borel measurable) function $F(x) = F(x_1, \ldots, x_d)$ on $\mathbb{R}^d$ varies regularly at $x = 0$ if

$$nF(L_n^{-1}x_n) \to \varphi(x) > 0 \quad \text{for all } x_n \to x \neq 0 \tag{6.104}$$

where

$$L_{[\lambda n]}L_n^{-1} \to \lambda^{-B} \quad \text{for all } \lambda > 0 \tag{6.105}$$

for some linear operator B whose eigenvalues all have negative real part. Then we also write $F \in \text{RV}_0(B)$. In this case, [146, Proposition 5.1.2] implies that

$$c\,\varphi(x) = \varphi(c^{-B}x) \quad \text{for all } c > 0. \tag{6.106}$$

If Y is normal and $\mathbb{E}[X] = 0$, then [146, Theorem 8.1.3] shows that (6.99) holds for some $b_n \in \mathbb{R}^d$ if and only if $F \in \text{RV}_0(-(1/2)I)$. In this case, (6.104) holds with $L_n^{-1} = A_n'$ and $\varphi(k) = k'Qk$, so that $B = -(1/2)I$, see [146, Corollary 8.1.8]. If we assume only that (6.104) holds for some sequence of invertible linear operators L_n such that $\|L_n\| \to 0$, then we can always choose L_n to be regularly varying, such that (6.105) holds, under some mild technical conditions, see [146, Theorem 5.2.16].

Remark 6.49. When Y is normal, $X \in \text{GDOA}(Y)$ implies that $\mu = \mathbb{E}[X]$ exists. The proof uses vector regular variation, see [146, Theorem 8.1.6]. In this case, we can apply Theorem 6.47 to the centered random vector $X - \mathbb{E}[X]$, and $F(k)$ is the truncated variance. Hence the assumption $\mathbb{E}[X] = 0$ entails no loss of generality. In fact, $X \in \text{GDOA}(Y)$ with Y normal implies that every one dimensional projection $X \cdot \theta$ belongs to the domain of attraction of a one dimensional normal law, see [146, Corollary 8.1.12].

Remark 6.50. If $F \in \text{RV}_0(-(1/2)I)$ and (6.103) holds, then [146, Theorem 5.3.4] implies that the truncated second moment $U_2(r, \theta) = \mathbb{E}\left[|X \cdot \theta|^2 I(|X \cdot \theta| \le r)\right]$ is slowly varying, uniformly in $\|\theta\| = 1$. That is, we have

$$\frac{U_2(\lambda r, \theta_r)}{U_2(r, \theta_r)} \to 1 \quad \text{as } r \to \infty$$

for all $\lambda > 0$ and all $\theta_r \to \theta$. Hahn and Klass [81] characterize the normal GDOA in terms of uniform slow variation of the truncated second moment.

If $A_n = a_n I$ in (6.99) then we say that X belongs to the *domain of attraction* of Y and we write $X \in \mathrm{DOA}(Y)$.

Remark 6.51. If $X \in \mathrm{DOA}(Y)$ and $Y \simeq [a, 0, \phi]$ then (6.101) reduces to

$$a_{[\lambda n]} a_n^{-1} \to \lambda^{-1/\alpha} \quad \text{for all } \lambda > 0 \tag{6.107}$$

for some $\alpha \in (0, 2)$, since $B = (1/\alpha)I$ for some $(1/\alpha) > 1/2$. Then a_n is $\mathrm{RV}(-1/\alpha)$ as in the case of random variables. Of course this must be true, for if (6.99) holds with $A_n = a_n I$, then every one dimensional projection $X \cdot \theta$ belongs to the domain of attraction of the random variable $Y \cdot \theta$ with the same sequence of norming constants.

The next result extends Theorem 4.5 to random vectors with the same power law tail behavior in every coordinate. It also shows that the scalar-normed limits of power law random walks in Theorem 6.17 cover all possible limits when $X \in \mathrm{DOA}(Y)$ and Y has no normal component. This verifies that Proposition 6.18 describes all stable random vectors with index $0 < \alpha < 2$, $\alpha \neq 1$.

Theorem 6.52. *If $X \in \mathrm{DOA}(Y)$, then Y is either normal, or stable with some index $0 < \alpha < 2$, and:*
(a) If Y is normal, then $\mu = E[X]$ exists and $X \in \mathrm{DOA}(Y)$ if and only if

$$nF(a_n k_n) \to k' Q k \quad \text{for all } k_n \to k \neq 0 \tag{6.108}$$

where $F(k) = \mathbb{E}[|(X - \mu) \cdot k|^2 I(|(X - \mu) \cdot k| \leq 1)]$;
(b) If Y is stable, then $X \in \mathrm{DOA}(Y)$ and (6.99) holds with $A_n = a_n I$ for some $b_n \in \mathbb{R}^d$ if and only if $V(r) = \mathbb{P}[\|X\| > r]$ is regularly varying with index $-\alpha$, and

$$P\left[\tfrac{X}{\|X\|} \in D \middle| \|X\| > r\right] = \frac{P[\|X\| > r, \frac{X}{\|X\|} \in D]}{V(r)} \to \frac{\Lambda(D)}{\Lambda(S)} \tag{6.109}$$

for some σ-finite Borel measure $\Lambda(d\theta)$ on the unit sphere. Then $Y \simeq S_\alpha(\Lambda, \mu)$ in the notation of Proposition 6.18, for some $\mu \in \mathbb{R}^d$ depending on (a_n).

Proof. Part (a) follows using Remark 6.49 and applying Theorem 6.47 (a) to $X - \mu$. Part (b) follows from Theorem 6.47 (b) with $B = (1/\alpha)I$. With this exponent, it follows from (6.102) and Remark 6.39 that

$$\phi\{t^{1/\alpha}\theta : t > r, \theta \in V\} = r^{-1}\Lambda(V) \tag{6.110}$$

for all $r > 0$ and all Borel subsets V of the unit sphere, where the spectral measure $\Lambda(V) = \phi\{r\theta : r > 1, \theta \in V\}$. Substitute $s = t^{1/\alpha}$ to see that

$$\phi\{s\theta : s > r, \theta \in V\} = r^{-\alpha}\Lambda(V). \tag{6.111}$$

Then Proposition 6.18 shows that $Y \simeq S_\alpha(\Lambda, \mu)$. A regular variation argument shows that (6.109) is equivalent to (6.100) with $A_n = a_n I$ and limit measure (6.111). The argument is similar to Proposition 4.15, see [146, Theorem 8.2.18]. □

Remark 6.53. When $X \in \mathrm{GDOA}(Y)$ and Y is normal, we can always choose A_n in (6.99) to be regularly varying with index $-(1/2)I$, as noted in Remark 6.48. Then we can also write $A_n = n^{-1/2}G_n$ where (G_n) is slowly varying, so that $G_{[\lambda n]}G_n^{-1} \to I$ as $n \to \infty$ for all $\lambda > 0$. If we write $G_n x = r_n \theta_n$ for $r_n > 0$ and $\|\theta_n\| = 1$, then r_n is slowly varying, and θ_n is *very slowly varying*, i.e., each coordinate of θ_{e^n} is slowly varying. Roughly speaking $G_n x$ grows like $\log n$, and rotates like $\log\log n$. The same is true for $X \in \mathrm{GDOA}(Y)$ and Y stable, except that we write $A_n = n^{-1/\alpha}G_n$. For more details, see [146, Corollary 8.1.14]. Hahn and Klass [80, 81] provide examples to show that the GDOA of a spherically symmetric normal or stable law is strictly larger than the DOA, i.e., there exist X such that the convergence (6.99) requires operator norming.

Suppose $X \in \mathrm{GDOA}(Y)$ and take (Y_n) iid with Y. The term *operator stable* comes from the fact [146, Theorem 7.2.1] that for all n, for some $b_n \in \mathbb{R}^d$, we have

$$Y_1 + \cdots + Y_n \simeq n^B Y + b_n \tag{6.112}$$

where B is any *exponent* of Y. That is, (6.99) holds with $S_n = Y_1 + \cdots + Y_n$, $A_n = n^{-B}$, and convergence in distribution is replaced by the stronger condition of equality in distribution. If $Y \simeq \mathcal{N}(a, Q)$, we can take $B = (1/2)I$. If Y is stable, then (6.112) holds with $B = (1/\alpha)I$. If Y is operator stable with no normal component and (6.102) holds, then (6.112) holds with the same exponent B [146, Corollary 8.2.11]. Exponents need not be unique, because of symmetry. For example, it follows by a computation similar to Example 6.35 that $B = (1/2)I + Q$ is an exponent of $Y \simeq \mathcal{N}(a, cI)$ for any skew-symmetric matrix Q. The exact relation between exponents and symmetries is given in [146, Theorem 7.2.11].

Example 6.54. A general operator stable law can have both a normal component and a non-normal Poissonian component. For example, suppose $X \in \mathrm{GDOA}(Y)$ and X has independent components, which are either Pareto with tail index $0 < \alpha < 2$, or have a finite variance. Then it follows from Theorems 3.36 and 3.37 that (6.99) holds with $A_n = \mathrm{diag}(n^{-1/\alpha_1}, \ldots, n^{-1/\alpha_d})$ where $\alpha_i = 2$ for the finite variance components, and $\alpha_i \in (0, 2)$ for the heavy tailed components. Make a simple change of coordinates so that $\alpha_1 \geq \ldots \geq \alpha_d$. Then we can write the norming operator in block-diagonal form

$$A_n = \begin{pmatrix} A_n^1 & 0 \\ 0 & A_n^2 \end{pmatrix}$$

with $A_n^2 = \mathrm{diag}(n^{-1/\alpha_{m+1}}, \ldots, n^{-1/\alpha_d})$ and $A_n^1 = n^{-1/2}I_m$, where I_m is the $m \times m$ identity matrix. The limit $Y = (Y^1, Y^2)'$ where Y^1 is an m dimensional normal random vector on a subspace of $\mathbb{R}^d$, Y^2 is a $d - m$ dimensional operator stable random vector with no normal component on another subspace, and the intersection of these two subspaces is the single point $x = 0$. The exponent of this operator stable random vector is

$$B = \begin{pmatrix} B^1 & 0 \\ 0 & B^2 \end{pmatrix}$$

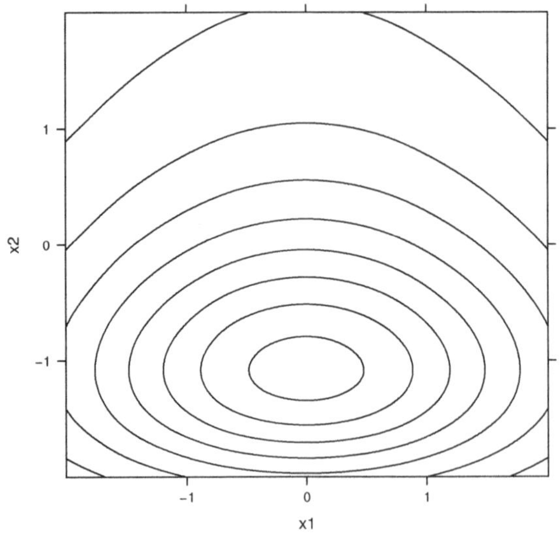

Fig. 6.10: Level sets of the solution $p(x, t)$ to the fractional diffusion equation (6.113) at time $t = 1$ in dimension $d = 2$, with $\alpha_1 = 2.0$, $\alpha_2 = 1.4$, and $D_1 = D_2 = 1$.

where $B^1 = (1/2)I_m$ and $B^2 = \text{diag}(1/\alpha_{m+1}, \ldots, 1/\alpha_d)$. The density $p(x, t)$ of the corresponding operator stable Lévy process $Z_t = (Z_t^1, Z_t^2)'$ with $Z_1 \simeq Y$ solves the fractional diffusion equation

$$\frac{\partial}{\partial t} p(x, t) = \sum_{j=1}^{d} \left[D_j \frac{\partial^{\alpha_j}}{\partial x_j^{\alpha_j}} p(x, t) \right] \tag{6.113}$$

where $\alpha_i = 2$ for $1 = 1, 2, \ldots, m$, $0 < \alpha_i < 2$ for $i > m$, $D_i < 0$ for $0 < \alpha_i < 1$, and $D_i > 0$ for $1 < \alpha_i \le 2$ (here we assume $\alpha_i \ne 1$). Figure 6.10 shows level sets of a typical solution $p(x, t)$ in $\mathbb{R}^2$ with $\alpha_1 = 2.0$ and $\alpha_2 = 1.4$, obtained using the R code from Figure 6.19.

The *spectral decomposition* takes Example 6.54 one step further. Suppose that Y is operator stable with exponent B and (6.112) holds. Theorem 7.2.1 in [146] shows that every eigenvalue $\lambda_j = a_j + ib_j$ of the exponent B has real part $a_i \ge 1/2$. Make a change of coordinates so that $a_1 \le \cdots \le a_d$ and write

$$B = \begin{pmatrix} B^1 & & \\ & \ddots & \\ & & B^p \end{pmatrix}$$

where p is the number of distinct values of a_i, every eigenvalue of the $m_j \times m_j$ matrix B^j has real part a_j, and $m_1 + \cdots + m_p = d$. Projecting (6.112) onto m_j-dimensional subspaces shows that $Y = (Y^1, \ldots, Y^p)'$, where each component Y^j is an m_j-dimensional operator stable random vector with exponent B^j, see [146, Theorem 7.2.9]. Since every eigenvalue of B^j has the same real part a_j, we say that the operator stable law Y^j is *spectrally simple*.

Furthermore, [146, Theorem 8.3.24] shows that we can take

$$A_n = \begin{pmatrix} A_n^1 & & \\ & \ddots & \\ & & A_n^p \end{pmatrix}$$

in these coordinates, where every A_n^j is RV$(-B^j)$. Then $X = (X^1, \ldots, X^p)'$ and we can apply Theorem 6.47 (a) to the normal component. Theorem 6.47 (b) describes each spectrally simple operator stable component, and (6.93) implies that the tails of X^j fall off like $r^{-\alpha_j}$ where $\alpha_j = a_j^{-1} \in (0, 2)$. The tails of a spectrally simple operator stable law need not be regularly varying, but they are R-O varying with the same upper and lower tail index, see [146, Theorem 6.4.15] for complete details. It follows from the Lévy representation (3.4) that the normal component is independent of the remaining components. The dependence of the remaining non-normal spectrally simple operator stable components is coded through the Lévy measure.

Suppose that (6.99) holds with $b_n = 0$. Then it follows as in the proof of Theorem 6.21 that we also get random walk convergence

$$A_n S_{[nt]} \Rightarrow Z_t$$

where Z_t is an operator stable Lévy motion with $Z_1 \simeq Y$. If every $a_i > 1$, then the density $p(x, t)$ of Z_t solves the operator scaling fractional diffusion equation

$$\frac{\partial}{\partial t} p(x, t) = C \nabla_M^B p(x, t) \tag{6.114}$$

for some $C < 0$, where the generalized fractional derivative ∇_M^B is given by (6.97). If every $a_i \in (1/2, 1)$, then $p(x, t)$ solves (6.114) for some $C > 0$, with ∇_M^B given by (6.98). Add a drift to see that the density $p(x, t)$ of $vt + Z_t$ solves the *generalized fractional advection-dispersion equation* (GADE)

$$\frac{\partial}{\partial t} p(x, t) = -v \cdot \nabla p(x, t) + C \nabla_M^B p(x, t). \tag{6.115}$$

Applications of operator stable laws and the GADE will be discussed in Section 7.12.

Details

Since X_j is tight for any fixed j, equation (6.29) holds with $X = X_j$. Write

$$\mathbb{P}[\|X_{nj}\| > \varepsilon] = \mathbb{P}[\|A_n X_j\| > \varepsilon] = \mathbb{P}[\|X_j\| \in A_n^{-1} B_\varepsilon]$$

where the set $B_\varepsilon = \{x \in \mathbb{R}^d : \|x\| > \varepsilon\}$. If $X \in \text{GDOA}(Y)$ and Y is full, then a simple argument with characteristic functions (a special case of [146, Lemma 3.3.3]) shows that $\|A_n\| \to 0$ as $n \to \infty$. Since $\|x\| = \|A_n A_n^{-1} x\| \le \|A_n\| \, \|A_n^{-1} x\|$, it follows that $\|A_n^{-1} x\| > \varepsilon / \|A_n\| \to \infty$ for all $x \in B_\varepsilon$, and then it follows that condition (6.22) holds.

```
library(lattice)
x = seq(-2,2,.01)
y = seq(-2,2,.01)
u <- dnorm(x, mean = 0.0, sd = 2.0)
v <- dnorm(x, mean = 0.0, sd = 2.0)
r <- as.vector(outer(u, v, FUN = "*"))
grid <- expand.grid(x=x, y=y)
grid$z <- r
levelplot(z~x*y, grid, cuts = 8,
region=FALSE, contour=TRUE, labels=FALSE)
```

Fig. 6.11: R code to plot the isotropic two dimensional Gaussian density with independent components in Figure 6.1.

```
library(lattice)
x = seq(-2,2,.01)
y = seq(-2,2,.01)
u <- dnorm(x, mean = 0.0, sd = 2.0)
v <- dnorm(x, mean = 0.0, sd = 1.0)
r <- as.vector(outer(u, v, FUN = "*"))
grid <- expand.grid(x=x, y=y)
grid$z <- r
levelplot(z~x*y, grid, cuts = 8,
region=FALSE, contour=TRUE, labels=FALSE)
```

Fig. 6.12: R code to plot the anisotropic two dimensional Gaussian density with independent components in Figure 6.2.

```
library(lattice)
library(stabledist)
D1=0.5 ; D2=0.5
v1=0.0 ; v2=0.0
a1=1.2 ; a2=1.2
q1=0.5 ; q2=0.5
t=5.0
mu1=v1*t ; mu2=v2*t
pi=3.1415927
g1=(D1*t*abs(cos(pi*a1/2)))^(1/a1)
g2=(D2*t*abs(cos(pi*a2/2)))^(1/a2)
b1=1-2*q1 ; b2=1-2*q2
x = seq(-2,2,.01)
y = seq(-2,2,.01)
u = dstable(x, alpha=a1, beta=b1, gamma = g1, delta = mu1, pm=1)
v = dstable(y, alpha=a2, beta=b2, gamma = g2, delta = mu2, pm=1)
r = as.vector(outer(u, v, FUN = "*"))
grid = expand.grid(x=x, y=y)
grid$z = r
levelplot(z~x*y, grid, cuts = 8, region=FALSE,
          contour=TRUE, labels=FALSE)
```

Fig. 6.13: R code to plot level curves of the solution $p(x, y, t)$ to the two dimensional fractional diffusion equation (6.16) shown in Figure 6.3.

```
library(lattice)
library(stabledist)
D1=0.5 ; D2=0.5
v1=0.0 ; v2=0.0
a1=0.8 ; a2=0.6
q1=0.0 ; q2=0.0
t=3.0
#
mu1=v1*t ; mu2=v2*t
pi=3.1415927
g1=(D1*t*abs(cos(pi*a1/2)))^(1/a1)
g2=(D2*t*abs(cos(pi*a2/2)))^(1/a2)
b1=1-2*q1 ; b2=1-2*q2
x1 = seq(0,4,.01)
x2 = seq(0,4,.01)
u <- dstable(x1, alpha=a1, beta=b1, gamma=g1, delta=mu1, pm=1)
v <- dstable(x2, alpha=a2, beta=b2, gamma=g2, delta=mu2, pm=1)
r <- as.vector(outer(u, v, FUN = "*"))
grid <- expand.grid(x1=x1, x2=x2)
grid$z <- r
levelplot(z~x1*x2, grid, cuts = 12, region=FALSE,
          contour=TRUE, labels=FALSE)
```

Fig. 6.14: R code to plot level curves of the solution $p(x, t)$ to the vector fractional diffusion equation (6.78) shown in Figure 6.5.

```
a1=0.7; a2=1.2
t=seq(0.001,10,.1)
x1=t^a1
x2=t^a2
plot(x1,x2,type="l",xlim=c(-3,3),ylim=c(-3,3))
lines(-x1,-x2,type="l")
lines(-x1,x2,type="l")
lines(x1,-x2,type="l")
lines(x1,0*x2,type="l")
lines(-x1,0*x2,type="l")
lines(0*x1,x2,type="l")
lines(0*x1,-x2,type="l")
theta=seq(0,6.29,.1)
x1=cos(theta)
x2=sin(theta)
lines(x1,x2,lty="dashed")
```

Fig. 6.15: R code to plot orbits $t \mapsto t^A x$ for Example 6.33.

```
a=0.5
t=seq(0.0001,10,.01)
x1=t^a*log(t)
x2=t^a
plot(x1,x2,type="l",xlim=c(-3,3),ylim=c(-3,3))
lines(-x1,-x2,type="l")
lines(3*x1,3*x2,type="l")
lines(-3*x1,-3*x2,type="l")
lines(x1,0*x2,type="l")
lines(-x1,0*x2,type="l")
t=seq(0,6.29,.1)
x1=cos(t)
x2=sin(t)
lines(x1,x2,lty="dashed")
```

Fig. 6.16: R code to plot orbits $t \mapsto t^A x$ for Example 6.34.

```
a=0.5
t=seq(-8,4,.05)
x1=exp(a*t)*cos(t)
x2=exp(a*t)*sin(t)
plot(x1,x2,type="l",xlim=c(-3,3),ylim=c(-3,3))
lines(-x1,-x2,type="l")
x1=-exp(a*t)*sin(t)
x2=exp(a*t)*cos(t)
lines(x1,x2,type="l")
lines(-x1,-x2,type="l")
t=seq(0,6.29,.1)
x1=cos(t)
x2=sin(t)
lines(x1,x2,lty="dashed")
```

Fig. 6.17: R code to plot orbits $t \mapsto t^A x$ for Example 6.35.

```
a1=0.7; a2=1.2
t=seq(0.001,10,.1)
x1=t^a1
x2=t^a2
plot(x1,x2,type="l",xlim=c(-3,3),ylim=c(-3,3))
lines(-x1,-x2,type="l")
lines(-x1,x2,type="l")
lines(x1,-x2,type="l")
lines(x1,0*x2,type="l")
lines(-x1,0*x2,type="l")
lines(0*x1,x2,type="l")
lines(0*x1,-x2,type="l")
theta=seq(0,6.3,.1)
x1=cos(theta)
x2=sin(theta)
lines(x1,x2,lty="dashed")
lines(2^a1*x1,2^a2*x2,lty="dashed")
lines(.5^a1*x1,.5^a2*x2,lty="dashed")
```

Fig. 6.18: R code to plot Jurek coordinates for Remark 6.40.

```
library(lattice)
library(stabledist)
D1=1.0 ; D2=1.0
v1=0.0 ; v2=0.0
a1=2.0 ; a2=1.4
q1=0.0 ; q2=0.0
t=1.0
#
mu1=v1*t ; mu2=v2*t
pi=3.1415927
g1=(D1*t*abs(cos(pi*a1/2)))^(1/a1)
g2=(D2*t*abs(cos(pi*a2/2)))^(1/a2)
b1=1-2*q1 ; b2=1-2*q2
x1 = seq(-2,2,.01)
x2 = seq(-2,2,.01)
u <- dstable(x1, alpha=a1, beta=b1, gamma=g1, delta=mu1, pm=1)
v <- dstable(x2, alpha=a2, beta=b2, gamma=g2, delta=mu2, pm=1)
r <- as.vector(outer(u, v, FUN = "*"))
grid <- expand.grid(x1=x1, x2=x2)
grid$z <- r
levelplot(z~x1*x2, grid, cuts = 8, region=FALSE,
          contour=TRUE, labels=FALSE)
```

Fig. 6.19: R code to plot level curves of the solution $p(x, t)$ to the fractional diffusion equation (6.113) shown in Figure 6.10.

7 Applications and Extensions

In this final chapter, we discuss a few of the many applications and extensions being developed today in the rapidly growing research area of fractional diffusion.

7.1 The fractional Poisson process

The classical Brownian motion and Poisson process are the most famous and useful continuous time stochastic processes in both theory and applications. In this section, we discuss the fractional Poisson process. It generalizes the classical Poisson process to allow long waiting times between events.

In Example 3.2 we showed that a Poisson random variable Y with mean $\mathbb{E}[Y] = \lambda > 0$ is infinitely divisible, with characteristic function $\mathbb{E}[e^{ikY}] = e^{\psi(k)}$ where $\psi(k) = \lambda[e^{ik} - 1]$. The classical Poisson process $N(t)$ is a Lévy process with $\mathbb{E}[e^{ikN(t)}] = e^{t\psi(k)}$, so that $N(t)$ has a Poisson distribution with mean λt. The family of linear operators $T_t f(x) = \mathbb{E}[f(x - N(t))]$ forms a semigroup on the Banach space $C_0(\mathbb{R})$. Given a discrete random variable X with probability mass function (pmf) $f(x)$, the semigroup $T_t f(x)$ gives the pdf of the discrete random variable $X + N(t)$, a Poisson process with random initial state. Let $p(n, t) = \mathbb{P}[N(t) = n]$ denote the pmf of the discrete random variable $N(t)$. Then $p(0, 0) = 1$ and $p(n, 0) = 0$ for $n = 1, 2, 3, \ldots$, and we can take FT to see that $\hat{p}(k, t) = \mathbb{E}[e^{-ikN(t)}] = e^{t\psi(-k)}$ for all $t \geq 0$. Use the convolution property (2.1) of the FT to see that

$$q(x, t) = T_t f(x) = \sum_{n=0}^{\infty} f(x - n)p(n, t)$$

has FT $\hat{q}(k, t) = e^{t\psi(-k)}\hat{f}(k)$ for all $t \geq 0$. It follows that

$$\frac{\partial}{\partial t}\hat{q}(k, t) = \psi(-k)\hat{q}(k, t) = \lambda(e^{-ik} - 1)\hat{q}(k, t) \tag{7.1}$$

for all $t \geq 0$. Invert (7.1) using the shift property (3.25) of the FT to see that

$$\frac{\partial}{\partial t}q(x, t) = \lambda\left[q(x - 1, t) - q(x, t)\right]; \quad q(x, 0) = f(x). \tag{7.2}$$

for all $t \geq 0$.

Equation (7.2) is a Cauchy problem on the Banach space $C_0(\mathbb{R})$: $\partial q/\partial t = Lq$; $q(0) = f$ where the generator $Lf(x) = \lambda[f(x - 1) - f(x)]$. By comparing the Fourier symbol $\psi(k) = \lambda[e^{ik} - 1]$ with the Lévy representation (3.4), it is easy to see that $N(1) \simeq [a, b, \phi]$ with $b = 0$, $\phi\{1\} = \lambda$ is a single point mass, and

$$a = \int \frac{y}{1 + y^2}\phi(dy) = \frac{\lambda}{2}.$$

Then the generator can also obtained from Theorem 3.17. If $f(0) = 1$, then $q(x, t) = T_t f(x) = p(x, t)$, and it is easy to check that the Poisson distribution

$$p(n, t) = \mathbb{P}[N(t) = n] = e^{-\lambda t}\frac{(\lambda t)^n}{n!} \tag{7.3}$$

solves the Cauchy problem (7.2) with initial condition $q(n, 0) = f(n)$.

The fractional Poisson process can be defined by $N_\beta(t) = N(E_t)$ where E_t is the inverse

$$E_t = \inf\{u > 0 : D(u) > t\} \tag{7.4}$$

of a standard stable subordinator $D(u)$, so that

$$\mathbb{E}[e^{-sD(u)}] = e^{-us^\beta} \tag{7.5}$$

for all $s > 0$, for some $0 < \beta < 1$. We assume that $D(u)$ and hence E_t are independent of $N(t)$, and then a simple conditioning argument shows that the pmf of the fractional Poisson process is given by

$$m(n, t) = \mathbb{P}[N_\beta(t) = n] = \int_0^\infty p(n, u)h(u, t)\, du \tag{7.6}$$

where $h(u, t)$ is the pdf of the inverse stable subordinator $u = E_t$. Then, using (4.47) and (4.48), we can write the pmf of the fractional Poisson process $N_\beta(t) = N(E_t)$ as

$$m(n, t) = \int_0^\infty p(n, u)\frac{t}{\beta}u^{-1-1/\beta}g_\beta(tu^{-1/\beta})\, du = \int_0^\infty p(n, (t/r)^\beta)g_\beta(r)\, dr \tag{7.7}$$

where $g_\beta(r)$ is the pdf of $D(1)$, and $p(n, t)$ is the Poisson pmf given by (7.3).

Take FT in (7.6) to see that

$$\hat{m}(k, t) = \mathbb{E}[e^{-ikN_\beta(t)}] = \int_0^\infty \hat{p}(k, u)h(u, t)\, du = \int_0^\infty e^{u\psi(-k)}h(u, t)\, du \tag{7.8}$$

and then take LT using (4.42) to see that the FLT

$$\begin{aligned}\bar{m}(k,s) &= \int_0^\infty e^{-st}\hat{m}(k,t)\,dt\\ &= \int_0^\infty e^{-st}\int_0^\infty e^{u\psi(-k)}h(u,t)\,du\,dt\\ &= \int_0^\infty\left(\int_0^\infty e^{-st}h(u,t)\,dt\right)e^{u\psi(-k)}\,du\\ &= \int_0^\infty s^{\beta-1}e^{-us^\beta}e^{u\psi(-k)}\,du\\ &= s^{\beta-1}\int_0^\infty e^{-u[s^\beta-\psi(-k)]}du = \frac{s^{\beta-1}}{s^\beta-\psi(-k)}\end{aligned} \tag{7.9}$$

by Fubini, a special case of (4.43). Rewrite (7.9) in the form

$$s^\beta\bar{m}(k,s) - s^{\beta-1} = \psi(-k)\bar{m}(k,s) = \lambda(e^{-ik}-1)\bar{m}(k,s)$$

and invert the LT to get

$$\partial_t^\beta\hat{m}(k,t) = \lambda(e^{-ik}-1)\hat{m}(k,t)$$

where ∂_t^β is the Caputo fractional derivative. Here we use the fact that $\hat{m}(k,0) = 1$. Then invert the FT, using the shift property (3.25) of the FT, to see that the pmf $m(n,t)$ of the fractional Poisson process solves the time-fractional equation

$$\partial_t^\beta m(n,t) = \lambda\,[m(n-1,t) - m(n,t)]\,. \tag{7.10}$$

Equation (7.10) explains why we call $N_\beta(t)$ a fractional Poisson process: Its pmf solves the time-fractional analogue (7.10) of the Cauchy problem (7.2) for the classical Poisson process.

The form $N(E_t)$ of the fractional Poisson process is quite similar to the CTRW limit $A(E_t)$ in Chapter 4. The only difference is that the outer process is now a Poisson process, instead of a normal or stable Lévy process. In fact, the fractional Poisson process is also a kind of CTRW limit. But in order to get the Poisson process as the limit of the jump process, we have to use triangular arrays. This is necessary in view of Theorem 4.5, which states that ordinary random walk limits are either normal or stable. Consider a triangular array $\{X_{nj} : j = 1,\ldots,k_n; n = 1,2,3,\ldots\}$ of CTRW jumps, where X_{nj} is Poisson with mean λ/n. The CTRW row sums

$$S_n(k_n) = X_{n1} + \cdots + X_{nk_n}$$

are then Poisson with mean $\lambda k_n/n$. Taking $k_n = [nt]$, it follows that

$$S_n([nt]) \Rightarrow N(t) \tag{7.11}$$

as $n \to \infty$ for any $t \geq 0$ (see details). Argue as in Section 4.3 that (7.11) holds in the sense of finite dimensional distributions, and then apply Theorem 3 in Bingham [38], as in Section 4.4, to conclude that (7.11) holds in $\mathbb{D}[0, \infty)$ with the Skorokhod J_1 topology.

Next assume iid waiting times $J_n \simeq J$ with a Pareto distribution $\mathbb{P}[J > t] = Bt^{-\beta}$ where $0 < \beta < 1$ and $B = 1/\Gamma(1-\beta)$. Let $T_n = J_1 + \cdots + J_n$ be the time of the nth CTRW jump, and let $R(t) = \max\{n \geq 0 : T_n \leq t\}$ denote the number of jumps by time $t \geq 0$. It follows from (4.30) with $c = n^{1/\beta}$ that

$$n^{-1}R(n^{1/\beta}t) \Rightarrow E_t \tag{7.12}$$

as $n \to \infty$ in the Skorokhod space $\mathbb{D}[0, \infty)$ with the J_1 topology, where where E_t is the inverse (7.4) of the standard stable subordinator $D(t)$ with index $0 < \beta < 1$. Since waiting times and jumps are independent, it follows that

$$(S_n([nt]), n^{-1}R(n^{1/\beta}t)) \Rightarrow (N(t), E_t)$$

in the J_1 topology. Since $N(t)$ and $D(t)$ are independent Lévy processes, they have almost surely no simultaneous jumps. Then it follows from the continuous mapping theorem and Theorem 13.2.4 in Whitt [219], as in Section 4.4, that

$$S_n(R(n^{1/\beta}t)) = S_n(n \cdot n^{-1}R(n^{1/\beta}t) \Rightarrow N(E_t)$$

in the Skorokhod M_1 topology. This shows that the fractional Poisson process is a CTRW limit.

Actually, the fractional Poisson process is itself a CTRW. The Poisson process $N(t)$ is a CTRW with iid exponential waiting times $\mathbb{P}[J_n > t] = e^{-\lambda t}$ and deterministic jumps $Y_n = 1$, see for example Ross [179, Proposition 5.1]. The fractional Poisson process $N_\beta(t)$ is a CTRW with the same deterministic jumps, and iid waiting times

$$\mathbb{P}[W_n > t] = E_\beta(-\lambda t^\beta) \tag{7.13}$$

for some $0 < \beta < 1$, using the Mittag-Leffler function (2.29). This definition of the fractional Poisson process is due to Laskin [115]. It can be motivated by the fact that the Mittag-Leffler function is a natural extension of the exponential, due to its power series definition. Mainardi and Gorenflo [129, Eq. (5.26)] show that the tail of the Mittag-Leffler function satisfies

$$E_\beta(-\lambda t^\beta) \sim \frac{t^{-\beta}}{\lambda\Gamma(1-\beta)} \quad \text{as } t \to \infty,$$

and it follows that the waiting times of the fractional Poisson process are heavy tailed.

To see that this CTRW is a fractional Poisson process, let

$$\tau_n = \sup\{t > 0 : N(E_t) < n\} \tag{7.14}$$

denote the time when the fractional Poisson process enters the state n. Then it suffices to show that $\tau_n = W_1 + \cdots + W_n$ where the waiting times W_n between state transitions are iid Mittag-Leffler distributed, as in (7.13). Note that $\{N(t) < n\} = \{T_n > t\}$ for the classical Poisson process. In other words, there have been less than n arrivals by time $t > 0$ if and only if the time of the nth arrival is greater than t. Apply this to (7.14), using the fact that E_t is independent of the Poisson process $N(t)$, to conclude that

$$\tau_n = \sup\{t > 0 : E_t < T_n\}. \tag{7.15}$$

The rest of the argument is delicate, involving sample paths properties of the process $t = D(u)$ and its inverse $u = E_t$. We will give a heuristic argument here, see [141, Theorem 2.1] for complete details: Since $u = E_t = T_n$ at $t = \tau_n$, we have $\tau_n = t = D(u) = D(T_n)$. Recall that the LT of the exponential distribution is

$$\mathbb{E}(e^{-sJ_n}) = \int_0^\infty e^{-st}\lambda e^{-\lambda t}dt = \frac{\lambda}{\lambda + s}$$

for any $\lambda > 0$ and $s > 0$. Then we can use (7.5) to write

$$\mathbb{E}[e^{-s\tau_1}] = \mathbb{E}[e^{-sD(T_1)}] = \mathbb{E}\left[\mathbb{E}\left(e^{-sD(J_1)}|J_1\right)\right] = \mathbb{E}\left[e^{-J_1 s^\beta}\right] = \frac{\lambda}{\lambda + s^\beta} \tag{7.16}$$

for any $\lambda > 0$ and $s > 0$.

Next we will show that the Mittag-Leffler random variable W_n in (7.13) has the same LT as $D(T_1)$. Recall from (2.31) that the function $G(t) = E_\beta(-\lambda t^\beta)$ has LT

$$\tilde{G}(s) = \frac{s^{\beta-1}}{s^\beta + \lambda}$$

for any $\lambda > 0$ and any $s^\beta > \lambda$. Let $f_\beta(t) = \partial_t[1 - G(t)]$ be the Mittag-Leffler pdf of W_n. Integrate by parts to see that

$$\begin{aligned}\mathbb{E}(e^{-sW_n}) &= \int_0^\infty e^{-st} f_\beta(t)\, dt \\ &= \int_0^\infty se^{-st}\,(1 - G(t))\, dt \\ &= s\left[\frac{1}{s} - \frac{s^{\beta-1}}{\lambda + s^\beta}\right] = \frac{\lambda}{\lambda + s^\beta}\end{aligned} \tag{7.17}$$

for any $\lambda > 0$ and any $s^\beta > \lambda$. Then the uniqueness theorem for LT (moment generating functions) implies that $W_1 \simeq \tau_1$. The general case $n > 1$ involves computing the joint LT of $\tau_1, \tau_2, \ldots, \tau_n$, see [141, Theorem 2.1] for complete details.

Remark 7.1. The heuristic formula $\tau_n = D(T_n)$ is not exactly true. Rather, we have $\tau_n = D(T_n-)$, which can be different if the process $D(u)$ has a jump at $u = T_n$. Then we will have $D(u) > D(u-)$ since $D(u)$ is a strictly increasing element of the Skorokhod space $\mathbb{D}[0, \infty)$, and hence is right-continuous with left-hand limits. However, we do have $\mathbb{P}[D(u-) = D(u)] = 1$ since $D(u)$ is a Lévy process, and therefore has no fixed points of discontinuity. Since T_n is independent of $D(u)$, it follows that $D(T_n) \simeq D(T_n-)$ and hence the LT of the pdf of $\tau_n = D(T_n-)$ is the same as that of $D(T_n)$, and in particular, (7.16) is the correct LT for τ_1. See [141, Theorem 2.1] for more details.

Remark 7.2. The CTRW representation of the fractional Poisson process can be useful for simulations. One only needs to compute the jump times $\tau_n = W_1 + \cdots + W_n$ by simulating iid Mittag-Leffler waiting times, e.g., using the `MittagLeffleR` package in R. This gives the exact sample paths of the fractional Poisson process. It can also be useful to consider a fractional compound Poisson process $S(N(E_t))$ where $S_n = X_1 + \cdots + X_n$ is the sum of iid random variables or vectors. This process jumps to the point $S(n)$ at time τ_n, and hence it can also be simulated exactly using the same approach.

Details

To show that (7.11) holds, note that the left-hand side is Poisson with mean $\lambda[nt]/n$, and hence its characteristic function

$$\mathbb{E}[e^{ikS_n([nt])}] = e^{\lambda\psi(k)[nt]/n} \to e^{\lambda t\psi(k)} = \mathbb{E}[e^{ikN(t)}],$$

since the limit $N(t)$ is Poisson with mean λt. Then it follows from Theorem 1.3 that (7.11) holds.

7.2 LePage series representation

As an application of the Poisson representation in Section 3.4, we now develop a very interesting *series representation* for stable laws and their domains of attraction. Suppose that (W_j) are iid Pareto with $\mathbb{P}[W_j > x] = Cx^{-\alpha}$ for some $0 < \alpha < 1$. Then Theorem 3.37 shows that

$$n^{-1/\alpha} \sum_{j=1}^{n} W_j \Rightarrow Y \tag{7.18}$$

where the α-stable limit Y has characteristic function

$$\mathbb{E}[e^{ikY}] = \exp\left[-C\Gamma(1-\alpha)(-ik)^\alpha\right]. \tag{7.19}$$

Suppose that (U_j) are iid uniform random variables on $(0, 1)$ with $\mathbb{P}[U_j \le x] = x$ for $0 \le x \le 1$. Then we can take

$$W_j = (U_j/C)^{-1/\alpha}$$

since

$$\mathbb{P}[(U_j/C)^{-1/\alpha} > x] = \mathbb{P}[U_j^{-1/\alpha} > C^{-1/\alpha}x]$$
$$= \mathbb{P}[U_j < Cx^{-\alpha}] = Cx^{-\alpha}$$

for all $x > C^{1/\alpha}$, as noted in Example 5.17. This is a special case of the inverse cdf method for simulating random variables: If $F(x) = \mathbb{P}[X \le x]$ and U is uniform on $(0,1)$, then $F^{-1}(U) \simeq X$ (e.g., see Press et al. [170, Chapter 7]).

Suppose N_t is a Poisson process with rate $\lambda = 1$. Take (E_n) iid with $\mathbb{P}[E_n > t] = e^{-t}$, the waiting times between jumps for this process, and let

$$\Gamma_n = E_1 + \cdots + E_n$$

be the time of the nth jump. Then we have the inverse process relation $\{N_t \ge n\} = \{\Gamma_n \le t\}$ as in Section 4.3. Now let

$$U_{(1)} \le \cdots \le U_{(n)}$$

denote the *order statistics* of the sequence $U_1, \ldots, U_n$. A standard result in extreme value theory (e.g., see Resnick [175, p. 322] or Bickel and Doksum [32]) states that

$$\left(\frac{\Gamma_1}{\Gamma_{n+1}}, \ldots, \frac{\Gamma_n}{\Gamma_{n+1}}\right) \simeq (U_{(1)}, \ldots, U_{(n)}).$$

That is, the first n arrival times are uniformly distributed in the interval $[0, \Gamma_{n+1}]$.

Now write

$$\begin{aligned}
n^{-1/\alpha}\sum_{j=1}^{n} W_j &= n^{-1/\alpha}\sum_{j=1}^{n}(U_j/C)^{-1/\alpha} \\
&= n^{-1/\alpha}\sum_{j=1}^{n}(U_{(j)}/C)^{-1/\alpha} \\
&\simeq C^{1/\alpha}n^{-1/\alpha}\sum_{j=1}^{n}(\Gamma_j/\Gamma_{n+1})^{-1/\alpha} \\
&= C^{1/\alpha}\left(\frac{\Gamma_{n+1}}{n}\right)^{1/\alpha}\sum_{j=1}^{n}\Gamma_j^{-1/\alpha} \Rightarrow Y
\end{aligned}$$

where Y is a stable random variable with characteristic function (7.19). The strong law of large numbers implies that

$$\frac{\Gamma_{n+1}}{n} = \frac{E_1 + \cdots + E_{n+1}}{n+1} \cdot \frac{n+1}{n} \to 1 \quad \text{almost surely, as } n \to \infty.$$

Then it follows using the Continuous Mapping Theorem 4.19 that

$$C^{1/\alpha}\sum_{j=1}^{n}\Gamma_j^{-1/\alpha} \Rightarrow Y.$$

In other words, the infinite series converges in distribution to a stable random variable with characteristic function (7.19):

$$C^{1/\alpha}\sum_{j=1}^{\infty}\Gamma_j^{-1/\alpha} \simeq S_\alpha(1,\sigma,0) \tag{7.20}$$

where $\sigma^\alpha = C\Gamma(1-\alpha)\cos(\pi\alpha/2)$. In fact, the series converges almost surely, see LePage, Woodroofe, and Zinn [124].

Remark 7.3. The argument above can be extended to any $W \in \mathrm{DOA}(Y)$ with Y stable. Suppose $a_n(W_1 + \cdots + W_n) \Rightarrow Y$, where (W_n) are iid with $W > 0$, and let $V_0(x) = \mathbb{P}[W > x]$. Then $nV_0(a_n^{-1}x) \to x^{-\alpha}$ for all $x > 0$, for some choice of a_n. A regular variation argument (a special case of Lemma 1.2 in Meerschaert and Scheffler [145]) shows that $a_n V_0^{-1}(n^{-1}y) \to y^{-1/\alpha}$. Roughly speaking, the argument equates

$$\begin{aligned} nV_0(a_n^{-1}x) &\approx x^{-\alpha} \\ V_0(a_n^{-1}x) &\approx n^{-1}x^{-\alpha} \\ a_n^{-1}x &\approx V_0^{-1}(n^{-1}x^{-\alpha}) \\ x &\approx a_n V_0^{-1}(n^{-1}x^{-\alpha}) \end{aligned}$$

and then substitutes $y = x^{-\alpha}$ to get $y^{-1/\alpha} \approx a_n V_0^{-1}(n^{-1}y)$. The *Skorokhod Theorem* (e.g., see Athreya and Lahiri [9, Theorem 9.4.1]) implies that $(W_1, \cdots, W_n) \simeq (V_0^{-1}(U_1), \ldots, V_0^{-1}(U_n))$, and then

$$a_n\sum_{j=1}^{n} W_j \simeq \sum_{j=1}^{n} a_n V_0^{-1}(\Gamma_j/\Gamma_{n+1}) \to \sum_{i=1}^{\infty}\Gamma_i^{-1/\alpha} \simeq Y.$$

See LePage, Woodroofe, and Zinn [124] for complete details.

Remark 7.4. The series representation (7.20) can be extended to Lévy processes. Suppose Z_t is an α-stable Lévy process with index $0 < \alpha < 1$ and Lévy measure (3.10). For V_j iid uniform on $(0, T)$, we have

$$Z_t \simeq (tC)^{1/\alpha}\sum_{j=1}^{\infty} I(V_j \le t)\Gamma_j^{-1/\alpha} \tag{7.21}$$

for all $0 < t < T$. Note that V_j is the exact time of the jth largest jump of the process Z_t in the interval $0 < t < T$. This representation extends to certain infinitely divisible Lévy processes $Z_t \simeq [0, 0, t\phi]$, with $\Gamma_j^{-1/\alpha}$ replaced by $G^{-1}(\Gamma_j)$, where $G(r,\infty) = \phi(r,\infty)$, see Rosiński [177].

To get a series representation for two-sided stable laws, assume $\mathbb{P}[W_j > x] = pCx^{-\alpha}$ and $\mathbb{P}[W_j < -x] = qCx^{-\alpha}$ for $x > C^{1/\alpha}$, for some $0 < \alpha < 1$ and $C > 0$, where $C > 0$, $p, q \ge 0$, and $p + q = 1$. We can construct this sequence of iid random variables by setting $W_j = \Theta_j X_j$ with iid random signs $\mathbb{P}[\Theta_j = +1] = p$, $\mathbb{P}[\Theta_j = -1] = q$, and $\mathbb{P}[X_j > x] = Cx^{-\alpha}$ iid Pareto independent of Θ_j. Now write

$$\begin{aligned} n^{-1/\alpha} \sum_{j=1}^{n} W_j &= n^{-1/\alpha} \sum_{j=1}^{n} \Theta_j (U_{(j)}/C)^{-1/\alpha} \\ &\simeq C^{1/\alpha} \left(\frac{\Gamma_{n+1}}{n} \right)^{1/\alpha} \sum_{j=1}^{n} \Theta_j \Gamma_j^{-1/\alpha} \Rightarrow Y. \end{aligned}$$

It follows using the strong law of large numbers that

$$C^{1/\alpha} \sum_{j=1}^{\infty} \Theta_j \Gamma_j^{-1/\alpha} \simeq S_\alpha(\beta, \sigma, 0) \tag{7.22}$$

with index $\beta = p - q$ and $\sigma^\alpha = C\Gamma(1-\alpha)\cos(\pi\alpha/2)$.

Remark 7.5. The series representation (7.22) was extended to operator stable laws by Hahn, Hudson, and Veeh [79]. There $\Gamma_j^{-1/\alpha}$ is replaced by Γ_j^{-B}, and Θ_j are iid according to the mixing measure $M(d\theta)$. The series representation for operator stable Lévy processes was modified and applied to operator stable laws in Cohen, Meerschaert and Rosiński [51] to provide a fast and accurate method for simulating operator stable sample paths.

Remark 7.6. The series representation for $\alpha > 1$ requires centering. In this case, $\Gamma_j^{-1/\alpha}$ has a finite mean, and the centering is needed to make the sum

$$\sum_{j=1}^{\infty} \left(\Theta_j \Gamma_j^{-1/\alpha} - \mathbb{E}[\Theta_j] \mathbb{E}\left[\Gamma_j^{-1/\alpha}\right] \right)$$

converge to a mean zero stable law, see LePage, Woodroofe, and Zinn [124]. If $p = q$, then $\mathbb{E}[\Theta_j] = 0$, and no centering is required. LePage, Podgórski, and Ryznar [123] proved almost sure convergence for stable series representations with centering. The centering is more delicate when $\alpha = 1$.

Example 7.7. Here we present a simple application of the LePage series representation to extreme value theory. Take W_j iid Pareto with index $0 < \alpha < 1$. Let $M_n = \max(W_1, \ldots, W_n) = W_{(n)}$. Then $M_n \simeq C^{1/\alpha}(\Gamma_1/\Gamma_{n+1})^{-1/\alpha}$ so that

$$\begin{aligned} n^{-1/\alpha} M_n &\simeq n^{-1/\alpha} C^{1/\alpha} (\Gamma_1/\Gamma_{n+1})^{-1/\alpha} \\ &= C^{1/\alpha} \left(\frac{\Gamma_{n+1}}{n} \right)^{1/\alpha} \Gamma_1^{-1/\alpha} \\ &\to C^{1/\alpha} \Gamma_1^{-1/\alpha} \end{aligned}$$

with probability one, by the strong law of large numbers. Then we have $n^{-1/\alpha}M_n \Rightarrow C^{1/\alpha}\Gamma_1^{-1/\alpha}$. This extreme value limit has the Frechét distribution:

$$\begin{aligned}\mathbb{P}[C^{1/\alpha}\Gamma_1^{-1/\alpha} \le x] &= \mathbb{P}[C^{1/\alpha}E_1^{-1/\alpha} \le x]\\ &= \mathbb{P}[E_1^{-1/\alpha} \le C^{-1/\alpha}x]\\ &= \mathbb{P}[E_1 \ge Cx^{-\alpha}] = \exp(-Cx^{-\alpha})\end{aligned}$$

for $x \ge 0$.

Example 7.8. The LePage series representation is also useful to compute the weak limit for self-normalized sums of heavy tailed random variables. Take $W_j = \Theta_j X_j$ as before: X_n iid Pareto with $0 < \alpha < 1$, and iid random signs. Then

$$\begin{aligned}\frac{\sum_{j=1}^n W_j}{\sqrt{\sum_{j=1}^n W_j^2}} &\simeq \frac{C^{1/\alpha}(\Gamma_{n+1}/n)^{1/\alpha}\sum_{j=1}^n \Theta_j\Gamma_j^{-1/\alpha}}{\sqrt{C^{2/\alpha}(\Gamma_{n+1}/n)^{2/\alpha}\sum_{j=1}^n \Gamma_j^{-2/\alpha}}}\\ &= \frac{\sum_{j=1}^n \Theta_j\Gamma_j^{-1/\alpha}}{\sqrt{\sum_{j=1}^n \Gamma_j^{-2/\alpha}}} \Rightarrow \frac{\sum_{j=1}^\infty \Theta_j\Gamma_j^{-1/\alpha}}{\sqrt{\sum_{j=1}^\infty \Gamma_j^{-2/\alpha}}} = \frac{Y_1}{\sqrt{Y_2}}\end{aligned} \tag{7.23}$$

so the weak limit of the self-normalized sum is a ratio of two *dependent* stable laws: Y_1 has index α, and Y_2 has index $\alpha/2$. In fact, we have

$$\left(n^{-1/\alpha}\sum_{j=1}^n W_j,\, n^{-2/\alpha}\sum_{j=1}^n W_j^2\right) \Rightarrow (Y_1, Y_2)$$

where the limit $Y = (Y_1, Y_2)'$ is operator stable with exponent $B = \operatorname{diag}(1/\alpha, 2/\alpha)$ and the Lévy measure ϕ of Y is concentrated on the set $\{y : y_2 = y_1^2\}$, see Meerschaert and Scheffler [146, Corollary 10.1.8]. The convergence (7.23) extends to arbitrary $X \in \mathrm{DOA}(Y_1)$ using the ideas in Remark 7.3, see Logan, Mallows, Rice and Shepp [125].

7.3 Tempered stable laws

Tempered stable laws reduce the probability of extremely large jumps, so that all moments exist. This can be preferable in applications where the moments have a physical meaning. Another motivation for considering a tempered power law comes from tail estimation. If $p = \mathbb{P}[X > x] \approx Cx^{-\alpha}$ as $x \to \infty$, then $\log p \approx \log C - \alpha \log x$, and a log-log plot of the upper order statistics fits a line with slope $-\alpha$. In many practical applications, this is true up to some point, beyond which the most extreme order statistics fall short of the power law model (e.g., see Aban, Meerschaert and Panorska [1]). For such applications, a tempered model may provide a better fit to real data.

For a general treatment of tempered stable laws and their governing equations, see Baeumer and Meerschaert [19]. To illustrate the basic idea, suppose $Y > 0$ is a stable random variable with index $0 < \alpha < 1$ and pdf $f(y)$ such that

$$\tilde{f}(s) = \mathbb{E}[e^{-sY}] = \int_0^\infty e^{-sy} f(y)\, dy = \exp\left[- Ds^\alpha \right] \tag{7.24}$$

for all $s > 0$, where $D > 0$. The exponentially tempered function $e^{-\lambda y} f(y)$ is not a pdf, since it will not integrate to 1. In fact, we have by (7.24) that

$$\int_0^\infty e^{-\lambda y} f(y)\, dy = \exp\left[- D\lambda^\alpha \right]$$

and it follows that $f_\lambda(y) = e^{-\lambda y} f(y) \exp\left[D\lambda^\alpha\right]$ is a pdf, called the (exponentially) *tempered stable* pdf. This pdf has LT

$$\tilde{f}_\lambda(s) = \int_0^\infty e^{-sy} e^{-\lambda y} f(y) \exp\left[D\lambda^\alpha\right] dy = \exp\left[- D\{(s+\lambda)^\alpha - \lambda^\alpha\}\right].$$

Zolotarev [228, Lemma 2.2.1] implies that (7.24) holds with $s = \lambda + ik$ for any $\lambda > 0$ and $k \in \mathbb{R}$, and then it follows by essentially the same argument that

$$\hat{f}_\lambda(k) = \int_0^\infty e^{-iky} e^{-\lambda y} f(y) \exp\left[D\lambda^\alpha\right] dy = \exp\left[- D\{(\lambda + ik)^\alpha - \lambda^\alpha\}\right]. \tag{7.25}$$

It is obvious from (7.25) that the tempered stable law with pdf $f_\lambda(y)$ is infinitely divisible with Fourier symbol $\psi_\lambda(-k) = -D\{(\lambda + ik)^\alpha - \lambda^\alpha\}$. Note that this reduces to the stable case when $\lambda = 0$. Now we will show that this infinitely divisible law comes from exponentially tempering the Lévy measure. It follows from Proposition 3.10 that the random variable Y with LT (7.24) has characteristic function $\mathbb{E}[e^{ikY}] = e^{\psi(k)}$ where

$$\psi(k) = \int \left(e^{iky} - 1\right) \phi(dy)$$

and $\phi(dy) = C\alpha y^{-\alpha-1} dy I(y > 0)$, where $D = C\Gamma(1-\alpha)$. Define the tempered Lévy measure

$$\phi_\lambda(dy) = e^{-\lambda y} C\alpha y^{-\alpha-1} dy I(y > 0).$$

Since $\int y\,I(0 < y \le R)\,\phi_\lambda(dy) < \infty$, it follows from Theorem 3.8 (a) that there exists a unique infinitely divisible law with characteristic function $e^{\psi_1(k)}$ where

$$\begin{aligned}
\psi_1(k) &= \int \left(e^{iky} - 1\right) \phi_\lambda(dy) \\
&= \int_0^\infty \left(e^{iky} - 1\right) e^{-\lambda y} C\alpha y^{-\alpha-1}\, dy \\
&= \int_0^\infty \left(e^{(ik-\lambda)y} - e^{-\lambda y}\right) C\alpha y^{-\alpha-1}\, dy \\
&= \int_0^\infty \left(e^{(ik-\lambda)y} - 1\right) C\alpha y^{-\alpha-1}\, dy - \int_0^\infty \left(e^{-\lambda y} - 1\right) C\alpha y^{-\alpha-1}\, dy \\
&= -C\Gamma(1-\alpha)(\lambda - ik)^\alpha + C\Gamma(1-\alpha)\lambda^\alpha = \psi_\lambda(k).
\end{aligned} \tag{7.26}$$

This shows that tempering a positive stable law is equivalent to tempering its Lévy measure.

Now suppose that Z_t is a tempered α-stable Lévy process whose pdf $p(x, t)$ has FT

$$\hat{p}(k, t) = \mathbb{E}[e^{-ikZ_t}] = \exp\left[t\psi_\lambda(-k)\right] = \exp\left[-Dt\{(\lambda + ik)^\alpha - \lambda^\alpha\}\right].$$

What is the governing equation of this process? Clearly $\hat{p}(k, t)$ solves the differential equation

$$\frac{d}{dt}\hat{p}(k, t) = -D\{(\lambda + ik)^\alpha - \lambda^\alpha\}\hat{p}(k, t)$$

and so we know that $p(x, t)$ solves

$$\frac{\partial}{\partial t}p(x, t) = Lp(x, t)$$

where $Lf(x)$ is the inverse FT of $\psi_\lambda(-k)\hat{f}(k)$. In order to understand the operator L, it is easiest to go back to the LT.

The pdf $p(x, t)$ of the tempered α-stable Lévy process Z_t has LT

$$\tilde{p}(s, t) = \int_0^\infty e^{-sx} p(x, t)\, dx = \exp\left[-Dt\{(\lambda + s)^\alpha - \lambda^\alpha\}\right].$$

This LT solves the differential equation

$$\frac{d}{dt}\tilde{p}(s, t) = -D\{(\lambda + s)^\alpha - \lambda^\alpha\}\tilde{p}(s, t)$$

and inverting the LT shows that $p(x, t)$ solves

$$\frac{\partial}{\partial t}p(x, t) = Lp(x, t)$$

where $Lf(x)$ is the inverse LT of $-D\{(\lambda + s)^\alpha - \lambda^\alpha\}\tilde{f}(s)$. Now we will use the fact that

$$\int_0^\infty e^{-sx} e^{\lambda x} f(x)\, dx = \tilde{f}(s - \lambda) \tag{7.27}$$

and the fact (proven in the details at the end of Section 2.3) that the Riemann-Liouville fractional derivative of order $0 < \alpha < 1$ has LT

$$\int_0^\infty e^{-sx} \frac{d^\alpha}{dx^\alpha}[f(x)]\, dx = s^\alpha \tilde{f}(s).$$

Putting these two facts together, we see that

$$\int_0^\infty e^{-sx} \frac{d^\alpha}{dx^\alpha}[e^{\lambda x} f(x)]\, dx = s^\alpha \tilde{f}(s - \lambda).$$

Using (7.27) one more time, we see that

$$\int_0^\infty e^{-sx} e^{-\lambda x} \frac{d^\alpha}{dx^\alpha}[e^{\lambda x} f(x)]\, dx = (s + \lambda)^\alpha \tilde{f}(s).$$

This shows that the generator of the tempered stable semigroup is defined (for suitable functions f) by

$$Lf(x) = e^{-\lambda x} \frac{d^\alpha}{dx^\alpha}[e^{\lambda x} f(x)] - \lambda^\alpha f(x).$$

We call

$$\partial_x^{\alpha,\lambda} f(x) = e^{-\lambda x} \frac{d^\alpha}{dx^\alpha}[e^{\lambda x} f(x)] - \lambda^\alpha f(x) \tag{7.28}$$

the (positive) *tempered fractional derivative* of order $0 < \alpha < 1$. With this notation, the pdf of the tempered fractional Lévy motion solves the *tempered fractional diffusion equation*

$$\frac{\partial}{\partial t} p(x, t) = -D\partial_x^{\alpha,\lambda} p(x, t).$$

Remark 7.9. A general theory of tempered stable laws in $\mathbb{R}^d$ has been developed by Rosiński [178]. Exponentially tempered stable processes were originally proposed by Koponen [110] as a model for turbulent velocity fluctuations, and developed further by Cartea and del Castillo-Negrete [45]. Tempered stable random variables (and Lévy processes) are the weak limits of triangular arrays where the row elements follow a power law jump distribution with exponential tempering, and the tempering strength tends to zero at a specific rate as $k_n \to \infty$, see Chakrabarty and Meerschaert [46]. Tempering can also be applied to the waiting times in a CTRW framework, and then a tempered fractional derivative in time replaces the usual first order time derivative, leading to a *tempered fractional Cauchy problem*. Tempered stable laws were applied in

Meerschaert, Zhang and Baeumer [157] to a variety of problems in geophysics. In those applications, the tempering is in the time variable. The tempered space-fractional diffusion is applied to hydrology in Zhang [227]. In a typical application, $\lambda > 0$ is very small, so that the pdf $f_\lambda(y)$ is indistinguishable from the stable pdf $f(y)$ until $|y|$ is quite large. A useful method for simulating tempered stable random variables is presented in Baeumer and Meerschaert [19, Section 4].

Now suppose that Y is stable with index $1 < \alpha < 2$ and Lévy measure $\phi(dy) = C\alpha y^{-\alpha-1}dyI(y > 0)$ as in Proposition 3.12. In this case, the pdf $f(y) > 0$ for all $y \in \mathbb{R}$, but the left tail $f(y) \to 0$ faster than $e^{-\lambda y}$ as $y \to \infty$ for any $\lambda > 0$, so the Laplace transform integral exists over the entire real line. In fact, Zolotarev [228, Lemma 2.2.1] shows that

$$\tilde{f}(\lambda + ik) = \mathbb{E}[e^{-(\lambda+ik)Y}] = \int_{-\infty}^{\infty} e^{-(\lambda+ik)y} f(y)\,dy = \exp\left[D(\lambda + ik)^\alpha\right] \tag{7.29}$$

for all $\lambda > 0$ and all $k \in \mathbb{R}$, where $D = C\Gamma(2-\alpha)/(\alpha-1)$. Then

$$\int_{-\infty}^{\infty} e^{-\lambda y} f(y)\,dy = \exp\left[D\lambda^\alpha\right]$$

and so $f_\lambda(y) = e^{-\lambda y} f(y) \exp\left[-D\lambda^\alpha\right]$ is a pdf on $-\infty < y < \infty$. Its FT is given by

$$\hat{f}_\lambda(k) = \exp\left[D\{(\lambda + ik)^\alpha - \lambda^\alpha\}\right], \tag{7.30}$$

the same form as $0 < \alpha < 1$ except for a change of sign.

Here it is also true that exponentially tempering the pdf is equivalent to tempering the Lévy measure, up to a shift: Define

$$\phi_\lambda(dy) = e^{-\lambda y} C\alpha y^{-\alpha-1} dyI(y > 0)$$

and note that, since $\int y\, I(y > R)\, \phi_\lambda(dy) < \infty$, Theorem 3.8 (b) implies that there exists a unique infinitely divisible random variable Y_0 with characteristic function $\mathbb{E}[e^{ikY_0}] = e^{\psi_2(k)}$ where

$$\begin{aligned}\psi_2(k) &= \int_0^\infty \left(e^{iky} - 1 - iky\right) e^{-\lambda y} C\alpha y^{-\alpha-1} dy \\ &= \int_0^\infty \left(e^{(ik-\lambda)y} - 1 - (ik-\lambda)y\right) C\alpha y^{-\alpha-1} dy \\ &\quad - \int_0^\infty \left(e^{-\lambda y} - 1 + \lambda y\right) C\alpha y^{-\alpha-1} dy - ik \int_0^\infty \left(e^{-\lambda y} - 1\right) y\, C\alpha y^{-\alpha-1} dy \\ &= C\frac{\Gamma(2-\alpha)}{\alpha-1}(\lambda - ik)^\alpha - C\frac{\Gamma(2-\alpha)}{\alpha-1}\lambda^\alpha - ika \end{aligned} \tag{7.31}$$

using Proposition 3.12 twice, where

$$
\begin{aligned}
a &= \frac{\alpha}{\alpha-1}\int_0^\infty \left(e^{-\lambda y}-1\right) C(\alpha-1)y^{-(\alpha-1)-1}\,dy \\
&= \frac{\alpha}{\alpha-1}\left[-C\Gamma(1-(\alpha-1))\lambda^{(\alpha-1)}\right] \\
&= -C\frac{\Gamma(2-\alpha)}{\alpha-1}\alpha\lambda^{\alpha-1}
\end{aligned} \tag{7.32}
$$

using Proposition 3.10 and noting that $\alpha-1\in(0,1)$. Then

$$\mathbb{E}[e^{ikY_0}] = \exp\left[D\{(\lambda-ik)^\alpha-\lambda^\alpha+ik\alpha\lambda^{\alpha-1}\}\right]$$

where $D = C\Gamma(2-\alpha)/(\alpha-1)$. Similar to Remark 3.38, it is not hard to check that

$$\mathbb{E}[Y_0] = (-i)\frac{d}{dk}\mathbb{E}[e^{ikY_0}]_{k=0} = 0.$$

If we define a tempered stable Lévy process Z_t with $Z_1 \simeq Y_0$, then $\mathbb{E}[Z_t] = 0$. Figure 7.1 illustrates the meaning of the truncation parameter λ, in the case $\alpha = 1.2$. The bottom left panel is almost indistinguishable from the corresponding stable Lévy motion, compare Figure 5.24. As λ grows, the large jumps diminish, and for large λ the sample path resembles a Brownian motion, compare Figure 5.18. The sample paths in Figure 7.1 were simulated using an exponential rejection scheme, see [19, Section 4] for details.

The density of Z_t has FT

$$\hat{p}(k,t) = \exp\left[Dt\{(\lambda+ik)^\alpha-\lambda^\alpha-ik\alpha\lambda^{\alpha-1}\}\right].$$

This FT solves the differential equation

$$\frac{d}{dt}\hat{p}(k,t) = D\{(\lambda+ik)^\alpha-\lambda^\alpha-ik\alpha\lambda^{\alpha-1}\}\hat{p}(k,t)$$

and inverting the FT shows that $p(x,t)$ solves

$$\frac{\partial}{\partial t}p(x,t) = Lp(x,t)$$

where $Lf(x)$ is the inverse FT of $D\{(\lambda+ik)^\alpha-\lambda^\alpha-ik\alpha\lambda^{\alpha-1}\}\hat{f}(k)$. An argument very similar to the case $0<\alpha<1$ shows that the generator is defined (for suitable functions f) by

$$Lf(x) = e^{-\lambda x}\frac{d^\alpha}{dx^\alpha}\left[e^{\lambda x}f(x)\right]-\lambda^\alpha f(x)-\alpha\lambda^{\alpha-1}f'(x).$$

We call

$$\partial_x^{\alpha,\lambda}f(x) = e^{-\lambda x}\frac{d^\alpha}{dx^\alpha}\left[e^{\lambda x}f(x)\right]-\lambda^\alpha f(x)-\alpha\lambda^{\alpha-1}f'(x) \tag{7.33}$$

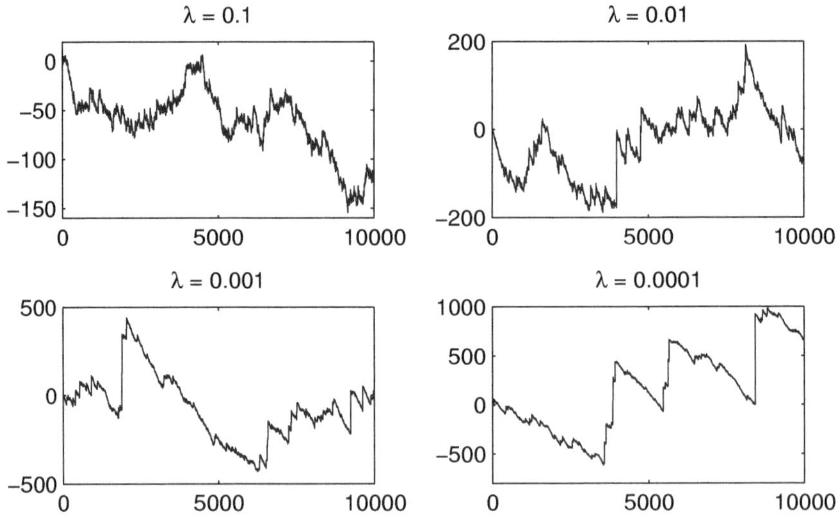

Fig. 7.1: Tempered stable Lévy motion Z_t with $\alpha = 1.2$, showing the effect of the tempering parameter λ, from Baeumer and Meerschaert [19].

the (positive) *tempered fractional derivative* of order $1 < \alpha < 2$. With this notation, the pdf of the tempered fractional Lévy motion with drift $Z_t + vt$ solves the *tempered fractional diffusion equation with drift*

$$\partial_t p(x, t) = -v\partial_x p(x, t) + D\partial_x^{\alpha,\lambda} p(x, t).$$

A two-sided tempered stable Lévy process has Lévy measure $\phi_\lambda(dy) = e^{-\lambda|y|}\phi(dy)$ where ϕ is the Lévy measure (3.30) of an arbitrary nonnormal stable law. Then we can write $Z_t = Z_t^+ - Z_t^-$ where Z_t^+ and Z_t^- are two independent one-sided tempered stable Lévy processes with the same index. If Z_t^+ has Lévy measure $\phi(dy) = pC\alpha y^{-\alpha-1}dyI(y > 0)$ and Z_t^- has Lévy measure $\phi(dy) = qC\alpha|y|^{-\alpha-1}dyI(y < 0)$, then it is not hard to check that the pdf $p(x, t)$ of Z_t solves the two-sided tempered fractional diffusion equation

$$\partial_t p(x, t) = qD\partial_{(-x)}^{\alpha,\lambda} p(x, t) + pD\partial_x^{\alpha,\lambda} p(x, t)$$

where the Fourier symbol of the negative tempered fractional derivative is obtained by substituting $-k$ for k in the Fourier symbol of the positive tempered fractional derivative. In the next section, we will consider alternative forms of the tempered fractional derivative, similar to our analysis of the fractional derivative in Chapter 2.

7.4 Tempered fractional derivatives

We first defined a fractional derivative $d^\alpha f(x)/dx^\alpha$ in Chapter 1 as the function with FT $(ik)^\alpha \hat{f}(k)$. Then in Chapters 2–3, we studied some alternative forms in terms of finite differences, convolution integrals, and the generator formula for a semigroup. Our present goal is to apply the same analysis to the tempered fractional derivative. For complete details, see Baeumer and Meerschaert [19].

Recall from Section 7.3 that a one-sided tempered stable Lévy process Z_t with index $0 < \alpha < 1$ has characteristic function

$$\mathbb{E}[e^{ikZ_t}] = e^{t\psi_\lambda(k)}$$

where $\psi_\lambda(k) = -D\{(\lambda - ik)^\alpha - \lambda^\alpha\}$ for some $\lambda > 0$ and $D > 0$. The pdf $p(x, t)$ of Z_t has FT

$$\hat{p}(k, t) = \mathbb{E}[e^{-ikZ_t}] = e^{t\psi_\lambda(-k)}$$

which solves

$$\frac{d}{dt}\hat{p}(k, t) = \psi_\lambda(-k)\hat{p}(k, t) = -D\{(\lambda + ik)^\alpha - \lambda^\alpha\}\hat{p}(k, t)$$

and so $p(x, t)$ solves the tempered fractional diffusion equation

$$\frac{\partial}{\partial t}p(x, t) = -D\partial_x^{\alpha,\lambda}p(x, t).$$

The tempered fractional derivative $\partial_x^{\alpha,\lambda}f(x)$ has FT $\{(\lambda + ik)^\alpha - \lambda^\alpha\}\hat{f}(k)$, and we know from (7.26) that

$$\int_0^\infty \left(e^{iky} - 1\right) e^{-\lambda y} C\alpha y^{-\alpha-1} dy = -C\Gamma(1-\alpha)[(\lambda - ik)^\alpha - \lambda^\alpha].$$

Set $C = 1/\Gamma(1-\alpha)$ to see that

$$-\int_0^\infty \left(e^{-iky} - 1\right) e^{-\lambda y} \frac{\alpha}{\Gamma(1-\alpha)} y^{-\alpha-1} dy = (\lambda + ik)^\alpha - \lambda^\alpha$$

and apply this formula to see that

$$\{(\lambda + ik)^\alpha - \lambda^\alpha\}\hat{f}(k) = \int_0^\infty \left(\hat{f}(k) - e^{-iky}\hat{f}(k)\right) e^{-\lambda y} \frac{\alpha}{\Gamma(1-\alpha)} y^{-\alpha-1} dy.$$

Inverting the FT shows that (for suitable functions f) the *generator form* of the (positive) tempered fractional derivative of order $0 < \alpha < 1$ is given by

$$\partial_x^{\alpha,\lambda}f(x) = \int_0^\infty (f(x) - f(x-y))e^{-\lambda y} \frac{\alpha}{\Gamma(1-\alpha)} y^{-\alpha-1} dy \qquad (7.34)$$

using Theorem 3.23 (a). This reduces to the generator form (2.17) of the fractional derivative when $\lambda = 0$. An alternative proof uses Theorem 3.17.

Can we also extend the Grünwald finite difference form (2.1) to tempered fractional derivatives? Recall from Section 2.1 that

$$\frac{d^\alpha f(x)}{dx^\alpha} = \lim_{h\to 0} \frac{\Delta^\alpha f(x)}{h^\alpha} \tag{7.35}$$

where

$$\Delta^\alpha f(x) = (I-B)^\alpha f(x) = \sum_{j=0}^{\infty} \binom{\alpha}{j} (-1)^j f(x-jh)$$

is written in terms of the shift operator $Bf(x) = f(x-h)$. In Section 2.2 we used the Grünwald form to motivate the generator form. Now we reverse that process, to explore one possible idea of a finite difference formula for the tempered fractional derivative. In the case of a fractional derivative, we can use the asymptotic expression (2.5) for the Grünwald weights,

$$w_j = (-1)^j \binom{\alpha}{j} \sim \frac{-\alpha}{\Gamma(1-\alpha)} j^{-\alpha-1} \quad \text{as } j \to \infty,$$

to write

$$\begin{aligned}
\frac{d^\alpha f(x)}{dx^\alpha} &= \int_0^\infty [f(x) - f(x-y)] \frac{\alpha}{\Gamma(1-\alpha)} y^{-\alpha-1} dy \\
&\approx \sum_{j=1}^{\infty} [f(x) - f(x-jh)] \frac{\alpha}{\Gamma(1-\alpha)} (jh)^{-\alpha-1} h \\
&= h^{-\alpha} \sum_{j=1}^{\infty} [f(x) - f(x-jh)] \frac{\alpha}{\Gamma(1-\alpha)} j^{-\alpha-1} \\
&\approx h^{-\alpha} \sum_{j=1}^{\infty} [f(x-jh) - f(x)]\, w_j \\
&= h^{-\alpha} \left[\sum_{j=1}^{\infty} f(x-jh) w_j - f(x) \sum_{j=1}^{\infty} w_j \right] = h^{-\alpha} \sum_{j=0}^{\infty} f(x-jh) w_j
\end{aligned}$$

since $\sum_{j=1}^{\infty} w_j = -w_0 = -1$ by (2.11). The Grünwald weights form a discrete approximation of the Lévy measure. (For more on this topic, see Meerschaert and Scheffler [148, Section 5].) Inspired by this heuristic argument, we may consider a kind of tempered finite difference operator

$$\Delta_\lambda^\alpha f(x) = \sum_{j=0}^{\infty} \binom{\alpha}{j} (-1)^j e^{-\lambda jh} f(x-jh) = \sum_{j=0}^{\infty} w_j e^{-\lambda jh} f(x-jh). \tag{7.36}$$

It follows from (2.2) that

$$\sum_{j=0}^{\infty} w_j e^{-\lambda jh} = \sum_{j=0}^{\infty} \binom{\alpha}{j} (-1)^j e^{-\lambda jh} = (1 - e^{-\lambda h})^\alpha \tag{7.37}$$

and, using this fact, we can write

$$
\begin{aligned}
\partial_x^{\alpha,\lambda} f(x) &= \int_0^\infty (f(x) - f(x-y))e^{-\lambda y}\frac{\alpha}{\Gamma(1-\alpha)}y^{-\alpha-1}dy \\
&\approx \sum_{j=1}^{\infty} [f(x) - f(x-jh)]\, e^{-\lambda jh}\frac{\alpha}{\Gamma(1-\alpha)}(jh)^{-\alpha-1}h \\
&= h^{-\alpha}\sum_{j=1}^{\infty} [f(x) - f(x-jh)]\, e^{-\lambda jh}\frac{\alpha}{\Gamma(1-\alpha)}j^{-\alpha-1} \\
&\approx h^{-\alpha}\sum_{j=1}^{\infty} [f(x-jh) - f(x)]\, e^{-\lambda jh}w_j \\
&= h^{-\alpha}\left[\sum_{j=1}^{\infty} f(x-jh)e^{-\lambda jh}w_j - f(x)\sum_{j=1}^{\infty} e^{-\lambda jh}w_j\right] \\
&= h^{-\alpha}\left[\sum_{j=0}^{\infty} f(x-jh)e^{-\lambda jh}w_j - f(x)(1-e^{-\lambda h})^{\alpha}\right] \qquad (7.38)
\end{aligned}
$$

where, in the last line, we add $w_0 f(x)$ to each term, and apply (7.37). This leads us to the following conjecture:

Proposition 7.10 (Baeumer's formula). *For a bounded function f, such that f and its derivatives up to some order $n > 1 + \alpha$ exist and are absolutely integrable, the tempered fractional derivative defined by* (7.34) *exists, and*

$$\partial_x^{\alpha,\lambda} f(x) = \lim_{h\to 0} \frac{\Delta^{\alpha,\lambda} f(x)}{h^{\alpha}} \qquad (7.39)$$

where the tempered finite difference operator

$$\Delta^{\alpha,\lambda} f(x) = \sum_{j=0}^{\infty}\binom{\alpha}{j}(-1)^j e^{-\lambda jh} f(x-jh) - (1-e^{-\lambda h})^{\alpha} f(x). \qquad (7.40)$$

Now we will prove this conjecture. Of course we would not have presented the discussion above, if it did not lead to a positive result! The rather informal presentation is intended to illustrate, for the beginning researcher, the thought process behind the result. In mathematical research, it is necessary (but not sufficient) to master the methods of mathematical proof. One also needs to guess, by some method, what result might be true, and then try to prove it. In this case, our first guess (7.36) had to be adjusted, by the second term in (7.40).

Proof. Write

$$h^{-\alpha}\Delta^{\alpha,\lambda} f(x) = h^{-\alpha}\left[\sum_{j=0}^{\infty}\binom{\alpha}{j}(-1)^j e^{-\lambda jh} f(x-jh) - (1-e^{-\lambda h})^{\alpha} f(x)\right]$$

and take FT to get

$$\begin{aligned}
&h^{-\alpha}\sum_{j=0}^{\infty}\binom{\alpha}{j}(-1)^j e^{-\lambda jh}e^{-ikjh}\hat{f}(k) - h^{-\alpha}(1-e^{-\lambda h})^{\alpha}\hat{f}(k)\\
&= h^{-\alpha}\sum_{j=0}^{\infty}\binom{\alpha}{j}(-1)^j e^{-(\lambda+ik)jh}\hat{f}(k) - h^{-\alpha}(1-e^{-\lambda h})^{\alpha}\hat{f}(k)\\
&= h^{-\alpha}(1-e^{-(\lambda+ik)h})^{\alpha}\hat{f}(k) - h^{-\alpha}(1-e^{-\lambda h})^{\alpha}\hat{f}(k)\\
&\to (\lambda+ik)^{\alpha}\hat{f}(k) - \lambda^{\alpha}\hat{f}(k)
\end{aligned}$$

by the same Taylor expansion argument as in the proof of Proposition 2.1. Apply the continuity theorem for FT to see that (7.39) holds for each $x \in \mathbb{R}$. Note that $z = e^{-(\lambda+ik)h}$ is a complex number with norm $|z| < 1$, so that the series in (7.40) converges absolutely, uniformly in x, in view of (2.2). The proof that $\partial_x^{\alpha,\lambda} f(x)$ exists as the inverse FT of $[(\lambda + ik)^{\alpha} - \lambda^{\alpha}]\hat{f}(k)$ is essentially identical to Proposition 2.5. □

Remark 7.11. The proof of Proposition 7.10 extends immediately to the case $1 < \alpha < 2$, with exactly the same proof, to show that

$$e^{-\lambda x}\frac{d^{\alpha}}{dx^{\alpha}}[e^{\lambda x}f(x)] - \lambda^{\alpha}f(x) = \lim_{h\to 0}\frac{\Delta^{\alpha,\lambda}f(x)}{h^{\alpha}} \tag{7.41}$$

for $1 < \alpha < 2$, where $\Delta^{\alpha,\lambda}f(x)$ is given by the same formula (7.40). In fact, (7.41) holds true, by the same proof, for any $\alpha > 0$. From this it is easy to derive a finite difference formula for the tempered fractional derivative (7.33) of order $1 < \alpha < 2$, see [19, Proposition 3]. Similar to Remark 2.2, a shifted version of the finite difference formula is useful for numerical solutions of the tempered fractional diffusion equation, see [19, Proposition 6].

Remark 7.12. The generator form of the negative tempered fractional derivative is

$$\partial_{(-x)}^{\alpha,\lambda}f(x) = \int_0^{\infty}(f(x)-f(x+y))e^{-\lambda y}\frac{\alpha}{\Gamma(1-\alpha)}y^{-\alpha-1}dy \tag{7.42}$$

using Theorem 3.23 (a). This reduces to the generator form (3.32) of the negative fractional derivative when $\lambda = 0$. This form is the generator of a Lévy process with Lévy measure

$$\phi(dy) = e^{-\lambda|y|}C\alpha|y|^{-\alpha-1}dyI(y<0).$$

We also have the obvious modification of Proposition 7.10: For a bounded function f, such that f and its derivatives up to some order $n > 1 + \alpha$ exist and are absolutely integrable, the negative tempered fractional derivative defined by (7.42) exists, and

$$\partial_{(-x)}^{\alpha,\lambda}f(x) = \lim_{h\to 0}\frac{\Delta_{(-x)}^{\alpha,\lambda}f(x)}{h^{\alpha}} \tag{7.43}$$

where the *tempered finite difference* operator

$$\Delta^{\alpha,\lambda}_{(-x)}f(x) = \sum_{j=0}^{\infty}\binom{\alpha}{j}(-1)^j e^{-\lambda jh}f(x+jh) - (1-e^{-\lambda h})^\alpha f(x). \tag{7.44}$$

Remark 7.13. The generator form for $1 < \alpha < 2$ can be obtained by inverting the FT

$$\psi_2(-k)\hat{f}(k) = \{(\lambda + ik)^\alpha - \lambda^\alpha - ik\alpha\lambda^{\alpha-1}\}\hat{f}(k)$$

of the positive tempered fractional derivative of order $1 < \alpha < 2$ in (7.33). Substitute $C = (\alpha - 1)/\Gamma(2-\alpha)$ in (7.31) and (7.32) to get

$$\int_0^\infty \left(e^{iky} - 1 - iky\right) e^{-\lambda y}\frac{\alpha(\alpha-1)}{\Gamma(2-\alpha)}y^{-\alpha-1}dy = (\lambda - ik)^\alpha - \lambda^\alpha + ik\alpha\lambda^{\alpha-1} = \psi_2(k).$$

Then the inverse FT of

$$\psi_2(-k)\hat{f}(k) = \int_0^\infty \left(e^{-iky}\hat{f}(k) - \hat{f}(k) + iky\hat{f}(k)\right) e^{-\lambda y}\frac{\alpha(\alpha-1)}{\Gamma(2-\alpha)}y^{-\alpha-1}dy$$

is

$$\partial_x^{\alpha,\lambda}f(x) = \int_0^\infty \left(f(x-y) - f(x) + yf'(x)\right) e^{-\lambda y}\frac{\alpha(\alpha-1)}{\Gamma(2-\alpha)}y^{-\alpha-1}dy \tag{7.45}$$

using Theorem 3.23 (b). Equation (7.45) is the generator form of the positive tempered fractional derivative of order $1 < \alpha < 2$. When $\lambda = 0$, (7.45) reduces to the generator form (2.20) for the positive fractional derivative of order $1 < \alpha < 2$. An alternative proof uses Theorem 3.17.

7.5 Distributed order fractional derivatives

The distributed order fractional derivative is defined by

$$\partial_t^{\nu(d\beta)}f(t) = \int_0^1 \partial_t^\beta f(t)\nu(d\beta) \tag{7.46}$$

where $\nu(d\beta)$ is a finite Borel measure on the unit interval $(0, 1)$, and ∂_t^β is the Caputo fractional derivative (2.33). The distributed order time-fractional diffusion equation

$$\partial_t^{\nu(d\beta)}p(x,t) = D\frac{\partial^2}{\partial x^2}p(x,t), \tag{7.47}$$

where $D > 0$, was introduced by Chechkin et al. [48, 49] in the physics literature.

If $\nu(d\beta)$ is a point mass at some $\beta \in (0, 1)$, then (7.47) reduces to the time-fractional diffusion equation (2.39), a model for subdiffusion where solutions spread at the rate

$t^{\beta/2}$ for all $t > 0$. If ν has point masses at $0 < \beta_1 < \cdots < \beta_n < 1$, we get a linear combination of fractional time derivatives of different orders. Chechkin et al. [48] show that in this case, the spreading rate is asymptotically equal to to $t^{\beta_1/2}$ as $t \to \infty$, so the smallest β dominates. If (7.47) includes fractional derivatives of arbitrarily small order $\beta > 0$, this suggests that solutions to (7.47) will spread at a rate slower than $t^{\beta/2}$ for any $\beta > 0$. Suppose for example that

$$\nu(d\beta) = \begin{cases} A\beta^{\alpha-1}d\beta & 0 < \beta < B \\ 0 & B < \beta < 1 \end{cases} \tag{7.48}$$

for some $\alpha > 0$, $A > 0$, and $B < 1$. We will show that this leads to ultraslow diffusion, where the solution to (7.47) spreads like $(\log t)^{\alpha/2}$ for all $t > 0$ sufficiently large, slower than any power law. The cutoff at $B < 1$ is required for technical reasons, see details.

Recall from Section 2.3 that the Caputo fractional derivative $\partial_t^\beta f(t)$ has Laplace transform $s^\beta \tilde{f}(s) - s^{\beta-1}f(0)$. Then for suitable functions $f(t)$ the Laplace transform of $\partial_t^{\nu(d\beta)} f(t)$ is

$$\int_0^\infty e^{-st}\partial_t^{\nu(d\beta)} f(t)\,dt = \int_0^1 \left[s^\beta \tilde{f}(s) - s^{\beta-1}f(0)\right]\nu(d\beta). \tag{7.49}$$

Now we will apply the alternative theory of infinitely divisible subordinators $Y > 0$ based on Laplace transforms, see the details at the end of Section 4.5. This Lévy representation takes the simplified form $\mathbb{E}[e^{-sY}] = e^{-\psi(s)}$, where $s > 0$ and

$$\psi(s) = as + \int_0^\infty \left(1 - e^{-sy}\right)\phi(dy)$$

for some $a \geq 0$, and some Lévy measure $\phi(dy)$. This Lévy representation is unique. The Lévy measure $\phi(dy)$ on $\{y : y > 0\}$ satisfies $\phi(R, \infty) < \infty$ and

$$\int_0^R y\phi(dy) < \infty \tag{7.50}$$

for all $R > 0$. Now use (7.49) to write

$$\int_0^\infty e^{-st}\partial_t^{\nu(d\beta)} f(t)\,dt = \psi(s)\tilde{f}(s) - s^{-1}\psi(s)f(0), \tag{7.51}$$

where

$$\psi(s) = \int_0^1 s^\beta \nu(d\beta).$$

Recall from (4.56), which is just from Proposition 3.10 with $s = -ik$, that

$$\int_0^\infty \left(1 - e^{-sy}\right) C\beta y^{-\beta-1}\,dy = C\Gamma(1-\beta)s^\beta$$

for all $s > 0$, for any $0 < \beta < 1$. Set $C = C(\beta) = 1/\Gamma(1-\beta)$ and substitute into (7.49) to see that (7.51) holds with

$$\psi(s) = \int_0^1 s^\beta \nu(d\beta) = \int_0^B \int_0^\infty (1 - e^{-sy})\, \beta y^{-\beta-1} dy\, C(\beta) A\beta^{\alpha-1}\, d\beta = \int_0^\infty (1 - e^{-sy})\, \phi(dy),$$

where

$$\phi(dy) = \int_0^B \beta y^{-\beta-1} dy\, C(\beta) A\beta^{\alpha-1}\, d\beta.$$

Let $p(\beta) = C(\beta)A\beta^{\alpha-1}$ for $0 < \beta < B$, and $p(\beta) = 0$ otherwise. Since $\Gamma(x)$ is a decreasing function for $0 < x < 1$, with $\Gamma(1) = 1$, we have $C(\beta) \le 1$ for all $0 < \beta < 1$. Then

$$M = \int_0^B p(\beta) d\beta = \int_0^B C(\beta) A\beta^{\alpha-1}\, d\beta \le \int_0^B A\beta^{\alpha-1}\, d\beta < \infty.$$

Since we can divide both sides of (7.47) by the constant M, it entails no loss of generality to assume that $M = 1$, and then $p(\beta)$ is a pdf. Now we can also write simply

$$\phi(dy) = \int_0^B \beta y^{-\beta-1} dy\, p(\beta)\, d\beta. \tag{7.52}$$

Next we want to show that $\phi(dy)$ is a Lévy measure. Take any $R > 0$ and write

$$\phi(R, \infty) = \int_R^\infty \int_0^B \beta y^{-\beta-1} p(\beta)\, d\beta\, dy = \int_0^B \int_R^\infty \beta y^{-\beta-1} dy\, p(\beta)\, d\beta = \int_0^B R^{-\beta} p(\beta)\, d\beta \tag{7.53}$$

using the Fubini-Tonelli Theorem. The last integral is bounded above by $\max\{1, R^{-B}\}$, since $p(\beta)$ is a pdf. Now we just need to check that (7.50) holds, see details. Then $\phi(dy)$ is a Lévy measure, and $\psi(s)$ is the Lévy symbol of some subordinator.

Let $D^\psi(u)$ be the Lévy subordinator with $\mathbb{E}[e^{-sD^\psi(u)}] = e^{-u\psi(s)}$ for all $s > 0$. Using some deep semigroup arguments, Corollary 2.1 in Kovács and Meerschaert [112] shows that $D^\psi(u)$ has a density $g(t,u)$, which is a smooth function of both $t > 0$ and $u > 0$. Define the inverse process $E_t^\psi = \inf\{u > 0 : D^\psi(u) > t\}$, so that

$$\{E_t^\psi \le u\} = \{D^\psi(u) \ge t\}. \tag{7.54}$$

Theorem 3.1 in [112] shows that E_t^ψ has a density $h(u,t)$ that is a smooth function of $u > 0$. Now we can argue as in Section 2.4, using (7.54), that

$$h(u,t) = \frac{d}{du}\mathbb{P}[E_t^\psi \le u] = \frac{d}{du}\mathbb{P}[D^\psi(u) \ge t] = \frac{d}{du}\left[1 - \int_0^t g(y,u)\,dy\right]$$

with LT

$$\begin{aligned}\tilde{h}(u,s) &= -\frac{d}{du}\left[s^{-1}\tilde{g}(s,u)\right]\\ &= -\frac{d}{du}\left[s^{-1}e^{-u\psi(s)}\right] = s^{-1}\psi(s)e^{-u\psi(s)}\end{aligned} \tag{7.55}$$

using the fact that integration corresponds to multiplication by s^{-1} in LT space. See Meerschaert and Scheffler [153, Theorem 3.1] for complete details.

Let $f(x,u)$ be the PDF of $B(u)$, independent of $D^\psi(u)$, and use a simple conditioning argument, as in Section 2.4, to see that $B(E_t^\psi)$ has PDF

$$p(x,t) = \int_0^\infty f(x,u)h(u,t)\,du.$$

Take FT and then LT to see that

$$\begin{aligned}\bar{p}(k,s) &= \int_0^\infty e^{-st}\int_{-\infty}^\infty e^{-ikx}p(x,t)\,dx\,dt\\ &= \int_0^\infty e^{-st}\int_{-\infty}^\infty e^{-ikx}\int_0^\infty f(x,u)h(u,t)\,du\,dx\,dt\\ &= \int_0^\infty\left(\int_{-\infty}^\infty e^{-ikx}f(x,u)\,dx\right)\left(\int_0^\infty e^{-st}h(u,t)\,dt\right)du\\ &= \int_0^\infty e^{-uDk^2}s^{-1}\psi(s)e^{-u\psi(s)}\,du = \frac{s^{-1}\psi(s)}{\psi(s)+Dk^2}.\end{aligned}$$

See [153, Theorem 3.6] for complete details. Rewrite in the form $\psi(s)\bar{p}(k,s) - s^{-1}\psi(s) = -Dk^2\bar{p}(k,s)$, invert the LT using (7.51) along with $\hat{p}(k,0) = 1$ to get

$$\partial_t^{\nu(d\beta)}\hat{p}(k,t) = -Dk^2\hat{p}(k,t),$$

then invert the FT to see that the pdf $p(x,t)$ of $B(E_t^\psi)$ solves the distributed order time-fractional diffusion equation (7.47).

Next we want to construct a CTRW model for $B(E_t^\psi)$. Take $S(n) = Y_1 + \cdots + Y_n$ iid with $\mathbb{E}[Y_i] = 0$ and $\mathbb{E}[Y_i^2] = 2D$. It follows from the Central Limit Theorem that $n^{-1/2}S(nt) \Rightarrow B(t)$ for any $t > 0$, see Example 3.31. This gives a suitable model for the CTRW jumps. A suitable model for the CTRW waiting times is more delicate. In view of Theorem 4.5, no random walk with iid waiting times can converge to $D^\psi(u)$, which is neither normal nor α-stable. Hence it is necessary to consider a triangular array.

Now we will use an idea from Meerschaert and Scheffler [152, Section 3]. Take $\{B_n : n = 1, 2, 3, \ldots\}$ iid with pdf $p(\beta)$. Then, for each $n = 1, 2, 3, \ldots$, define iid waiting times $\{J_{nj} : j = 1, 2, 3, \ldots, k_n\}$ with distribution

$$P\{J_{nj} > t | B_n = \beta\} = \begin{cases} 1 & 0 \le t < n^{-1/\beta} \\ n^{-1}t^{-\beta} & t \ge n^{-1/\beta} \end{cases} . \tag{7.56}$$

Given any $t > 0$, let $k_n = [nt]$, and consider the triangular array of waiting times $\{J_{nj} : j = 1, 2, \ldots, k_n; n = 1, 2, 3, \ldots\}$. At any time scale n, the waiting times J_{nj} are iid conditionally Pareto distributed, conditional on the random power law indices B_n. The time of the kth CTRW jump is given by the row sum

$$T_k^n = \sum_{j=1}^{k} J_{nj}$$

for any $k = 1, 2, 3, \ldots$ at any time scale n.

Remark 7.14. If we define (J_j) iid Pareto with $\mathbb{P}[J_j > t] = t^{-\beta}$ then

$$\mathbb{P}[n^{-1/\beta}J_j > t] = \mathbb{P}[J_j > n^{1/\beta}t] = (n^{1/\beta}t)^{-\beta} = n^{-1}t^{-\beta},$$

which is the same as (7.56). This shows that, conditional on $B_n = \beta$, the norming for this triangular array is the same as in Theorem 3.37.

Next we want to show that $T_{[nt]}^n \Rightarrow D^\psi(t)$ for any $t > 0$, using the convergence criteria for triangular arrays, Theorem 3.33. The proof is quite similar to Theorem 3.37. To show that condition (*i*) holds with $k_n = [nt]$, suppose $u > 0$ and note that for any n sufficiently large we have

$$\begin{aligned} [nt]\mathbb{P}[J_{nj} > u] &= [nt] \int_0^B \mathbb{P}[J_{nj} > u | B_1 = \beta] p(\beta)\, d\beta \\ &= \frac{[nt]}{n} \int_0^B u^{-\beta} p(\beta)\, d\beta \\ &\to t \int_0^B u^{-\beta} p(\beta)\, d\beta = t\,\phi(u, \infty), \end{aligned}$$

using (7.53). Hence condition (*i*) holds. See details for the proof that condition (*ii*) holds. Then it follows from Theorem 3.33 that $T^n_{[nt]} - a_n \Rightarrow D^\psi(t)$ for some centering constants a_n. Finally we want to argue that a_n can be made as small as we like by choosing $R > 0$ sufficiently small. Again, the proof is quite similar to Theorem 3.37, see details. Then we have shown that $T^n_{[nt]} \Rightarrow D^\psi(t)$.

Now consider a triangular array of CTRW with jumps (Y_j) and waiting times (J_{nj}). We have already shown that $n^{-1/2}S(nt) \Rightarrow B(t)$ and $T^n_{[nt]} \Rightarrow D^\psi(t)$. The number of jumps by time $t > 0$ at time scale n is defined by $N^n_t = \max\{k \geq 0 : T^n_k \leq t\}$, and then we can argue as in Section 4.3 that $n^{-1}N_n(t) \Rightarrow E^\psi_t$, see details. The CTRW particle position at time $t \geq 0$ and scale $n = 1, 2, 3, \dots$ is given by $S(N^n_t)$. Now argue as in Section 4.3, assuming the jumps (Y_j) are independent of the waiting times (J_{nj}), that

$$(n^{-1/2}S(nt), n^{-1}N^n_t) \Rightarrow (B(t), E^\psi_t)$$

in the sense of finite dimensional distributions. Then it follows as in Section 4.4 that

$$n^{-1/2}S(N^n_t) = n^{-1/2}S(n \cdot n^{-1}N^n_t) \Rightarrow B(E^\psi_t)$$

in the Skorokhod M_1 topology, see [153, Corollary 2.4] for complete details. Convergence in the J_1 topology follows from Straka and Henry [210, Theorem 3.6]. This establishes a CTRW model for the distributed order fractional diffusion equation (7.47).

Finally we explain how the distributed order fractional diffusion equation (7.47) with $\nu(d\beta)$ given by (7.48) models ultraslow diffusion. The Lévy symbol

$$\psi(s) = \int_0^1 s^\beta \nu(d\beta) = \int_0^1 e^{\beta \log s} \nu(d\beta) = \tilde{\nu}(-\log s)$$

where the LT

$$\tilde{\nu}(r) = \int_0^1 e^{-r\beta}\nu(d\beta) = \int_0^B e^{-r\beta}A\beta^{\alpha-1}\,d\beta = Ar^{-\alpha}\int_0^{rB} e^{-x}x^{\alpha-1}\,dx = Ar^{-\alpha}\Gamma(\alpha, rB)$$

using the incomplete gamma function. As $r \to \infty$ we have $\tilde{\nu}(r) \sim Ar^{-\alpha}\Gamma(\alpha)$, and hence

$$\psi(s) \sim A\Gamma(\alpha)(-\log s)^{-\alpha} \quad \text{as } s \to 0.$$

The CTRW limit process $B(E^\psi_t)$ has mean zero and, since $\mathbb{E}[B(t)^2] = 2Dt$, a simple conditioning argument shows that $\mathrm{Var}[B(E^\psi_t)] = 2D\mathbb{E}[E^\psi_t]$. Let

$$m(t) = \mathbb{E}[E^\psi_t] = \int_0^\infty \mathbb{P}[E^\psi_t \geq u]\,du = \int_0^\infty \mathbb{P}[D^\psi_u \leq t]\,du$$

and take LT as in (7.55) to get

$$\begin{aligned}\tilde{m}(s) &= \int_0^\infty e^{-st}\int_0^\infty \mathbb{P}[D_u^\psi \le t]\,du\,dt\\ &= \int_0^\infty\Big(\int_0^\infty e^{-st}\mathbb{P}[D_u^\psi \le t]\,dt\Big)\,du\\ &= \int_0^\infty s^{-1}e^{-u\psi(s)}\,du = s^{-1}\psi(s)^{-1} \sim Cs^{-1}(-\log s)^\alpha\end{aligned}$$

as $s \to 0$, where $C^{-1} = A\Gamma(\alpha)$. Example 2.10 shows that the function $f(t) = t^p$ on $t \ge 0$ has LT $\tilde{f}(s) = s^{-p-1}\Gamma(p+1)$ for any $p > -1$. An extension of this argument yields the Karamata Tauberian Theorem:

$$f(t) \sim t^p L(t) \text{ as } t \to \infty \quad \Longleftrightarrow \quad \tilde{f}(s) \sim s^{-p-1}\Gamma(p+1)L(1/s) \text{ as } s \to 0, \tag{7.57}$$

assuming that $p > -1$, $L(t)$ is slowly varying, and $f(t)$ is monotone for $t > 0$ sufficiently large (see Theorem 4, p. 446 in Feller [68]). Since E_t^ψ is nondecreasing, the moment function $m(t) = \mathbb{E}[E_t^\psi]$ is clearly monotone. Note that $-\log s = \log(1/s)$, and that the function $L(t) = C(\log t)^\alpha$ is slowly varying. Then we can apply (7.57) with $p = 0$ to see that $m(t) \sim C(\log t)^\alpha$ as $t \to \infty$. Hence the stochastic process $B(E_t^\psi)$ models ultraslow diffusion, since its variance $\mathrm{Var}[B(E_t^\psi)] \sim 2DC(\log t)^\alpha$ for some $\alpha > 0$. Remark 3.2 in [153] shows that $\mathbb{P}[J_{nj} > t]$ is also slowly varying as $t \to \infty$. Hence the ultraslow diffusion results from very long waiting times.

Details

To show that (7.52) defines a Lévy measure, we also need to check that (7.50) holds. Write

$$\begin{aligned}\int_0^R y\phi(dy) = \int_0^R y\int_0^B \beta y^{-\beta-1}dy\,p(\beta)\,d\beta &= \int_0^B\int_0^R y\beta y^{-\beta-1}dy\,p(\beta)\,d\beta\\ &= \int_0^B \frac{\beta}{1-\beta}R^{1-\beta}p(\beta)\,d\beta.\end{aligned} \tag{7.58}$$

Then the integral (7.58) is bounded above by $B\max\{R, R^{1-B}\}/(1-B)$. This along with upper bound on (7.53) shows that $\phi(dy)$ is a Lévy measure. Equation (7.58) also shows the reason for the cutoff at $B < 1$: If we integrate to $\beta = 1$, then (7.58) diverges.

To show that condition (*ii*) holds, first note that the conditional density of J_{nj} given $B_n = \beta$ is

$$\psi_n(u|\beta) = \begin{cases}0 & 0 \le u < n^{-1/\beta}\\ n^{-1}\beta u^{-\beta-1} & u \ge n^{-1/\beta}\end{cases}. \tag{7.59}$$

Then for any n sufficiently large we can write

$$
\begin{aligned}
[nt]\,\mathrm{Var}[J_{nj}^{\varepsilon}] &\le [nt]\int_0^{\varepsilon} u^2 \int_0^{B} \psi_n(u|\beta)p(\beta)\,d\beta\,du \\
&= [nt]\int_0^{B}\int_{n^{-1/\beta}}^{\varepsilon} u^2 n^{-1}\beta u^{-\beta-1}\,du\,p(\beta)\,d\beta \\
&= \frac{[nt]}{n}\int_0^{B} \varepsilon^{2-\beta}\frac{\beta}{2-\beta}p(\beta)\,d\beta - \frac{[nt]}{n}\int_0^{B} n^{1-2/\beta}\frac{\beta}{2-\beta}p(\beta)\,d\beta.
\end{aligned}
$$

Since $\beta/(2-\beta) \le 1$ and $1 - 2/\beta < -1$, the second integral

$$
\int_0^{B} n^{1-2/\beta}\frac{\beta}{2-\beta}p(\beta)\,d\beta \le n^{-1}\int_0^{1} p(\beta)\,d\beta = \frac{1}{n},
$$

and then it follows that the second term tends to zero as $n \to \infty$. Since $\varepsilon^{2-\beta} < \varepsilon$ for $0 < \varepsilon < 1$ and $0 < \beta < 1$, the first integral is bounded above by

$$
\varepsilon\int_0^{B} \frac{\beta}{2-\beta}p(\beta)\,d\beta \le \varepsilon.
$$

Then it follows that

$$
\lim_{\varepsilon\to 0}\limsup_{n\to\infty}[nt]\,\mathrm{Var}[J_{nj}^{\varepsilon}] \le \lim_{\varepsilon\to 0} t\varepsilon = 0.
$$

This shows that condition *(ii)* holds.

Next we show that a_n can be made as small as we like by choosing $R > 0$ sufficiently small. By (3.37) we can take

$$
\begin{aligned}
a_n = k_n\mathbb{E}[J_{nj}^{R}] &= [nt]\int_0^{R}\int_0^{B} u\psi_n(u|\beta)\,du\,p(\beta)\,d\beta \\
&= [nt]\int_0^{B}\int_{n^{-1/\beta}}^{R} un^{-1}\beta u^{-\beta-1}\,du\,p(\beta)\,d\beta \\
&= \frac{[nt]}{n}\int_0^{B} \frac{\beta}{1-\beta}R^{1-\beta}p(\beta)\,d\beta - \frac{[nt]}{n}\int_0^{B} n^{1-1/\beta}\frac{\beta}{1-\beta}p(\beta)\,d\beta.
\end{aligned}
$$

Since $n^{1-1/\beta} \le n^{1-1/B}$ and $1 - 1/B < 0$, it follows that the second term tends to zero as $n \to \infty$. Since $R^{1-\beta} < R^{1-B}$ for $0 < R < 1$, the first integral tends to zero as $R \to 0$. Hence the lim sup of the first term as $n \to \infty$ can be made arbitrarily small by choosing

$R > 0$ sufficiently small. As in the proof of Theorem 3.37, this implies that $T^n_{[nt]} \Rightarrow D^\psi(t)$ without any centering.

To show that $n^{-1}N^n_t \Rightarrow E^\psi_t$, let $\lceil x \rceil$ denote the smallest integer greater than or equal to $x \geq 0$, and note that as in Section 4.3 we have $\{N^n_t \geq x\} = \{T^n_{\lceil x \rceil} \leq t\}$. Next observe that $T^n_{\lceil nt \rceil} \Rightarrow D^\psi(t)$ for any $t > 0$. This is easy to check using characteristic functions, as in (3.64). Since both $D^\psi(x)$ and E^ψ_t have a density, it follows that

$$\mathbb{P}[D^\psi(u) > t] = \mathbb{P}[D^\psi(u) \geq t] = \mathbb{P}[E^\psi_t \leq u] = \mathbb{P}[E^\psi_t < u].$$

Now we can argue as in Section 4.3 that

$$\begin{aligned}\mathbb{P}[n^{-1}N^n_t < x] = \mathbb{P}[N^n_t < nx] &= \mathbb{P}[T^n_{\lceil nx \rceil} > t] \\ &\to \mathbb{P}[D^\psi(x) > t] = \mathbb{P}[E^\psi_t < x]\end{aligned}$$

using (7.54) and $T^n_{\lceil nt \rceil} \Rightarrow D^\psi(t)$. This shows that $n^{-1}N^n_t \Rightarrow E^\psi_t$.

7.6 Pearson diffusions

The diffusion equation with constant coefficients

$$\frac{\partial}{\partial t}p = -\frac{\partial}{\partial x}[vp] + \frac{\partial^2}{\partial x^2}[Dp] \tag{7.60}$$

from (1.9) governs the scaling limit of a random walk with finite variance jumps. In this section, we consider *Pearson diffusions* governed by (7.60) with space-variable coefficients

$$D(x) = d_0 + d_1 x + d_2 x^2 \quad \text{and} \quad v(x) = a_0 + a_1 x. \tag{7.61}$$

A Pearson diffusion is a Markov process that can tend to steady state: $X(t) \Rightarrow X(\infty)$ as $t \to \infty$. Then the density $\mathbf{m}(x)$ of the limit variable $X(\infty)$ is a time-invariant solution to equation (7.60): $p(x, t) = \mathbf{m}(x)$ for all $t \geq 0$. The steady state density of a Pearson diffusion follows one of the six classes of *Pearson distributions*: normal, gamma, beta, Student-t, inverse gamma, or F-distribution.

A Pearson diffusion is a *time-homogeneous Markov process* whose *transition density* $\mathbf{p}(x, t; y)$ is the conditional pdf of $x = X_t$ given $X_0 = y$. For any initial state $X_0 = y$, the function $p = p(x, t) = \mathbf{p}(x, t; y)$ solves the *forward equation* (7.60) with the point source initial condition $p(x, 0) = \delta(x - y)$. Then the *forward semigroup*

$$T_t f(x) = \int \mathbf{p}(x, t; y) f(y)\, dy \tag{7.62}$$

gives the pdf of X_t, given that X_0 has pdf $f(x)$. The function $p(x, t) = T_t f(x)$ solves a Cauchy problem

$$\frac{\partial}{\partial t}p(x, t) = \mathcal{L}p(x, t); \quad p(x, 0) = f(x) \tag{7.63}$$

where the generator of the forward equation is

$$\mathcal{L}f(x) = -\frac{\partial}{\partial x}\left[v(x)f(x)\right] + \frac{\partial^2}{\partial x^2}\left[D(x)f(x)\right]. \tag{7.64}$$

The forward equation is sometimes called the *Fokker-Planck equation*, especially in applications to physics. For Markov processes, it is common for technical reasons to first consider the *backward semigroup*

$$T_t^* g(y) = \mathbb{E}[g(X_t)|X_0 = y] = \int \mathbf{p}(x, t; y) g(x)\, dx. \tag{7.65}$$

If $g(y) = I(y \in B)$ for some Borel set B, then $T_t^* g(y) = \mathbb{P}[X_t \in B|X_0 = y]$, the probability of finding a particle in the set B after time $t > 0$, given that it started at location y at time $t = 0$. The function $p(y, t) = T_t^* g(y)$ solves the backward equation

$$\frac{\partial}{\partial t} p(y, t) = v(y) \frac{\partial}{\partial y} p(y, t) + D(y) \frac{\partial^2}{\partial y^2} p(y, t) \tag{7.66}$$

with initial condition $p(y, 0) = g(y)$. The backward equation is simpler, because the coefficients v and D appear outside the derivatives.

If a steady-state solution $p = p(x, t) = \mathbf{m}(x)$ to (7.60) exists, it satisfies:

$$0 = -\frac{\partial}{\partial x}\left[v(x)\mathbf{m}(x)\right] + \frac{\partial^2}{\partial x^2}\left[D(x)\mathbf{m}(x)\right]. \tag{7.67}$$

Integrating (7.67) once yields

$$\frac{d}{dx}\left[D(x)\mathbf{m}(x)\right] - v(x)\mathbf{m}(x) = C_1. \tag{7.68}$$

With $C_1 = 0$, equation (7.68) reduces to

$$\frac{\mathbf{m}'(x)}{\mathbf{m}(x)} = \frac{v(x) - D'(x)}{D(x)} = \frac{(a_0 - d_1) + (a_1 - 2d_2)x}{d_0 + d_1 x + d_2 x^2}. \tag{7.69}$$

Equation (7.69) is the famous *Pearson equation*, introduced by K. Pearson [166] in 1914 to unify the six classes of Pearson distributions.

The six types of Pearson diffusions will be described in Remark 7.19 at the end of this section. The study of Pearson diffusions began with Kolmogorov [108] and Wong [220], and continued in Forman and Sørensen [72], Avram, Leonenko and Rabehasaina [11], Leonenko and Šuvak [121, 120], and Avram, Leonenko and Šuvak [10]. For the remainder of this section, we will restrict our attention to Pearson diffusions of type (1–3), where the steady state density $\mathbf{m}(x)$ is normal, gamma, or beta. Then the backward equation (7.66) can be solved by *separation of variables*. See Leonenko, Meerschaert and Sikorskii [122, Theorem 3.2 and Remark 3.5] for a complete and detailed proof. Next we will sketch the main ideas of the proof. Write (7.66) in the form of a Cauchy problem

$$\frac{\partial}{\partial t} p(y, t) = \mathcal{G} p(y, t); \quad p(y, 0) = g(y) \tag{7.70}$$

where

$$\mathcal{G} g(y) = v(y) \frac{\partial g(y)}{\partial y} + D(y) \frac{\partial^2 g(y)}{\partial y^2} \tag{7.71}$$

is the generator of the backward semigroup. Suppose that $p(y,t) = S(t)\varphi(y)$ solves (7.70), where the functions S and φ may depend on x. Then

$$\frac{\partial}{\partial t}[S(t)\varphi(y)] = \mathcal{G}[S(t)\varphi(y)],$$

which is equivalent for non-zero functions to

$$\frac{1}{S(t)}\frac{dS(t)}{dt} = \frac{\mathcal{G}\varphi(y)}{\varphi(y)}. \tag{7.72}$$

Equation (7.72) can hold only if both sides are equal to a constant. Denote this constant by $-\lambda$, and consider the two resulting equations: The *Sturm-Liouville equation*

$$\mathcal{G}\varphi = -\lambda\varphi \tag{7.73}$$

and the time equation

$$\frac{dS(t)}{dt} = -\lambda S(t). \tag{7.74}$$

Recall from Section 2.3 that φ is an *eigenfunction* of $\mathcal{G}$ if (7.73) holds for some complex number λ. Write the Sturm-Liouville equation (7.73) using (7.71) and (7.61) to get

$$(d_0 + d_1x + d_2x^2)\varphi'' + (a_0 + a_1x)\varphi' + \lambda\varphi = 0. \tag{7.75}$$

The steady state solutions $\mathbf{m}(x)$ for Pearson diffusions of type (1–3) are the normal, gamma, and beta probability density functions. In these three cases, (7.75) is solved by the Hermite, Laguerre, or Jacobi polynomials, respectively (see Remark 7.19). Each of these families of polynomials forms an *orthogonal system*:

$$\int Q_n(x)Q_m(x)\mathbf{m}(x)dx = \begin{cases} c_n^2 > 0 & \text{if } n = m, \\ 0 & \text{if } n \neq m \end{cases} \tag{7.76}$$

such that $\mathcal{G}Q_n(x) = -\lambda_n Q_n(x)$ for all n, where $Q_0(x) \equiv 1$ and $0 = \lambda_0 < \lambda_1 < \lambda_2 < \cdots$ with $\lambda_n \to \infty$. The corresponding solutions to the time equation (7.74) have the form

$$S_n(t) = e^{-\lambda_n t}$$

since the exponential functions are the eigenfunctions of the first derivative. Then $p(y,t) = e^{-\lambda_n t}Q_n(y)$ solves the Cauchy problem (7.70) with initial condition $p(y,0) = Q_n(y)$. Since any finite linear combination of these functions will also solve equation (7.70), is it reasonable to consider the infinite sum

$$p(y,t) = \sum_{n=0}^{\infty} b_n e^{-\lambda_n t} Q_n(y). \tag{7.77}$$

If

$$g(x) = \sum_{n=0}^{\infty} b_n Q_n(x) \tag{7.78}$$

where the series converges uniformly on compact sets, then some analytic estimates show that the series (7.77) can be differentiated term-by-term, so that the function $p(y,t)$ in (7.77) solves equation (7.66). If the polynomials $Q_n(x)$ are normalized so that $c_n^2 = 1$ in (7.76), then

$$\int g(x)Q_n(x)\mathbf{m}(x)\,dx = b_n \quad \text{for all } n.$$

Then it follows from (7.78) that (7.77) solves the backward equation (7.66) with initial condition $p(y,0) = g(y)$.

Equating (7.65) to (7.77) we see that

$$\begin{aligned} p(y,t) = T_t^* g(y) &= \sum_{n=0}^{\infty} b_n e^{-\lambda_n t} Q_n(y) \\ &= \sum_{n=0}^{\infty} \left(\int g(x)Q_n(x)\mathbf{m}(x)\,dx \right) e^{-\lambda_n t} Q_n(y) \\ &= \int \left(\mathbf{m}(x) \sum_{n=0}^{\infty} e^{-\lambda_n t} Q_n(x)Q_n(y) \right) g(x)\,dx. \end{aligned} \tag{7.79}$$

Then

$$\mathbf{p}(x,t;y) = \mathbf{m}(x) \sum_{n=0}^{\infty} e^{-\lambda_n t} Q_n(x)Q_n(y) \tag{7.80}$$

is the transition density for the Pearson diffusion X_t. This heuristic argument is made rigorous in Proposition 7.21 in the details at the end of this section, which proves that (7.80) is the point source solution to the forward equation (7.60) and the backward equation (7.70) for Pearson diffusions of type (1–3).

A very similar separation of variables argument shows that

$$T_t f(x) = \int \mathbf{p}(x,t;y) f(y)\,dy$$

solves the forward equation (7.60) with initial condition $p(x,0) = f(x)$, where $\mathbf{p}(x,t;y)$ is given by (7.80), for any initial function such that

$$\frac{f(x)}{\mathbf{m}(x)} = \sum_{n=0}^{\infty} b_n Q_n(x) \tag{7.81}$$

where the series converges uniformly on compact sets. See [122, Theorem 3.3 and Remark 3.5] for details.

Example 7.15. A type (1) Pearson diffusion has $D(x)$ constant. Suppose that $D(x) = 1$, and $v(x) = -x$. Then equation (7.69) becomes

$$\frac{\mathbf{m}'(x)}{\mathbf{m}(x)} = -x,$$

and it is easy to check that the normal density

$$\mathbf{m}(x) = \frac{1}{\sqrt{2\pi}} e^{-x^2/2}$$

is a solution to this equation. The eigenfunction equation (7.75) becomes

$$\varphi'' - x\varphi' + \lambda\varphi = 0, \tag{7.82}$$

The eigenfunctions are the Hermite polynomials $H_n(x)$ and the corresponding eigenvalues are $\lambda_n = n$ for $n \geq 0$. The first three Hermite polynomials are $H_0 = 1$, $H_1(x) = x$, $H_2(x) = x^2 - 1$. Check that each of these functions solves (7.82) with $\lambda = n$.

Remark 7.16. Since we always have $Q_0(x) \equiv 1$, and since $\lambda_n > 0$ for all $n > 0$, it follows from (7.80) that $\mathbf{p}(x, t; y) \to \mathbf{m}(x)$ as $t \to \infty$ for any y, i.e., the Pearson diffusion X_t tends to the same steady state distribution $\mathbf{m}(x)$ regardless of the initial state $X_0 = y$. See [122, Theorems 4.6–4.8] for details.

Remark 7.17. The forward equation (7.60) can be derived from the backward equation (7.66) using integration by parts. Since the Pearson diffusion X_t is a Markov process, its transition densities satisfy the *Chapman-Kolmogorov equation*

$$\mathbf{p}(x, t+s; y) = \int \mathbf{p}(x, s; z)\mathbf{p}(z, t; y)\, dz \tag{7.83}$$

which adds up the probabilities of all the paths that transition from $X(0) = y$ to $X(t + s) = x$ through some point $X(t) = z$ in between (e.g., see Karlin and Taylor [102, p. 286]). Equation (7.83) can be established by an argument similar to (3.29). Let $p(x, t) = T_t f(x)$, and use (7.83) to write

$$p(x, s+t) = T_s T_t f(x) = T_s p(x, t) = \int \mathbf{p}(x, s; y) p(y, t)\, dy$$

for all $s,\ t > 0$. Observe that

$$\frac{\partial p(x, t+s)}{\partial t} = \frac{\partial p(x, t+s)}{\partial s},$$

and assuming that the derivative can be taken inside the integral, arrive at

$$\frac{\partial p(x, s+t)}{\partial t} = \int p(y, t) \frac{\partial \mathbf{p}(x, s; y)}{\partial s}\, dy.$$

Apply the backward equation (7.66) to get

$$\frac{\partial p(x, s+t)}{\partial t} = \int p(y, t) \left[v(y) \frac{\partial \mathbf{p}(x, s; y)}{\partial y} + D(y) \frac{\partial^2 \mathbf{p}(x, s; y)}{\partial y^2} \right] dy.$$

Integrate by parts twice to get

$$\frac{\partial p(x, s+t)}{\partial t} = \int \left(\frac{\partial^2}{\partial y^2} [D(y)p(y, t)] - \frac{\partial}{\partial y} [v(y)p(y, t)] \right) \mathbf{p}(x, s; y)\, dy,$$

assuming that the boundary terms vanish. Then let $s \to 0$, and use the fact that $\mathbf{p}(x, s; y) \to \delta(x - y)$ as $s \to 0$ to get the forward equation

$$\frac{\partial p(x, t)}{\partial t} = \frac{\partial^2}{\partial x^2} [D(x)u(x, t)] - \frac{\partial}{\partial x} [v(x)u(x, t)].$$

See [102, p. 219] and the details at the end of this section for more information.

Remark 7.18. We saw in Section 1.1 that the diffusion equation (1.9) governs a Brownian motion with drift, the scaling limit of a random walk $S_n = W_1 + \cdots + W_n$ with iid finite variance jumps. The forward equation (7.60) with parameters $v(x)$ and $D(x)$ governs the scaling limit of a Markov process $X_t = W_1 + \cdots + W_{N(t)}$, where $N(t)$ is a standard Poisson process with $\mathbb{E}[N(t)] = t$, and the jump distribution depends on the current state: Given $X_t = x$, the next jump has mean $v(x)$ and variance $2D(x)$. Then a suitably normalized version of the Markov process X_t converges to the Pearson diffusion with these coefficients. See Barkai, Metzler and Klafter [22] and Kolokoltsov [109] for additional details. In applications, this model is useful when the particle velocity $v(x)$ and dispersivity $D(x)$ vary in space.

Details

A Pearson diffusion is a Markov process defined on the state space $E = (a, b) \subseteq \mathbb{R}^1$, where we allow infinite endpoints. The interval (a, b) is chosen so that $D(x) > 0$ for $x \in (a, b)$. A *Markov process* on the state space E is a stochastic process on E with the *Markov property*:

$$\mathbb{P}[X_{t+s} \in B | X_t = y, X_{t_1} = y_1 \ldots, X_{t_n} = y_n] = \mathbb{P}[X_{t+s} \in B | X_t = y]$$

for any Borel set $B \subseteq E$, $s > 0$, $0 < t_1 < \cdots < t_n < t$, and $y, y_1, \ldots, y_n \in E$. We say that a Markov process X_t is time-homogeneous if

$$\mathbb{P}[X_{t+s} \in B | X_s = y] = \mathbb{P}[X_t \in B | X_0 = y].$$

Then the Markov process has stationary increments. A Lévy process is one example of a time-homogeneous Markov process, with independent increments.

The existence of a Markov process X_t on E whose backward semigroup has the generator (7.71) follows from Ikeda and Watanabe [95, Theorem 6.1]. That theorem proves the existence of a Markov process with generator (7.71), where $v(x)$ and $D(x)$ are continuous functions of x. Further, when the coefficients satisfy a local Lipschitz condition, the Markov process is unique. The local Lipschitz condition holds for the coefficients (7.61), and the corresponding Markov process is called a *Pearson diffusion*. The proof of [95, Theorem 6.1] is based on the theory of *stochastic differential equations*. The process X_t is defined as the solution to the stochastic differential equation

$$dX_t = v(X_t)dt + \sigma(X_t)dB_t$$

where $D(x) = \sigma^2(x)/2$ and B_t is a standard Brownian motion, see [95] for more details.

The general solution to (7.67) can be obtained as in Karlin and Taylor [102, p. 221]. Multiply both sides of (7.68) by the integrating factor

$$s(x) = \exp\left\{ -\int_{a_0}^{x} \frac{v(y)}{D(y)} dy \right\},$$

where a_0 is an arbitrary point in (a, b), and note that $s'(x) = -v(x)s(x)/D(x)$. Then (7.68) reduces to

$$\frac{d}{dx}\left(s(x)D(x)\mathbf{m}(x)\right) = C_1 s(x). \tag{7.84}$$

Equation (7.84) is solved by another integration

$$\mathbf{m}(x) = C_1\frac{S(x)}{s(x)D(x)} + C_2\frac{1}{s(x)D(x)}, \tag{7.85}$$

where

$$S(x) = \int_{a_0}^{x} s(y)dy$$

is called the *scale function* of the diffusion. The constants C_1 and C_2 are chosen so that $\mathbf{m}(x) > 0$ for $x \in E$, and $\int \mathbf{m}(x)dx = 1$. If a non-negative integrable solution of equation (7.67) does not exist, the stationary density does not exist. If a non-negative integrable solution of (7.67) exists, then it can be normalized so that it integrates to one [102, p. 221]. If the distribution of $X(0)$ has this density $\mathbf{m}(x)$, then $X(t)$ has the same density for all $t > 0$ (e.g., see [102, p. 220]). For Pearson diffusions, we choose $C_1 = 0$.

To prove that (7.80) is the transition density of a type (1–3) Pearson diffusion, use [122, Remark 3.4] to see that any smooth function $g(y)$ with compact support in E can be written in the form (7.78), where the series converges uniformly on compact sets. Since the indicator function of any compact interval $B \in E$ can be approximated arbitrarily closely by such functions, it follows that

$$\mathbb{P}[g(X_t)|x_0 = y] = \int_{x\in B} \mathbf{p}(x, t; y)g(x)\,dx$$

for all such intervals. Then it follows that $\mathbf{p}(x, t; y)$ is the conditional density of X_t given $X_0 = y$. The Fubini argument in (7.79) can be justified using Lemma 7.28.

In Remark 7.17, we outlined the derivation of the forward equation from the backward equation, following the brief discussion in [102, p. 219]. Here we provide some additional detail. As discussed in [69, 70, 102], the backward equation plays a central role in the theory of diffusion processes. Some analytical difficulties arise when considering the forward equation. Suppose that $f, g : E \to \mathbb{R}$ are twice continuously differentiable and have compact support in E. Then it is easy to check, using integration by parts, that $\int [\mathcal{L}f(x)]g(x)\,dx = \int f(y)[\mathcal{G}g(y)]\,dy$. That is, we have $\langle \mathcal{L}f, g\rangle = \langle f, \mathcal{G}g\rangle$ where the inner product $\langle f, g\rangle = \int f(x)g(x)\,dx$. The derivation of Remark 7.17 also assumed that the derivative can be passed through the integral. Since this does not always hold in general, the forward equation does not follow directly from the backward equation without additional assumptions.

There are six types of Pearson diffusions, corresponding to six classes of solutions to the ordinary differential equation (7.75). The solutions vary depending on the degree of polynomial $D(x)$ in (7.61) (zero, one, or two) and, if $D(x)$ has degree two, on the discriminant of $D(x)$ (zero, positive, or negative), see [11, 72]. These solution classes also vary in terms of the spectrum of the operator $\mathcal{G}$: The *spectrum* of a linear operator A defined on a Banach space $\mathbb{B}$ is the set of complex numbers λ such that $A - \lambda I$ has no bounded inverse. If $\mathbb{B} = \mathbb{R}^n$ and A is an $n \times n$ matrix, then the spectrum is the set of eigenvalues of the matrix A. If $\mathbb{B}$ is a space of functions, and if $Af = \lambda f$ for some $f \neq 0$ in $\mathbb{B}$, then f is an eigenfunction of A with eigenvalue λ, and λ belongs to the spectrum of A. For the first three types of Pearson diffusions, the spectrum of the operator $\mathcal{G}$ is purely discrete, and the sequence of eigenvalues increases to infinity. For the three remaining types of Pearson diffusions, the spectrum has a (possibly empty) finite discrete part, and a continuous part called the *essential spectrum*. A complete description of all six classes of Pearson diffusions is included below.

If the spectrum of the generator $\mathcal{G}$ is purely discrete, the Sturm-Liouville problem (7.73) is solved by an infinite system of classical orthogonal polynomials $\{Q_n\}$. This system is called *orthonormal* if $c_n^2 = 1$ for all n in (7.76). Then this system of polynomials forms an *orthonormal basis* for the space of functions $L^2(E, \mathbf{m}(x)\,dx)$ consisting of all Borel measurable functions $f : E \to \mathbb{R}$ such that $\int |f(x)|^2\mathbf{m}(x)dx < \infty$, with the inner product $\langle f, g\rangle_{\mathbf{m}} = \int f(x)g(x)\mathbf{m}(x)\,dx$. Any function $g \in L^2(E, \mathbf{m}(x)\,dx)$ can be written in the form (7.78) for some constants b_n, where the series on the right hand side of (7.78) converges in the L^2 sense:

$$\int \left| g(x) - \sum_{n=0}^{N} b_n Q_n(x) \right|^2 \mathbf{m}(x)\,dx \to 0 \quad \text{as } N \to \infty.$$

The coefficients in (7.78) are computed using the inner product: $g_n = \langle g, Q_n\rangle_{\mathbf{m}}$ for all n. Some additional technical conditions (see Szegő [212, pp. 245–248] and [186, p. 381]) are needed to assert that (7.78) holds point-wise.

Remark 7.19. In this remark, we catalog the six types of Pearson diffusions in terms of their invariant densities, and their polynomial families of eigenfunctions. Types (1–3) have a purely discrete spectrum, and types (4–6) have a mixture of discrete and continuous spectrum.

(1) The Ornstein-Uhlenbeck (OU) process is obtained when $D(x)$ in (7.61) is a constant. The traditional parametrization for the process is

$$v(x) = -\theta(x - \mu), \qquad D(x) = \theta\sigma^2.$$

The convenience of this parametrization is in separating the distributional and covariance parameters. For a stationary OU process, θ is a correlation function parameter, and μ and σ are distribution parameters. The state space is $E = \mathbb{R}^1$ and the stationary distribution is normal:

$$\mathbf{m}(x) = \frac{1}{\sigma\sqrt{2\pi}} \exp\left[-\frac{(x-\mu)^2}{2\sigma^2}\right], \qquad x \in \mathbb{R}. \tag{7.86}$$

When $\theta < 0$, the process is transient. When $\theta > 0$, the diffusion is a stationary OU process when the initial distribution has density $\mathbf{m}(x)$. The eigenvalues of $(-\mathcal{G})$ are $\lambda_n = \theta n$, $n \geq 0$. The corresponding eigenfunctions are Hermite polynomials. The first three Hermite polynomials are $H_0 = 1$, $H_1(x) = x$, $H_2(x) = x^2 - 1$. The *Rodrigues formula*

$$H_n(x) = (-1)^n \left[\mathbf{m}(x)\right]^{-1} \frac{d^n}{dx^n}\mathbf{m}(x), \quad x \in \mathbb{R}, \; n = 0, 1, 2, \ldots$$

can be used to compute the remaining polynomials.

(2) The Cox-Ingersoll-Ross (CIR) process is obtained when $D(x)$ is a first degree polynomial $D(x) = d_1 x + d_0$. We may suppose $d_0 = 0$ (after normalizing, which would change a_0 to $a_0 - a_1 d_0/d_1$). If $d_1 > 0$ then the process is a CIR (square root Feller) diffusion on the state space $E = (0, \infty)$, see Cox, Ingersoll and Ross [52]. If $d_1 < 0$, then the state space is $E = (-\infty, 0)$, where $D(x)$ is positive. This case can be reduced to the case $d_1 > 0$ by a simple reparametrization. The traditional parametrization of the CIR process is

$$v(x) = -\theta\left(x - \frac{b}{a}\right), \quad D(x) = \frac{\theta}{a}x.$$

The invariant density is gamma:

$$\mathbf{m}(x) = \frac{a^b}{\Gamma(b)} x^{b-1} e^{-ax} \quad x > 0. \tag{7.87}$$

The eigenvalues are $\lambda_n = \theta n$, $n \geq 0$. The orthogonal polynomials are the Laguerre polynomials $L_n^{(b-1)}(ax)$ for $n = 0, 1, 2, \ldots$ where

$$L_n^{(\gamma)}(x) = \frac{1}{n!} x^{-\gamma} e^x \frac{d^n}{dx^n}\left[x^{n+\gamma} e^{-x}\right].$$

(3) The Jacobi diffusion process is obtained when $D(x)$ is a second degree polynomial with positive discriminant. Suppose $D(x) = d_2(x - x_1)(x - x_2)$, and $d_2 < 0$. The state space is $E = (x_1, x_2)$ with $x_1 < x_2$. After rescaling we may assume $d_2 = -1$, and after a linear change of variables $\tilde{x} = 2x - (x_1 + x_2)/(x_2 - x_1)$, we can take

$$v(x) = -(a + b + 2)x + b - a, \quad D(x) = 1 - x^2.$$

In the recurrent case $a, b > -1$, we obtain the Beta density:

$$\mathbf{m}(x) = (1 - x)^a (1 + x)^b \frac{\Gamma(a + b + 2)}{\Gamma(b + 1)\Gamma(a + 1)2^{a+b+1}}, \quad x \in (-1, 1). \tag{7.88}$$

The eigenvalues are $\lambda_n = n(n + a + b + 1)$, $n \geq 0$. The orthogonal polynomials are Jacobi polynomials given by the formula:

$$2^n n! P_n^{(a,b)}(x) = (-1)^n (1 - x)^{-a} (1 + x)^{-b} \frac{d^n}{dx^n}\left\{(1 - x)^{a+n} (1 + x)^{b+n}\right\}.$$

(4) The Student diffusion process is obtained when $D(x)$ is a second degree polynomial with negative discriminant, and $d_2 > 0$. The state space is $E = \mathbb{R}$, and the traditional parametrization is

$$\nu(x) = -\theta(x-\mu), \quad D(x) = \theta a\left[(x-\mu')^2 + \delta^2\right]$$

The invariant density is

$$\mathbf{m}(x) = c(\mu, \mu', a, \delta)\frac{\exp\left[\left(\frac{\mu-\mu'}{a\delta}\right)\operatorname{Arctan}\left(\frac{x-\mu'}{\delta}\right)\right]}{\left[1+\left(\frac{x-\mu'}{\delta}\right)^2\right]^{1+1/(2a)}},$$

where $x \in \mathbb{R}$, $1 + 1/(2a) > 1/2$, and

$$c(\mu, \mu', a, \delta) = \frac{\Gamma\left(1+\frac{1}{2a}\right)}{\delta\sqrt{\pi}\,\Gamma\left(\frac{1}{2}+\frac{1}{2a}\right)}\prod_{k=0}^{\infty}\left[1+\left(\frac{\frac{\mu-\mu'}{2a\delta}}{1+\frac{1}{2a}+k}\right)^2\right]^{-1}.$$

Note that in the symmetric case ($\mu = \mu'$) the density function is

$$\mathbf{m}(x) = c(\mu, a, \delta)\frac{1}{\left[1+\left(\frac{x-\mu}{\delta}\right)^2\right]^{1+1/(2a)}}, \quad x \in \mathbb{R}.$$

In the classical parametrization for the Student distribution, with degrees of freedom $\nu = 1 + (1/a)$, this reduces to

$$\mathbf{m}(x) = \frac{\Gamma\left(\frac{\nu+1}{2}\right)}{\delta\sqrt{\pi}\,\Gamma\left(\frac{\nu}{2}\right)}\frac{1}{\left[1+\left(\frac{x-\mu}{\delta}\right)^2\right]^{(\nu+1)/2}}, \quad x \in \mathbb{R}.$$

Only a finite number of central moments of the invariant distribution exist; the nth central moment exists if $n < \nu$. Also, the invariant density has heavy tails that decrease like $|x|^{-(1+\nu)}$.

In this case, there are only finitely many simple eigenvalues in $[0, \Lambda]$, where $\Lambda = \theta\nu^2/(4(\nu-1))$, $\nu > 1$, and the absolutely continuous spectrum of multiplicity two is in (Λ, ∞), see Leonenko and Šuvak [121]. The simple eigenvalues are

$$\lambda_n = \frac{\theta}{\nu-1}n(\nu-n), \quad 0 \le n \le \left\lfloor\frac{\nu}{2}\right\rfloor.$$

The orthogonal polynomials are generalized Romanovski polynomials given by the Rodrigues formula:

$$R_0(x) = 1,$$

$$R_n(x) = \delta^n\left[1+\left(\frac{x-\mu}{\delta}\right)^2\right]^{(\nu+1)/2}\frac{d^n}{dx^n}\left[1+\left(\frac{x-\mu}{\delta}\right)^2\right]^{n-(\nu+1)/2}$$

$$n = 1, \ldots, \left\lfloor\frac{\nu}{2}\right\rfloor.$$

(5) The reciprocal gamma diffusion is obtained in the case of zero discriminant, with the polynomial $D(x)$ proportional to x^2 (after a change of variables). The coefficients are

$$v(x) = -\theta\left(x - \frac{a}{b-1}\right), \quad D(x) = \frac{\theta}{b-1}x^2,$$

where $\theta > 0$, $a > 0$, $b > 1$. The invariant density is the inverse gamma:

$$\mathbf{m}(x) = \frac{a^b}{\Gamma(b)}x^{-b-1}e^{-a/x}, \quad x > 0.$$

This is a heavy tailed diffusion, whose moments of order n exist only for $n < b$. The discrete part of the spectrum of $(-\mathcal{G})$ consists of finitely many eigenvalues given by

$$\lambda_n = \frac{n\theta(b-n)}{b-1}, \quad 0 \le n \le \left\lfloor \frac{b}{2} \right\rfloor.$$

These eigenvalues lie within $[0, \Lambda]$, and the continuous part of the spectrum has multiplicity one and lies inside (Λ, ∞), see Leonenko and Šuvak [120], where the cut-off

$$\Lambda = \frac{\theta b^2}{4(b-1)}.$$

The orthogonal polynomials in the point spectrum case are generalized Bessel polynomials:

$$\widetilde{B}_0(x) = 1,$$
$$\widetilde{B}_n(x) = x^{b+1}e^{(a/x)}\frac{d^n}{dx^n}\left[x^{2n-(b+1)}e^{-(a/x)}\right], \quad n \in \left\{1, \dots, \left\lfloor \frac{b}{2} \right\rfloor\right\}, \quad b > 1.$$

(6) The Fisher-Snedecor diffusion is obtained when $D(x)$ is a second degree polynomial with positive discriminant. After transformations, we can assume that the first root of $D(x)$ is negative, and the second is zero, so the state space is $E = (0, \infty)$. The coefficients are

$$v(x) = -\theta\left(x - \frac{b}{b-2}\right), \quad D(x) = \frac{\theta}{a(b-2)}x(ax+b)$$

where the parameters $a \ge 2$, $b > 2$, and $\theta > 0$. The invariant density is the density of F-distribution (also known as Fisher-Snedecor distribution):

$$\mathbf{m}(x) = \frac{x^{(a/2)-1}(ax+b)^{-(a+b)/2}}{a^{-(a/2)}b^{-b/2}B(a/2,\, b/2)}, \quad x > 0,$$

where B is the beta function

$$B(a, b) = \int_0^1 x^{a-1}(1-x)^{b-1}dx = \frac{\Gamma(a)\Gamma(b)}{\Gamma(a+b)}, \quad a > 0,\ b > 0.$$

This is another heavy-tailed diffusion. The moments of order $2n$ exist when $2n < b$. Only finitely many discrete eigenvalues exist, and they are given by the formula

$$\lambda_n = \frac{\theta}{b-2} n(b-2n), \quad 0 \le n \le \left\lfloor \frac{b}{4} \right\rfloor, \quad b > 2.$$

The cut-off for the discrete spectrum is

$$\Lambda = \frac{\theta b^2}{8(b-2)}, \quad b > 2,$$

so that the essential spectrum lies in (Λ, ∞). The essential spectrum has multiplicity one, see Avram, Leonenko and Šuvak [10]. The orthogonal polynomials have no common name in this case; in [10] they are called Fisher-Snedecor polynomials since they are orthogonal with respect to the Fisher-Snedecor density. These polynomials $\{\widetilde{F}_n(x),\ n = 0, 1, \ldots, \lfloor b/4 \rfloor\}$ are given by the Rodrigues formula:

$$\begin{aligned}
\widetilde{F}_0(x) &= 1, \\
\widetilde{F}_n(x) &= x^{1-(a/2)}(ax+b)^{(a+b)/2} \frac{d^n}{dx^n} \left[2^n x^{(a/2)+n-1} (ax+b)^{n-(a+b)/2} \right], \\
n &\in \left\{ 1, \ldots, \left\lfloor \frac{b}{4} \right\rfloor \right\}.
\end{aligned}$$

Remark 7.20. The heavy tailed Pearson diffusions (4–6) have only a finite number N of orthogonal polynomials, because only a finite number of moments exist for the invariant distribution. Since $Q_n(x)$ is the polynomial of degree n, $Q_n^2(x)$ has degree $2n$. For case (4), moments of order $2n$ exists only for $2n < \nu$, so $N = \lfloor \nu/2 \rfloor$. For case (5), these moments exists only for $2n < b$, so $N = \lfloor b/2 \rfloor$. For case (6), moments of order $2n < b/2$ exist, so that $N = \lfloor b/4 \rfloor$.

Proposition 7.21. *For the Pearson diffusions (1–3) the series*

$$\mathbf{p}(x, t; y) = \mathbf{m}(x) \sum_{n=0}^{\infty} e^{-\lambda_n t} Q_n(y) Q_n(x), \tag{7.89}$$

where $\{Q_n,\ n \ge 0\}$ are Hermite, Laguerre or Jacobi polynomials, converges for fixed $t > 0$, $x, y \in E$. Equation (7.89) *can be differentiated term by term on any finite intervals $t \in [t_1, t_2]$, $0 < t_1 < t_2$, $x, y \in [x_1, x_2] \subset E$, and hence the function $\mathbf{p}(x, t; y)$ in* (7.89) *satisfies the backward and forward equations* (7.70) *and* (7.60).

Proof. Recall that the eigenvalues are $\lambda_n = \theta n$ in the Hermite and Laguerre cases (1–2), and $\lambda_n = n(n + a + b + 1)$ in the Jacobi case (3). In the rest of the proof, we will assume without loss of generality that $\mu = 0$ and $\sigma = 1$ in the OU case (1), and $a = 1$ in the CIR case (2).

The *orthonormal* Hermite polynomials

$$\bar{H}_n(x) = H_n(x)/\|H_n(x)\| = \frac{1}{\sqrt{n!}} H_n(x), \quad n = 0, 1, 2, \ldots..$$

satisfy (7.76) with $Q_n = \bar{H}_n$ and $c_n^2 = 1$ for all n. For orthonormal Hermite polynomials (e.g., see Sansone [186, p. 369])

$$\bar{H}_n(x) \le K e^{x^2/4} n^{-1/4}(1 + |x/\sqrt{2}|^{5/2}), \tag{7.90}$$

where K is a constant that does not depend on x.

To make the system of Laguerre polynomials orthonormal, we use the fact that

$$\int_0^\infty |L_n^{(b-1)}(x)|^2 x^{b-1} e^{-x} dx = \frac{\Gamma(b+n)}{n!}.$$

The orthonormal system of polynomials in this case is given by

$$\bar{L}_n^{(b-1)}(x) = \frac{L_n^{(b-1)}(ax)}{\sqrt{\Gamma(b+n)/(\Gamma(b)n!)}}.$$

For orthonormal Laguerre polynomials [186, p. 348]

$$\bar{L}_n^{(b-1)}(x) = O\left(\frac{e^{x/2}}{x^{(2b-1)/4}} n^{-1/4}\right), \tag{7.91}$$

uniformly for x in finite intervals $[x_1, x_2]$.

Finally, for Jacobi polynomials using the fact that

$$\int_{-1}^1 (P_n^{(a,b)}(x))^2 (1-x)^a (1+x)^b dx = c_n^2 = \frac{2^{a+b+1}}{2n+a+b+1} \frac{\Gamma(n+a+1)\Gamma(n+b+1)}{\Gamma(n+1)\Gamma(n+a+b+1)},$$

we obtain the orthonormal system

$$\bar{P}_n^{(a,b)}(x) = \frac{P_n^{(a,b)}(x)}{c_n}. \tag{7.92}$$

From [212, p. 196] we have

$$\bar{P}_n^{(a,b)}(x) = C(x, a, b)\cos(N\theta + \gamma) + O(n^{-1}), \tag{7.93}$$

where $x = \cos\theta$, $N = n + 1/2(a + b + 1)$, and $\gamma = -(a + 1/2)\pi/2$.

Convergence of the series (7.89) for fixed $x, y \in E$ and $t > 0$ follows from the above relations. Specifically, in the Hermite case,

$$|e^{-\lambda_n t} Q_n(y) Q_n(x)| \le \frac{C(x, y)e^{-n\theta t}}{n^{1/2}}.$$

In the Laguerre case,

$$|e^{-\lambda_n t} Q_n(y) Q_n(x)| \le \frac{C(x, y)e^{-n\theta t}}{n^{1/2}}.$$

In the Jacobi case

$$|e^{-\lambda_n t} Q_n(y) Q_n(x)| \le C(x, y) e^{-n(n+a+b+1)t}.$$

Above and later in the proof, we use notation $C(x, y)$ for constants not all equal but not dependent on n. These constants may also depend on the parameters of the distributions (i.e., the coefficients $v(x)$ and $D(x)$ in (7.61)).

Now we show that the series in (7.89) can be differentiated term by term, and in view of standard results in analysis (e.g., see Rudin [181, Theorem 7.16, p. 151; Theorem 7.17, p. 152]) this would follow from absolute and uniform convergence on finite intervals of the series that involve the derivatives:

$$\sum_{n=0}^{\infty} \frac{\partial}{\partial t} e^{-\lambda_n t} Q_n(y) Q_n(x),$$

$$\sum_{n=0}^{\infty} e^{-\lambda_n t} Q_n'(y) Q_n(x),$$

$$\sum_{n=0}^{\infty} e^{-\lambda_n t} Q_n''(y) Q_n(x).$$

For the derivative with respect to t, we have

$$\left| \frac{\partial}{\partial t} e^{-\lambda_n t} Q_n(y) Q_n(x) \right| \le C(x, y) \theta n^{1/2} e^{-n\theta t}$$

for the Hermite and Laguerre cases, and for the Jacobi case

$$\left| \frac{\partial}{\partial t} e^{-\lambda_n t} Q_n(y) Q_n(x) \right| \le C(x, y) n(n + a + b + 1) e^{-n(n+a+b+1)t}.$$

The terms on the right hand side of the two inequalities above form series that converge uniformly for $t \in [t_1, t_2]$, $x, y \in [x_1, x_2] \subset E$. For the derivatives with respect to y, we use the properties of Hermite, Laguerre, and Jacobi polynomials. For the Hermite series involving derivatives in y, we use the relation (e.g., see Abramowitz and Stegun [3, p. 783]):

$$\frac{d}{dx} H_n(x) = n H_{n-1}(x).$$

For orthonormal Hermite polynomials,

$$\frac{d}{dx} \bar{H}_n(x) = \frac{n}{\sqrt{n!}} H_{n-1}(x) = \sqrt{n} \bar{H}_{n-1}(x),$$

and so in this case

$$\left| e^{-\lambda_n t} Q_n'(y) Q_n(x) \right| \le C(x, y) e^{-n\theta t} \left(\frac{n}{n-1} \right)^{1/4}.$$

For the second derivative in space, use the differential equation (7.75):

$$H_n''(y) = y H_n'(y) - n H_n(y).$$

The series involving the first derivative in space was treated above, and for the second term

$$\left| e^{-\lambda_n t} n \bar{H}_n(y) \bar{H}_n(x) \right| \le C(x, y) e^{-n\theta t} \sqrt{n},$$

and this upper bound leads to the series that converge uniformly for $t \in [t_1, t_2]$.

For Laguerre polynomials (e.g., see Szegő [212, p. 102])

$$\frac{d}{dx} L_n^{(b-1)}(x) = -L_{n-1}^{(b)}(x),$$

and for orthonormal Laguerre polynomials

$$\frac{d}{dx} \bar{L}_n^{(b-1)}(x) = -\frac{(n-1)^{b/2}}{n^{(b-1)/2}} \bar{L}_{n-1}^{(b)}(x).$$

The last quantity behaves like $C(x, b)n^{1/4}$ uniformly on finite intervals (see Sansone [186, p. 348]). Therefore in this case

$$\left| e^{-\lambda_n t} Q_n'(y) Q_n(x) \right| \le C(x, y) e^{-n\theta t}$$

and the rest of the argument for the series involving the first derivative in space is the same as for Hermite polynomials. The same argument also applies to the second derivative in space because, for Laguerre polynomials, equation (7.75) has the form

$$y \frac{d^2}{dy^2} L_n^{(b-1)} = (y - b) \frac{d}{dy} L_n^{(b-1)}(y) - n L_n^{(b-1)}(y).$$

For Jacobi polynomials,

$$(2n + a + b)(1 - x^2) \frac{d}{dx} P_n^{(a,b)}(x) = n(a - b - (2n + a + b)x) P_n^{(a,b)}(x) +$$
$$2(n + a)(n + b) P_{n-1}^{(a,b)}(x)$$

and for orthonormal Jacobi polynomials

$$\frac{d}{dx} \bar{P}_n^{(a,b)}(x) = \frac{n(a - b - (2n + a + b)x)}{(2n + a + b)(1 - x^2)} \bar{P}_n^{(a,b)}(x) + \frac{2(n + a)(n + b)}{(2n + a + b)(1 - x^2)} \sqrt{n/(n-1)} \bar{P}_{n-1}^{(a,b)}(x).$$

The first term in the last relation leads to the series

$$\sum_n n e^{-\lambda_n t} \bar{P}_n^{(a,b)}(y) \bar{P}_n^{(a,b)}(x)$$

that converges because it is dominated by the absolutely convergent series

$$C(x, y) \sum_n n e^{-n(n+a+b+1)t}.$$

The second term in the expression for the derivative of the normalized Jacobi polynomial behaves in the same way as the first, and finally, the expression for the second derivative from (7.75) is

$$(1 - y^2) \frac{d^2}{dy^2} P_n^{(a,b)}(y) = -((b - a) - (a + b - 2)y) \frac{d}{dy} P_n^{(a,b)}(y) - n(n + a + b + 1) P_n^{(a,b)}(y).$$

The term with the first derivative was treated above. The second term leads to the series

$$\sum_n e^{-\lambda_n t} n(n+a+b+1)\bar{P}_n^{(a,b)}(y)\bar{P}_n^{(a,b)}(x)$$

which is dominated by the series $C(x,y)\sum_n n^2 e^{-n(n+a+b+1)t}$. This completes the proof of term by term differentiation of (7.89).

It remains to note that each term of the series in (7.89) satisfies the backward and forward equations. For the backward equation, with operator $\mathcal{G}$ acting on y,

$$\mathcal{G}\mathbf{m}(x)\,e^{-\lambda_n t}Q_n(y)Q_n(x) = -\lambda_n\mathbf{m}(x)\,e^{-\lambda_n t}Q_n(y)Q_n(x) = \frac{\partial}{\partial t}\mathbf{m}(x)\,e^{-\lambda_n t}Q_n(y)Q_n(x),$$

since $Q_n(y)$ is an eigenfunction of $(-\mathcal{G})$ with the eigenvalue λ_n. For the forward (Fokker-Planck) equation, the left hand side is

$$\frac{\partial}{\partial t}[\mathbf{m}(x)\,e^{-\lambda_n t}Q_n(y)Q_n(x)] = -\lambda_n\mathbf{m}(x)\,e^{-\lambda_n t}Q_n(y)Q_n(x).$$

For the right-hand side, use the fact that $\mathbf{m}(x)$ satisfies time-independent Fokker-Planck equation, and therefore

$$\frac{1}{2}\frac{d^2}{dx^2}[\sigma^2(x)\mathbf{m}(x)]e^{-\lambda_n t}Q_n(y)Q_n(x) - \frac{d}{dx}[\mu(x)\mathbf{m}(x)]e^{-\lambda_n t}Q_n(y)Q_n(x) = 0.$$

Then the right-hand side of the equation is

$$\begin{aligned}
&\frac{1}{2}\frac{\partial^2}{\partial x^2}[\sigma^2(x)\mathbf{m}(x)e^{-\lambda_n t}Q_n(y)Q_n(x)] - \frac{\partial}{\partial x}[\mu(x)\mathbf{m}(x)e^{-\lambda_n t}Q_n(y)Q_n(x)]\\
&\quad = \frac{1}{2}\sigma^2(x)\mathbf{m}(x)\,e^{-\lambda_n t}Q_n(y)Q_n''(x) + \frac{d}{dx}[\sigma^2(x)\mathbf{m}(x)]e^{-\lambda_n t}Q_n(y)Q_n'(x)\\
&\qquad + \frac{1}{2}\frac{d^2}{dx^2}[\sigma^2(x)\mathbf{m}(x)]e^{-\lambda_n t}Q_n(y)Q_n(x)\\
&\quad - \frac{d}{dx}[\mu(x)\mathbf{m}(x)]e^{-\lambda_n t}Q_n(y)Q_n(x) - \mu(x)\mathbf{m}(x)\,e^{-\lambda_n t}Q_n(y)Q_n'(x)\\
&\quad = \frac{1}{2}\sigma^2(x)\mathbf{m}(x)\,e^{-\lambda_n t}Q_n(y)Q_n''(x) + \frac{d}{dx}[\sigma^2(x)\mathbf{m}(x)]e^{-\lambda_n t}Q_n(y)Q_n'(x)\\
&\quad - \mu(x)\mathbf{m}(x)\,e^{-\lambda_n t}Q_n(y)Q_n'(x).
\end{aligned}\tag{7.94}$$

Using the fact that $\mathbf{m}(x)$ satisfies (7.68) with $C_1 = 0$, i.e,

$$\frac{d}{dx}[\sigma^2(x)\,\mathbf{m}(x)] = 2\,\mathbf{m}(x)\,\mu(x),$$

equation (7.94) reduces to

$$\begin{aligned}
&\frac{1}{2}\sigma^2(x)\mathbf{m}(x)\,e^{-\lambda_n t}Q_n(y)Q_n''(x) + 2\,\mathbf{m}(x)\,\mu(x)e^{-\lambda_n t}Q_n(y)Q_n'(x)\\
&\quad - \mu(x)\mathbf{m}(x)\,e^{-\lambda_n t}Q_n(y)Q_n'(x)\\
&\quad = \frac{1}{2}\sigma^2(x)\mathbf{m}(x)\,e^{-\lambda_n t}Q_n(y)Q_n''(x) - \mathbf{m}(x)\,\mu(x)e^{-\lambda_n t}Q_n(y)Q_n'(x)\\
&\quad = \mathbf{m}(x)\,e^{-\lambda_n t}Q_n(y)[\mathcal{G}Q_n(x)] = -\lambda_n\mathbf{m}(x)\,e^{-\lambda_n t}Q_n(y)Q_n(x)
\end{aligned}$$

which finishes the proof. □

Remark 7.22. As discussed in Feller [69], it follows from Proposition 7.21 that the solutions of Cauchy problems for the forward equation (7.63) and the backward equation (7.66) are given by

$$T_t^* f(x) = \int \mathbf{p}(x,t;y) f(y)\,dy$$

and

$$T_t f(x) = \int \mathbf{p}(x,t;y) g(x)\,dx$$

respectively, where the transition density $\mathbf{p}(x,t;y)$ is given by (7.89).

7.7 Fractional Pearson diffusions

The time-fractional diffusion equation with constant coefficients

$$\partial_t^\beta p(x,t) = -\frac{\partial}{\partial x}[vp(x,t)] + \frac{\partial^2}{\partial x^2}[Dp(x,t)] \tag{7.95}$$

from (4.52) governs the scaling limit of a CTRW with finite variance jumps and power law waiting times, see Remark 4.23. The CTRW scaling limit process $A'(E_t)$ whose probability densities $p(x,t)$ solve this time-fractional diffusion equation is a Brownian motion with drift, where the time variable t has been replaced by an independent inverse stable subordinator E_t. In this section, we allow the coefficients of the time-fractional forward equation (7.95) to vary in space, thereby extending the results of Section 7.6 to the case of a time-fractional derivative. First we consider a time-fractional backward equation

$$\partial_t^\beta p = \mathcal{G}p(y,t) = v(y)\frac{\partial}{\partial y}p(y,t) + D(y)\frac{\partial^2}{\partial y^2}p(y,t) \tag{7.96}$$

with initial condition $p(y,0) = g(y)$. Note that x is a constant in this equation. The Caputo fractional derivative ∂_t^β of order $0 < \beta \le 1$ in (7.96) is defined by (2.33). Equation (7.96) is the time-fractional analog of the backward equation (7.66) considered in Section 7.6.

The fractional backward equation (7.96) governs a stochastic process that is not Markovian. Let D_t be a standard stable subordinator with Laplace transform

$$E[e^{-sD_t}] = \exp\{-ts^\beta\},\ s \ge 0. \tag{7.97}$$

As in Section 2.3, we define the inverse (hitting time, first passage time) process

$$E_t = \inf\{x > 0 : D_x > t\}. \tag{7.98}$$

Let $X_1(t)$ be a Pearson diffusion whose transition densities $\mathbf{p}_1(x,t;y)$ solve the backward Kolmogorov (7.66) and forward Fokker-Planck equation (7.60) with the point source initial condition $\mathbf{p}_1(x, 0; y) = \delta(x - y)$. Define the *fractional Pearson diffusion process*

$$X_\beta(t) = X_1(E_t), \ t \geq 0. \tag{7.99}$$

Since E_t rests for periods of time whose distribution is not exponential, $X_\beta(t)$ is not a Markov process.

Given a C_0 semigroup T_t on some Banach space $\mathbb{B}$, Theorem 3.16 shows that $q(t) = T_t f$ solves the Cauchy problem

$$\frac{d}{dt} q = Lq; \ q(0) = f \tag{7.100}$$

for any $f \in \mathrm{Dom}(L)$. If we replace the first derivative d/dt in (7.100) by a Caputo fractional derivative of order $0 < \beta < 1$, we obtain the *fractional Cauchy problem*

$$\partial_t^\beta p = Lp; \ p(0) = f. \tag{7.101}$$

Then a general result on semigroups, Baeumer and Meerschaert [18, Theorem 3.1], shows that

$$p(t) = S_t f = \int_0^\infty T_{(t/r)^\beta} f \, g_\beta(r) \, dr \tag{7.102}$$

solves the fractional Cauchy problem (7.101) for any $f \in \mathrm{Dom}(L)$. Here $g_\beta(r)$ is the probability density function of a standard stable subordinator D_1 with Laplace transform (7.97). A simple change of variable $u = (t/r)^\beta$ in (7.102) leads to an equivalent form

$$S_t f = \int_0^\infty T_u f \, \frac{t}{\beta} u^{-1-1/\beta} g_\beta(t u^{-1/\beta}) \, du. \tag{7.103}$$

The main ideas behind the proof of [18, Theorem 3.1] were illustrated in the derivation of (4.48). In particular, using equation (4.47) we can see that

$$S_t f = \int_0^\infty T_u f \, h(u, t) \, du \tag{7.104}$$

where $h(u, t)$ is the pdf of the inverse stable subordinator (7.98).

Remark 7.23. The mathematical study of fractional Cauchy problems was initiated by Kochubei [105, 106] and Schneider and Wyss [192]. Fractional Cauchy problems were also invented independently by Zaslavsky [222] as a model for Hamiltonian chaos, see also Saichev and Zaslavsky [183].

Now we apply (7.104) to the time-fractional backward equation (7.96) of a Pearson diffusion. Proposition 7.27, in the details at the end of this section, shows that

$$T_t^* g(y) = E[g(X_1(t)) | X_1(0) = y]$$

is a C_0 semigroup, and then it follows from Theorem 3.16 that

$$q(y,t) = T_t^* g(y) = E[g(X_1(t))|X_1(0) = y] = \int \mathbf{p}_1(x,t;y)g(x)dx$$

solves the Cauchy problem

$$\frac{\partial q}{\partial t} = \mathcal{G}q, \quad q(y,0) = g(y) \tag{7.105}$$

for any $g \in \mathrm{Dom}(\mathcal{G})$, where the transition density $\mathbf{p}_1(x,t;y)$ is given by (7.80). Then (7.104) implies that

$$p(y,t) = S_t g(y) = \int_0^\infty T_u g(y)\, h(u,t)\, du \tag{7.106}$$

solves the fractional Cauchy problem (7.96) for any $g \in \mathrm{Dom}(\mathcal{G})$. Now write

$$\begin{aligned} S_t g(y) &= \int_0^\infty T_u g(y)\, h(u,t)\, du \\ &= \int_0^\infty \mathbb{E}[g(X_1(u))|X_1(0) = y]\, h(u,t)\, du \\ &= \mathbb{E}[g(X_1(E_t))|X_1(0) = y] \\ &= \mathbb{E}[g(X_\beta(t))|X_\beta(0) = y] \end{aligned} \tag{7.107}$$

since $E_0 = 0$ almost surely. This shows that the fractional Pearson diffusion $X_\beta(t) = X_1(E_t)$ is governed by the time-fractional backward equation (7.96).

We will say that the non-Markovian Pearson diffusion process $X_\beta(t)$ has a *transition density* $\mathbf{p}_\beta(x,t;y)$ if

$$\mathbb{P}[X_\beta(t) \in B|X_\beta(0) = y] = \int_B \mathbf{p}_\beta(x,t;y)\, dx$$

for any Borel subset B of the state space E. That is, the transition density is the conditional probability density of $X_\beta(t)$, given $X_\beta(0) = y$. Since $\mathbf{p}_1(x,t;y)$ is the transition density of the Pearson diffusion $X_1(t)$, a simple conditioning argument shows that the transition density of the fractional Pearson diffusion $X_1(E_t)$ is

$$\mathbf{p}_\beta(x,t;y) = \int_0^\infty \mathbf{p}_1(x,u;y)h(u,t)\, du \tag{7.108}$$

where $h(u,t)$ is the pdf (4.47) of the inverse stable subordinator (7.98). Then we can write (7.107) in the form

$$S_t g(y) = \mathbb{E}[g(X_\alpha(t))|X_0 = y] = \int \mathbf{p}_\beta(x,t;y)g(x)\, dx. \tag{7.109}$$

The transition density $\mathbf{p}_\beta(x, t; y)$, along with the initial distribution of the random variable $X_\beta(0) = X_1(0)$, determine the distribution of $X_\beta(t)$ for any single $t > 0$.

An explicit formula for the transition density (7.108) can be obtained by separation of variables. Here we sketch the argument. For complete details, see Leonenko, Meerschaert and Sikorskii [122, Theorem 3.2]. Suppose that $p(y, t) = S(t)\varphi(y)$ solves the fractional backward equation (7.96), where the functions S and φ may depend on x and β. Write

$$\partial_t^\beta S(t)\varphi(y) = S(t)\mathcal{G}\varphi(y) \quad \text{or} \quad \frac{1}{S(t)}\partial_t^\beta S(t) = \frac{\mathcal{G}\varphi(y)}{\varphi(y)}.$$

Set both sides equal to a constant to obtain the Sturm-Liouville equation $\mathcal{G}\varphi = -\lambda\varphi$ and the fractional time equation

$$\partial_t^\beta S(t) = -\lambda S(t). \tag{7.110}$$

Recall from Section 2.3 that solutions to equation (7.110) have the form

$$S(t) = E_\beta\left(-\lambda t^\beta\right) = \sum_{j=0}^{\infty} \frac{\left(-\lambda t^\beta\right)^j}{\Gamma(1+\beta j)} \tag{7.111}$$

for any $\lambda > 0$, where $S(0) = 1$, and $E_\beta(\cdot)$ is the Mittag-Leffler function (2.29). For Pearson diffusions of type (1–3), the Sturm-Liouville equation has polynomial solutions $\mathcal{G}Q_n(x) = -\lambda_n Q_n(x)$ for all n, where $0 = \lambda_0 < \lambda_1 < \lambda_2 < \cdots$ and $\lambda_n \to \infty$. For each n, we also have that

$$\partial_t^\beta S_n(t) = -\lambda_n S_n(t)$$

where the Mittag-Leffler eigenfunctions $S_n(t) = E_\beta(-\lambda_n t^\beta)$ solve the fractional time equation. Then $p(y, t) = E_\beta(-\lambda_n t^\beta)Q_n(y)$ solves the time-fractional backward equation (7.96) with initial condition $p(y, 0) = Q_n(y)$. Since any finite linear combination of functions of this form will also solve the backward equation, is it reasonable to consider the infinite sum

$$p(y, t) = \sum_{n=0}^{\infty} b_n E_\beta(-\lambda_n t^\beta)Q_n(y). \tag{7.112}$$

If $g(y)$ is a function such that (7.78) holds, where the series converges uniformly on compact intervals $y \in [c, d]$, then the Caputo fractional derivative and the generator $\mathcal{G}$ can be applied to the series (7.77) term-by-term, so that the function $p(y, t)$ in (7.112) solves (7.96). If the polynomials Q_n are normalized so that $c_n^2 = 1$ for all n in (7.76), then (7.112) solves the backward equation (7.70) with the initial condition $p(y, 0) = g(y)$ given by (7.78).

Equating (7.109) to (7.112) we see that

$$p(y,t) = S_t g(y) = \sum_{n=0}^{\infty} b_n e^{-\lambda_n t} Q_n(y)$$
$$= \sum_{n=0}^{\infty} \left(\int g(x) Q_n(x) \mathbf{m}(x)\,dx \right) E_\beta(-\lambda_n t^\beta) Q_n(y)$$
$$= \int \left(\mathbf{m}(x) \sum_{n=0}^{\infty} E_\beta(-\lambda_n t^\beta) Q_n(x) Q_n(y) \right) g(x)\,dx.$$

It follows that the transition density of the fractional Pearson diffusion is

$$\mathbf{p}_\beta(x,t;y) = \mathbf{m}(x) \sum_{n=0}^{\infty} E_\beta\left(-\lambda_n t^\beta\right) Q_n(x) Q_n(y). \tag{7.113}$$

Since we always have $Q_0(x) \equiv 1$, and since $\lambda_n > 0$ for all $n > 0$, it follows from (7.113) that $\mathbf{p}_\beta(x,t;y) \to \mathbf{m}(x)$ as $t \to \infty$ for any y, i.e., the fractional Pearson diffusion $X_\beta(t)$ tends to the same steady state distribution $\mathbf{m}(x)$ regardless of the initial state $X_\beta(0) = y$. See [122] for complete details.

A very similar separation of variables argument shows that

$$T_t f(x) = \int \left(\mathbf{m}(x) \sum_{n=0}^{\infty} E_\beta(-\lambda_n t^\beta) Q_n(x) Q_n(y) \right) f(y)\,dy$$

solves the time-fractional forward equation (7.60) with initial condition $p(x,0) = f(x)$, for any initial function such that (7.81) holds uniformly on compact intervals $x \in [c, d]$. See [122, Theorem 3.3] for details.

Remark 7.24. If $\beta = 1$, then (7.113) becomes

$$\mathbf{p}_1(x,t;y) = \mathbf{m}(x) \sum_{n=0}^{\infty} e^{-\lambda_n t} Q_n(x) Q_n(y),$$

which agrees with (7.89).

Remark 7.25. The transition density (7.108) for a fractional Pearson diffusion $X_\beta(t)$ of type (1–3) can also be obtained by a different argument. Use (7.80) to write

$$\mathbf{p}_1(x,t;y) = \mathbf{m}(x) \sum_{n=0}^{\infty} e^{-\lambda_n t} Q_n(x) Q_n(y) \tag{7.114}$$

Bingham [38] and Bondesson, Kristiansen, and Steutel [40] show that the inverse stable subordinator E_t has a Mittag-Leffler distribution with

$$E[e^{-sE_t}] = \int_0^\infty e^{-su} h(u,t)\,du = E_\beta(-st^\beta). \tag{7.115}$$

Write

$$
\begin{aligned}
S_t g(y) &= \int_0^\infty T_u g(y)\, h(u,t)\, du \\
&= \int_0^\infty \left(\int \mathbf{p}_1(x,u;y) g(x)\, dx \right) h(u,t)\, du \\
&= \int_0^\infty \left(\int \mathbf{m}(x) \sum_{n=0}^\infty e^{-\lambda_n u} Q_n(x) Q_n(y) g(x)\, dx \right) h(u,t)\, du \\
&= \int \mathbf{m}(x) \sum_{n=0}^\infty \left(\int_0^\infty e^{-\lambda_n u}\, h(u,t)\, du \right) Q_n(x) Q_n(y) g(x)\, dx \\
&= \int \left(\mathbf{m}(x) \sum_{n=0}^\infty E_\beta(-s t^\beta) Q_n(x) Q_n(y) \right) g(x)\, dx.
\end{aligned} \tag{7.116}
$$

It follows that (7.108) is the transition density of $X_\beta(t)$. See [122, Lemma 4.1] for complete details.

Remark 7.26. For more on the connection between Lévy-type Markov processes, semigroups, and generators, see for example Schilling [191]. When the Lévy characteristics $[a, Q, \phi]$ in (6.21) vary with x, the resulting generator is called a *pseudo-differential operator*, see Jacob [96].

Details

The backward semigroup (7.65) can be defined on the Banach space $C_0(E)$ of bounded continuous real-valued functions on E, such that the limits

$$A = \lim_{x \downarrow a} f(x) \quad \text{and} \quad B = \lim_{x \uparrow b} f(x)$$

exist, with $A = 0$ if $a = -\infty$, and $B = 0$ if $b = +\infty$, with the supremum norm. The semigroup property $T_t^* T_s^* = T_{t+s}^*$ follows from the Chapman-Kolmogorov equation (7.83). The backward semigroup is bounded, and in fact $\|T_t^* f\| \le \|f\|$ for all $f \in C_0(E)$ and all $t \ge 0$: We say that $\{T_t^*\}$ is a *contraction semigroup*. In the terminology of Rogers and Williams [176, Definition 6.5, p. 241], this is also called a *Feller-Dynkin semigroup*.

Proposition 7.27. *The backward semigroup defined in* (7.65), *where* $\mathbf{p}(x,t;y)$ *is the transition density* (7.80) *of the Pearson diffusion process with diffusion coefficients* $v(x)$ *and* $D(x)$ *defined in* (7.61), *is strongly continuous on* $C_0(E)$. *That is,* $\|T_t^* g - g\| \to 0$ *in the supremum norm as* $t \to 0$ *for any* $g \in C_0(E)$.

Proof. In view of Friedman [73, Theorem 3.4, p. 112], the operators $\{T_t^* : t \geq 0\}$ form a uniformly bounded semigroup on $C_0(E)$. In addition, for any fixed $y \in E$ we have

$$\begin{aligned} T_t^* g(y) - g(y) &= \int \mathbf{p}(x, t; y)(g(x) - g(y))dx \\ &= \int\limits_{|x-y|\leq\varepsilon} \mathbf{p}(x, t; y)(g(x) - g(y))dx \\ &\quad + \int\limits_{|x-y|>\varepsilon} \mathbf{p}(x, t; y)(g(x) - g(y))dx \\ &\leq \sup_{|x-y|\leq\varepsilon} |g(x) - g(y)| \int\limits_{|x-y|\leq\varepsilon} \mathbf{p}(x, t; y)dx \\ &\quad + C \int\limits_{|x-y|>\varepsilon} \mathbf{p}(x, t; y)dx, \end{aligned}$$

where $C = \sup_{x,y} |g(x) - g(y)|$ is finite since function g is bounded. It follows from the form of the generator of the semigroup $\{T_t^*\}$ that

$$\int\limits_{|x-y|>\varepsilon} \mathbf{p}(x, t; y)dx \to 0$$

as $t \to 0$ for any $\varepsilon > 0$ (see Feller [70]), therefore the second term in the above expression tends to zero as $t \to 0$. The first term is bounded by

$$\sup_{|x-y|\leq\varepsilon} |g(x) - g(y)|,$$

which tends to zero as $\varepsilon \to 0$. This proves point-wise continuity of the semigroup: For every fixed y, $T_t^* g(y) \to g(y)$ as $t \to 0$. Then Rogers and Williams [176, Lemma 6.7, p. 241] yields strong continuity of the semigroup: $\|T_t^* g - g\| \to 0$ as $t \to 0$ in the Banach space (supremum) norm. □

To prove that (7.113) is the transition density of a type (1–3) fractional Pearson diffusion, use [122, Remark 3.4] to see that any smooth function $g(y)$ with compact support in E can be written in the form (7.78), where the series converges uniformly on compact sets. Since the indicator function of any compact interval $B \in E$ can be approximated arbitrarily closely by such functions, it follows that

$$\mathbb{P}[g(X_t)|x_0 = y] = \int\limits_{x\in B} \mathbf{p}(x, t; y)g(x)\, dx$$

for all such intervals. Then it follows that $\mathbf{p}(x, t; y)$ is the conditional density of X_t, given $X_0 = y$. The Fubini argument in (7.79) can be justified using Lemma 7.28.

Lemma 7.28. *For the three classes of fractional Pearson diffusions with discrete spectrum (OU, CIR, Jacobi) and* $0 < \beta \le 1$, *The series*

$$p_\beta(x,t;y) = \mathbf{m}(x) \sum_{n=0}^{\infty} E_\beta\left(-\lambda_n t^\beta\right) Q_n(y) Q_n(x) \tag{7.117}$$

converges for fixed $t > 0$, $x,\ y \in E$.

Proof. For a Mittag-Leffler function with $0 < \beta < 1$ (see Mainardi and Gorenflo [129, Eq. (5.26)])

$$E_\beta(-\lambda_n t^\beta) \sim \frac{1}{\Gamma(1-\beta)\lambda_n t^\beta}$$

as the argument $\lambda_n t^\beta \to \infty$. The eigenvalues are $\lambda_n = \theta n$ in the Hermite and Laguerre cases, and $\lambda_n = n(n+a+b+1)$ in the Jacobi case. In the rest of the proof, we will assume without loss of generality that $\mu = 0$ and $\sigma = 1$ in the OU case, and $a = 1$ in the CIR case. For orthonormal Hermite polynomials (Sansone [186], p. 369)

$$\bar{H}_n(x) \le K e^{x^2/4} n^{-1/4} (1 + |x/\sqrt{2}|^{5/2}),$$

where K is a constant that does not depend on x.

For orthonormal Laguerre polynomials ([186], p. 348)

$$\bar{L}_n^{(b-1)}(x) = O\left(\frac{e^{x/2}}{x^{(2b-1)/4}} n^{-1/4}\right),$$

uniformly for x in finite intervals $[x_1,\ x_2]$.

For orthonormal Jacobi polynomials

$$\bar{P}_n^{(a,b)}(x) = C(x,a,b)\cos(N\theta + \gamma) + O(n^{-1}),$$

where $x = \cos\theta$, $N = n + 1/2(a+b+1)$, and $\gamma = -(a+1/2)\pi/2$.

Convergence of the series (7.117) for fixed x, y, t follows from the above relations. Specifically, in the Hermite case,

$$|E_\beta\left(-\lambda_n t^\beta\right) Q_n(y) Q_n(x)| \le \frac{C(x,y,t,\beta)}{n^{1+1/2}}.$$

In the Laguerre case,

$$|E_\beta\left(-\lambda_n t^\beta\right) Q_n(y) Q_n(x)| \le \frac{C(x,y,t,\beta)}{n^{1+1/2}}.$$

In the Jacobi case

$$|E_\beta\left(-\lambda_n t^\beta\right) Q_n(y) Q_n(x)| \le \frac{C(x,y,t,\beta)\cos(N\theta + \gamma)}{n^2}.$$

When $\beta = 1$, we have $E_\beta(-\lambda_n t^\beta) = e^{-\lambda_n t}$, and the proof is similar. □

7.8 Correlation structure of fractional processes

In many applications [49, 204, 97], it is useful to compute second order properties of the process used to model a particular phenomenon. In this section we develop explicit computational formulae for the correlation function of fractional Pearson diffusions discussed in Section 7.7 and time changed Lévy processes such as the fractional Poisson process discussed in Section 7.1. A time changed Lévy process can also arise as the limit of CTRW considered in Chapter 4. We show that the random time change in Pearson diffusions and in Lèvy process, using the inverse of the standard stable subordinator, introduces a long-range dependence in the corresponding fractional processes.

The consideration of the correlation function is premised on the existence of the second moment, and only processes with finite second moment are considered in this section. For three Pearson diffusions with purely discrete spectrum (Ornstein-Uhlenbeck, Cox-Ingersol-Ross, and Jacobi) all moments exist. The conditions for the existence of moments for three heavy-tailed Pearson diffusions are in Remark 7.20.

If the time-homogeneous Markov process $X_1(t)$ is in steady state, then its probability density $\mathbf{m}(x)$ stays the same over all time. The stationary Pearson diffusion has correlation function

$$\operatorname{corr}[X_1(t), X_1(t+s)] = \exp(-\theta s), \quad t \geq 0,\ s \geq 0, \tag{7.118}$$

where the correlation parameter $\theta = \lambda_1$ is the smallest positive eigenvalue of the backward generator [122, 117]. Thus the Pearson diffusion exhibits short range dependence, meaning that the correlation function falls off rapidly (exponentially in this case), so that it is integrable at infinity.

Recall that the fractional Pearson diffusion was defined in Section 7.7 as $X_\beta(t) = X_1(E_t)$, where E_t is the inverse or first passage time

$$E_t = \inf\{u > 0 : D(u) > t\}$$

of the standard β-stable stable subordinator $D(u)$ with

$$\mathbb{E}[e^{-sD(u)}] = e^{-us^\beta}.$$

We will say that a fractional Pearson diffusion is in steady state if it starts with the distribution $\mathbf{m}(x)$. The fractional Pearson diffusion in steady state is first order stationary, i.e., $X_\beta(t)$ has the same probability density $\mathbf{m}(x)$ for all $t > 0$. Indeed, in view of [18, Theorem 3.1]

$$\int_0^\infty \mathbf{m}(x) f_t(u)\, du = \mathbf{m}(x),$$

where f_t is the density of E_t. Thus the fractional Pearson diffusion in steady state has mean $\mathbb{E}[X_\beta](t) = \mathbb{E}[X_1(t)] = m_1$ and variance $\operatorname{Var}[X_\beta(t)] = \operatorname{Var}[X_1(t)] = m_2^2$ which do not vary over time. The next result gives an explicit formula for the correlation function of a fractional Pearson diffusion in steady state.

Theorem 7.29. *Suppose that $X_1(t)$ is a Pearson diffusion in steady state, so that its correlation function is given by (7.118). Then the correlation function of the corresponding fractional Pearson diffusion $X_\beta(t)$ is given by*

$$\operatorname{corr}[X_\beta(t), X_\beta(s)] = E_\beta(-\theta t^\beta) + \frac{\theta \beta t^\beta}{\Gamma(1+\beta)} \int_0^{s/t} \frac{E_\beta(-\theta t^\beta (1-z)^\beta)}{z^{1-\beta}} dz \tag{7.119}$$

for $t \geq s > 0$, where $E_\beta(z)$ is the Mittag-Leffler function.

See details at the end of this section for proof.

Remark 7.30. When $t = s$, it must be true that $\operatorname{corr}[X_\beta(t), X_\beta(s)] = 1$. To see that this follows from (7.119), recall the formula for the beta density

$$\int_0^x y^{a-1}(x-y)^{b-1}\, dy = B(a,b)x^{a+b-1} \quad \text{where} \quad B(a,b) := \frac{\Gamma(a)\Gamma(b)}{\Gamma(a+b)}$$

for $a > 0$ and $b > 0$, and write

$$\begin{aligned}
&\frac{\theta \beta t^\beta}{\Gamma(1+\beta)} \int_0^1 \frac{E_\beta(-\theta t^\beta (1-z)^\beta)}{z^{1-\beta}} dz \\
&= \frac{\theta \beta t^\beta}{\Gamma(1+\beta)} \int_0^1 \sum_{j=0}^\infty \frac{(-\theta t^\beta (1-z)^\beta)^j}{\Gamma(1+\beta j)} z^{1-\beta} dz \\
&= \frac{\theta \beta t^\beta}{\Gamma(1+\beta)} \sum_{j=0}^\infty \frac{(-\theta t^\beta)^j}{\Gamma(1+\beta j)} \int_0^1 (1-z)^{\beta j} z^{\beta-1} dz \\
&= \frac{\theta \beta t^\beta}{\Gamma(1+\beta)} \sum_{j=0}^\infty \frac{(-\theta t^\beta)^j}{\Gamma(1+\beta j)} B(\beta j + 1, \beta) \\
&= \frac{\theta t^\beta}{\Gamma(1+\beta)} \sum_{j=0}^\infty \frac{\beta \Gamma(\beta)(-\theta t^\beta)^j}{\Gamma(1+\beta(j+1))} \\
&= -\sum_{j=0}^\infty \frac{(-\theta t^\beta)^{j+1}}{\Gamma(1+\beta(j+1))} = 1 - E_\beta(-\theta t^\beta).
\end{aligned}$$

Then it follows from (7.119) that $\operatorname{corr}[X_\beta(t), X_\beta(s)] = 1$.

Remark 7.31. Stationary Pearson diffusions exhibit short range dependence, since their correlation function (7.118) falls off exponentially fast. However, the correlation function of a fractional Pearson diffusion falls off like a power law with exponent $\beta \in$

(0, 1). When s is fixed and $t \to \infty$, the correlation function is not integrable, so this process exhibits *long range dependence*. To see this, fix $s > 0$ and recall [129, Eq. (5.26)] that

$$E_\beta(-\theta t^\beta) \sim \frac{1}{\Gamma(1-\beta)\theta t^\beta} \quad \text{as } t \to \infty.$$

Then

$$E_\beta(-\theta t^\beta(1-sy/t)^\beta) \sim \frac{1}{\Gamma(1-\beta)\theta t^\beta(1-sy/t)^{-\beta}}$$

as $t \to \infty$ for any $y \in [0, 1]$. In addition from [114]

$$|E_\beta(-\theta t^\beta(1-sy/t)^\beta)| \le \frac{c}{1+\theta t^\beta(1-sy/t)^\beta}$$

for all $t > 0$, and using the dominated convergence theorem we get

$$\begin{aligned}
&\frac{\theta\beta t^\beta}{\Gamma(1+\beta)} \int_0^{s/t} \frac{E_\beta(-\theta t^\beta(1-z)^\beta)}{z^{1-\beta}} dz \\
&= \left(\frac{s}{t}\right)^\beta \frac{\theta\beta t^\beta}{\Gamma(1+\beta)} \int_0^1 y^{\beta-1} E_\beta\left(-\theta t^\beta(1-sy/t)^\beta\right) dy \\
&\sim \left(\frac{s}{t}\right)^\beta \frac{\beta}{\Gamma(1+\beta)\Gamma(1-\beta)} \int_0^1 y^{\beta-1} dy = \left(\frac{s}{t}\right)^\beta \frac{1}{\Gamma(1+\beta)\Gamma(1-\beta)}
\end{aligned}$$

as $t \to \infty$. It follows from (7.119) that for any fixed $s > 0$ we have

$$\operatorname{corr}(X_\beta(t), X_\beta(s)) \sim \frac{1}{t^\beta \Gamma(1-\beta)} \left(\frac{1}{\theta} + \frac{s^\beta}{\Gamma(\beta+1)}\right) \quad \text{as } t \to \infty. \tag{7.120}$$

We now consider a general time change in a Lévy process $Z(t) = X(Y(t))$ where X is a Lèvy process, X, Y are independent, and in general $Y(t)$ may be non-Markovian with non-stationary and non-independent increments. For example, it might be an inverse subordinator considered earlier in this section to time-change Pearson diffusions. Then $Z(t)$ may also be also non-Markovian with non-stationary and non-independent increments. The next result gives an explicit expression for the correlation function of this time-changed process.

Theorem 7.32. *Suppose that $X(t)$, $t \ge 0$ is a homogeneous Lévy process with $X(0) = 0$ and finite variance, and $Y(t)$ is a non-decreasing process independent of X, with $\mathbb{P}[Y(t) > 0] = 1$ for $t > 0$, finite mean $U(t) = EY(t)$ and finite variance. Then the mean of the process $Z = X(Y(t))$ is*

$$\mathbb{E}[Z(t)] = U(t)\mathbb{E}[X(1)], \tag{7.121}$$

the variance is

$$\operatorname{Var}[Z(t)] = [\mathbb{E}X(1)]^2 \operatorname{Var}[Y(t)] + U(t) \operatorname{Var}[X(1)], \tag{7.122}$$

and the covariance function of the process $Z = X(Y(t))$ is given by

$$\mathrm{Cov}[Z(t), Z(s)] = \mathrm{Var}[X(1)]U(\min(t, s)) + [\mathbb{E}X(1)]^2 \,\mathrm{Cov}[Y(t), Y(s)]. \tag{7.123}$$

Proof is included in the details at the end of this section.

Remark 7.33. When $EX(1) = 0$, then

$$\mathrm{Var}[Z(t)] = U(t)\,\mathrm{Var}[X(1)],$$

the covariance function is

$$\mathrm{Cov}[Z(t), Z(s)] = \mathrm{Var}[X(1)]U(\min(t, s)),$$

and the correlation function is

$$\mathrm{corr}[Z(t), Z(s)] = \frac{U(\min(t, s))}{\sqrt{U(t)U(s)}} = \sqrt{\frac{U(\min(t, s))}{U(\max(t, s))}}.$$

When the random time change is to the inverse or hitting time of a Lévy subordinator L with the Laplace exponent ϕ so that

$$\mathbb{E}\left[e^{-sL(t)}\right] = e^{-t\phi(s)}, \; s \geq 0,$$

the inverse process

$$Y(t) = \inf\{u \geq 0 : L(u) > t\}, t \geq 0 \tag{7.124}$$

is non-decreasing, and its sample paths are almost surely continuous if L is strictly increasing. For any Lévy subordinator L, Veillette and Taqqu [213] show that the renewal function $U(t) = \mathbb{E}[Y(t)]$ of its inverse (7.124) has Laplace transform $\tilde{U}$ given by:

$$\tilde{U}(s) = \int_0^\infty U(t)e^{-st}dt = \frac{1}{s\phi(s)}, \tag{7.125}$$

where ϕ is Laplace exponent of L. Thus, U characterizes the inverse process Y, since ϕ characterizes L. For example, from [213, Theorem 4.2] the second moment of Y is

$$\mathbb{E}Y^2(t) = \int_0^t 2U(t-\tau)dU(\tau) \tag{7.126}$$

and the covariance function of Y is given by [213, Eq. (18)]:

$$\mathrm{Cov}[Y(t_1), Y(t_2)] = \int_0^{t_1 \wedge t_2} (U(t_1-\tau) + U(t_2-\tau))dU(\tau) - U(t_1)U(t_2). \tag{7.127}$$

For many inverse subordinators, the Laplace exponent ϕ can be written explicitly, but then the Laplace transform (7.125) has to be inverted to obtain the renewal function.

Numerical methods for the inversion were proposed in [213]. We consider one example where the Laplace transform can be inverted analytically and its asymptotic behavior can be found in order to describe the behavior of the correlation function of the time changed process. For more examples, see [118].

When L is standard β-stable subordinator with index $0 < \beta < 1$, and the Laplace exponent $\phi(s) = s^\beta$ for all $s > 0$, the inverse stable subordinator has the Laplace transform

$$\mathbb{E}\left[e^{-sY(t)}\right] = \sum_{n=0}^{\infty} \frac{(-st^\beta)^n}{\Gamma(\beta n + 1)} = E_\beta(-st^\beta),$$

using the Mittag-Leffler function (2.29). When the outer process $X(t)$ is a homogeneous Poisson process, the time changed process $X(Y(t))$ is fractional Poisson process [141] discussed in Section 7.1. More generally, for any Lévy process $X(t)$, the time changed process $X(Y(t))$ is a CTRW limit where the waiting times between particle jumps belong to the domain of attraction of the stable subordinator $L(t)$, see [153].

Since

$$\tilde{U}(s) = \frac{1}{s^{\beta+1}} \tag{7.128}$$

the renewal function

$$U(t) = \mathbb{E}[Y(t)] = \frac{t^\beta}{\Gamma(1+\beta)}. \tag{7.129}$$

The renewal function (7.129) can also be obtained from a result of Bingham [38], who showed that for all $0 < t_1 < \cdots < t_k$

$$\frac{\partial \mathbb{E}[Y(t_1)\cdots Y(t_k)]}{\partial t_1 \cdots \partial t_k} = \frac{1}{\Gamma(\beta)^k} \frac{1}{[t_1(t_2 - t_1)\cdots(t_k - t_{k-1})]^{1-\beta}}. \tag{7.130}$$

Apply (7.130) with $k = 1$ to see that

$$\frac{d}{dt} U(t) = \frac{t^{\beta-1}}{\Gamma(\beta)},$$

integrate once, and use $\Gamma(\beta + 1) = \beta\Gamma(\beta)$.

For $0 < s \le t$, substitute (7.128) into (7.127) to see that the covariance function of the inverse stable subordinator is

$$\begin{aligned}
\mathrm{Cov}[Y(t), Y(s)] &= \frac{\beta}{\Gamma(1+\beta)^2} \int_0^s \left((t-\tau)^\beta + (s-\tau)^\beta\right) \tau^{\beta-1} d\tau \\
&\quad - \frac{(ts)^\beta}{\Gamma(1+\beta)^2} \\
&= \frac{\beta t^{2\beta}}{\Gamma(1+\beta)^2} \int_0^{s/t} (1-u)^\beta u^{\beta-1} du \\
&\quad + \frac{\beta s^{2\beta}}{\Gamma(1+\beta)^2} B(\beta, \beta+1) - \frac{(ts)^\beta}{\Gamma(1+\beta)^2} \\
&= \frac{1}{\Gamma(1+\beta)^2} \Big[\beta t^{2\beta} B(\beta, \beta+1; s/t) \\
&\qquad + \beta s^{2\beta} B(\beta, \beta+1) - (ts)^\beta\Big],
\end{aligned} \tag{7.131}$$

using a substitution $u = \tau/t$, where

$$B(a, b; x) := \int_0^x u^{a-1}(1-u)^{b-1} du$$

is the incomplete beta function. An equivalent form of the covariance function in terms of the hypergeometric function was obtained in [213, Eq. (74)]. Apply the Taylor series expansion $(1-u)^{b-1} = 1 + (1-b)u + O(u^2)$ as $u \to 0$ to see that

$$B(a, b; x) = \frac{x^a}{a} + (1-b)\frac{x^{a+1}}{a+1} + O(x^{a+2}) \quad \text{as } x \to 0.$$

Then it follows that for $s > 0$ fixed and $t \to \infty$ we have

$$\begin{aligned}
F(\beta; s, t) &:= \beta t^{2\beta} B(\beta, \beta+1; s/t) - (ts)^\beta \\
&= \beta t^{2\beta} \frac{(s/t)^\beta}{\beta} - \beta \frac{(s/t)^{\beta+1}}{\beta+1} + O((s/t)^{\beta+2}) - (ts)^\beta \\
&= -\beta \frac{(s/t)^{\beta+1}}{\beta+1} + O((s/t)^{\beta+2}),
\end{aligned}$$

so that

$$\mathrm{Cov}[Y(t), Y(s)] = \frac{1}{\Gamma(1+\beta)^2} \left[\beta s^{2\beta} B(\beta, \beta+1) + F(\beta; s, t)\right] \tag{7.132}$$

as $t \to \infty$, where $F(\beta; s, t) \to 0$ as $t \to \infty$. Hence

$$\mathrm{Cov}[Y(t), Y(s)] \to \frac{\beta s^{2\beta} B(\beta, \beta+1)}{\Gamma(1+\beta)^2} = \frac{s^{2\beta}}{\Gamma(2\beta+1)} \quad \text{as } t \to \infty. \tag{7.133}$$

Letting $s = t$ it follows from (7.131) that

$$\begin{aligned}\operatorname{Var}[Y(t)] &= \frac{1}{\Gamma(1+\beta)^2}\left[2t^{2\beta}\frac{\beta\Gamma(\beta)\Gamma(\beta+1)}{\Gamma(2\beta+1)} - t^{2\beta}\right] \\ &= t^{2\beta}\left[\frac{2}{\Gamma(2\beta+1)} - \frac{1}{\Gamma(1+\beta)^2}\right], \end{aligned} \tag{7.134}$$

which agrees with the computation in [15, Section 5.1]. From (7.132) and (7.134) it follows that for $0 < s \le t$ the inverse stable subordinator has correlation function

$$\operatorname{corr}[Y(s), Y(t)] = \frac{\left[\beta s^{2\beta} B(\beta, \beta+1) + F(\beta; s, t)\right]}{(st)^\beta\left[\frac{2\Gamma(1+\beta)^2}{\Gamma(2\beta+1)} - 1\right]}$$

where $F(\beta; s, t) \to 0$ as $t \to \infty$, and hence

$$\operatorname{corr}[Y(s), Y(t)] \sim \left(\frac{s}{t}\right)^\beta \left[2 - \frac{\Gamma(2\beta+1)}{\Gamma(1+\beta)^2}\right]^{-1} \quad \text{as } t \to \infty.$$

This power law decay of the correlation function is a kind of long range dependence for the inverse stable subordinator $Y(t)$.

From (7.121) and (7.129) we can see that the time-changed process $Z(t) = X(Y(t))$ has mean

$$\mathbb{E}[Z(t)] = \frac{t^\beta \mathbb{E}[X(1)]}{\Gamma(1+\beta)}.$$

Substituting (7.134) into (7.122) yields the variance of the time-changed process:

$$\operatorname{Var}[Z(t)] = \frac{t^\beta \operatorname{Var}[X(1)]}{\Gamma(1+\beta)} + \frac{t^{2\beta}[\mathbb{E}X(1)]^2}{\beta}\left(\frac{1}{\Gamma(2\beta)} - \frac{1}{\beta\Gamma(\beta)^2}\right). \tag{7.135}$$

It follows from (7.123), (7.129), and (7.132) that for $0 < s \le t$ the covariance function of $Z(t) = X(Y(t))$ is

$$\begin{aligned}\operatorname{Cov}[Z(t), Z(s)] &= \frac{s^\beta \operatorname{Var}[X(1)]}{\Gamma(1+\beta)} \\ &\quad + \frac{[\mathbb{E}X(1)]^2}{\Gamma(1+\beta)^2}\left[\beta s^{2\beta} B(\beta, \beta+1) + F(\beta; s, t)\right] \end{aligned} \tag{7.136}$$

where $F(\beta; s, t) \to 0$ as $t \to \infty$, hence

$$\operatorname{Cov}[Z(t), Z(s)] \to \frac{s^\beta \operatorname{Var}[X(1)]}{\Gamma(1+\beta)} + \frac{s^{2\beta}[\mathbb{E}X(1)]^2}{\Gamma(1+2\beta)} \quad \text{as } t \to \infty. \tag{7.137}$$

For $0 < s \le t$, the time changed process $Z(t) = X(Y(t))$ has correlation

$$\operatorname{corr}[Z(t), Z(s)] = \frac{\operatorname{Cov}[Z(s), Z(t)]}{\sqrt{\operatorname{Var}[Z(s)]\operatorname{Var}[Z(t)]}}$$

where $\operatorname{Cov}[Z(s), Z(t)]$ is given by (7.136) and the remaining terms are specified in (7.135).

The asymptotic behavior of the correlation depends on whether the outer process has zero mean. If $\mathbb{E}[X(1)] \neq 0$, then for any $s > 0$ fixed we have

$$\mathrm{Var}[Z(t)] \sim \frac{t^{2\beta}[\mathbb{E}X(1)]^2}{\beta}\left(\frac{1}{\Gamma(2\beta)} - \frac{1}{\beta\Gamma(\beta)^2}\right) \quad \text{as } t \to \infty,$$

and so we have

$$\mathrm{corr}[Z(t), Z(s)] \sim t^{-\beta} C(\beta, s) \quad \text{as } t \to \infty,$$

where

$$C(\beta, s) = \left(\frac{s^{\beta}\,\mathrm{Var}[X(1)]}{\Gamma(1+\beta)} + \frac{s^{2\beta}[\mathbb{E}X(1)]^2}{\Gamma(1+2\beta)}\right) \times \left(\sqrt{\frac{\beta}{\Gamma(2\beta)} - \frac{1}{\Gamma(\beta)^2}}\,|\mathbb{E}[X(1)]|\,\sqrt{\mathrm{Var}[Z(s)]}\right)^{-1},$$

where $\mathrm{Var}[Z(s)]$ is given by (7.135).

On the other hand, if $\mathbb{E}[X(1)] = 0$, then the covariance function of the time-changed process for $0 < s \leq t$ simplifies to

$$\mathrm{Cov}[Z(t), Z(s)] = \mathrm{Var}[X(1)]\frac{s^{\beta}}{\Gamma(1+\beta)}. \tag{7.138}$$

and the correlation function is

$$\mathrm{corr}[Z(t), Z(s)] = \left(\frac{s}{t}\right)^{\beta/2},$$

a formula obtained by Janczura and Wyłomańska [97] for the special case when the outer process $X(t)$ is a Brownian motion.

In summary, the correlation function of $Z(t)$ falls off like a power law $t^{-\beta}$ when $\mathbb{E}[X(1)] \neq 0$, and even more slowly, like the power law $t^{-\beta/2}$ when $\mathbb{E}[X(1)] = 0$. In either case, the non-stationary time-changed process $Z(t)$ exhibits long range dependence. If $\mathbb{E}[X(1)] = 0$, the time-changed process $Z(t) = X(Y(t))$ also has uncorrelated increments: Since $\mathrm{Cov}[Z(t), Z(s)]$ does not depend on t, we have $\mathrm{Var}[Z(s)] = \mathrm{Cov}[Z(s), Z(s)] = \mathrm{Cov}[Z(s), Z(t)]$ and hence, since the covariance is additive, we have $\mathrm{Cov}[Z(s), Z(t) - Z(s)] = 0$ for $0 < s < t$. Uncorrelated increments together with long range dependence is a hallmark of financial data [188], and hence this process can be useful to model such data. Since the outer process $X(t)$ can be any Lévy process, the distribution of the time-changed process $Z(t) = X(Y(t))$ can take many forms.

Details

The proof of Theorem 7.29 is presented below.

Proof. Write

$$\operatorname{corr}[X_\beta(t), X_\beta(s)] = \operatorname{corr}[X_1(E_t), X_1(E_s)] = \int_0^\infty \int_0^\infty e^{-\theta|u-v|} H(du, dv), \tag{7.139}$$

a Lebesgue-Stieltjes integral with respect to $H(u, v) := \mathbb{P}[E_t \le u, E_s \le v]$, the bivariate distribution function of the process E_t.

To compute the integral in (7.139), we use the bivariate integration by parts formula [77, Lemma 2.2]

$$\begin{aligned}\int_0^a \int_0^b F(u, v) H(du, dv) = & \int_0^a \int_0^b H([u, a] \times [v, b]) F(du, dv) + \\ & + \int_0^a H([u, a] \times (0, b]) F(du, 0) \\ & + \int_0^b H((0, a] \times [v, b]) F(0, dv) \\ & + F(0, 0) H((0, a] \times (0, b]).\end{aligned} \tag{7.140}$$

with $F(u, v) = e^{-\theta|u-v|}$, and the limits of integration a and b are infinite:

$$\begin{aligned}\int_0^\infty \int_0^\infty F(u, v) H(du, dv) = & \int_0^\infty \int_0^\infty H([u, \infty] \times [v, \infty]) F(du, dv) \\ & + \int_0^\infty H([u, \infty] \times (0, \infty]) F(du, 0) \\ & + \int_0^\infty H((0, \infty] \times [v, \infty]) F(0, dv) \\ & + F(0, 0) H((0, \infty] \times (0, \infty]) \\ = & \int_0^\infty \int_0^\infty \mathbb{P}[E_t \ge u, E_s \ge v] F(du, dv) \\ & + \int_0^\infty \mathbb{P}[E_t \ge u] F(du, 0) \\ & + \int_0^\infty \mathbb{P}[E_s \ge v] F(0, dv) + 1,\end{aligned} \tag{7.141}$$

since $E_t > 0$ with probability 1 for all $t > 0$. Note that $F(du, v) = f_v(u)du$ for all $v \geq 0$, where

$$f_v(u) = -\theta e^{-\theta(u-v)} I\{u > v\} + \theta e^{-\theta(v-u)} I\{u \leq v\}. \tag{7.142}$$

Integrate by parts to get

$$\begin{aligned} \int_0^\infty \mathbb{P}[E_t \geq u] F(du, 0) &= \int_0^\infty (1 - \mathbb{P}[E_t < u]) \left(-\theta e^{-\theta u}\right) du \\ &= \left[e^{-\theta u} \mathbb{P}[E_t \geq u]\right]_0^\infty + \int_0^\infty e^{-\theta u} f_t(u) du \\ &= E_\beta(-\theta t^\beta) - 1. \end{aligned} \tag{7.143}$$

Similarly

$$\int_0^\infty \mathbb{P}[E_s \geq v] F(0, dv) = E_\beta(-\theta s^\beta) - 1,$$

and hence (7.141) reduces to

$$\int_0^\infty \int_0^\infty F(u, v) H(du, dv) = I + E_\beta(-\theta t^\beta) + E_\beta(-\theta s^\beta) - 1 \tag{7.144}$$

where

$$I = \int_0^\infty \int_0^\infty \mathbb{P}[E_t \geq u, E_s \geq v] F(du, dv).$$

Assume (without loss of generality) that $t \geq s$. Then $E_t \geq E_s$, so that $\mathbb{P}[E_t \geq u, E_s \geq v] = P[E_s \geq v]$ for $u \leq v$. Write $I = I_1 + I_2 + I_3$ where

$$\begin{aligned} I_1 &:= \int_{u<v} \mathbb{P}[E_t \geq u, E_s \geq v] F(du, dv) = \int_{u<v} \mathbb{P}[E_s \geq v] F(du, dv) \\ I_2 &:= \int_{u=v} \mathbb{P}[E_t \geq u, E_s \geq v] F(du, dv) = \int_{u=v} \mathbb{P}[E_s \geq v] F(du, dv) \\ I_3 &:= \int_{u>v} \mathbb{P}[E_t \geq u, E_s \geq v] F(du, dv). \end{aligned}$$

Since $F(du, dv) = -\theta^2 e^{-\theta(v-u)} du\, dv$ for $u < v$, we may write

$$
\begin{aligned}
I_1 &= -\theta^2 \int_{v=0}^{\infty} \int_{u=0}^{v} \mathbb{P}[E_s \geq v] e^{\theta(u-v)} du dv \\
&= -\theta \int_{v=0}^{\infty} \mathbb{P}[E_s \geq v](1 - e^{-\theta v}) dv \\
&= -\theta\, \mathbb{E}[E_s] - \theta \int_{v=0}^{\infty} \mathbb{P}[E_s \geq v] e^{-\theta v} dv \\
&= -\frac{\theta s^\beta}{\Gamma(1-\beta)} - (E_\beta(-\theta s^\beta) - 1)
\end{aligned} \tag{7.145}
$$

using the well known formula $\mathbb{E}[X] = \int_0^\infty \mathbb{P}[X \geq x] dx$ for any positive random variable, the relation (7.143), and the formula $\mathbb{E}[E_t] = t^\beta / \Gamma(1+\beta)$ for the mean of the standard inverse β-stable subordinator [15, Eq. (9)].

Since $F(du, v) = f_v(u) du$, where the function (7.142) has a jump of size 2θ at the point $u = v$, we also have

$$
I_2 = 2\theta \int_0^\infty \mathbb{P}[E_s \geq v] dv = 2\theta\, \mathbb{E}[E_s] = \frac{2\theta s^\beta}{\Gamma(1+\beta)}.
$$

Since $F(du, dv) = -\theta^2 e^{-\theta(u-v)} du\, dv$ for $u > v$ as well, we have

$$
I_3 = -\theta^2 \int_{v=0}^{\infty} \mathbb{P}[E_t \geq u,\ E_s \geq v] \int_{u=v}^{\infty} e^{-\theta(u-v)} du\, dv. \tag{7.146}
$$

Next, we obtain an expression for $\mathbb{P}[E_t \geq u,\ E_s \geq v]$. Since the process E_t is inverse to the stable subordinator D_u, we have $\{E_t > u\} = \{D_u < t\}$ and since E_t has a density, it follows that $\mathbb{P}[E_t \geq u,\ E_s \geq v] = \mathbb{P}[D_u < t, D_v < s]$. Since $D(u)$ has the same distribution as $u^{1/\beta} D(1)$, the random variable $D(u)$ has the density function $g_\beta(x, u) = u^{-1/\beta} g_\beta(x u^{-1/\beta})$, and

$$
\frac{x}{\beta} g_\beta(x, u) = u f_x(u),
$$

where $f_t(u)$ is the probability density of $u = E_t$. Since D_u has stationary independent increments, it follows that

$$
\begin{aligned}
\mathbb{P}[E_t \geq u, E_s \geq v] &= \mathbb{P}[D_u < t, D_v < s] \\
&= \mathbb{P}[(D_u - D_v) + D_v < t, D_v < s] \\
&= \int_{y=0}^{s} g_\beta(y, v) \int_{x=0}^{t-y} g_\beta(x, u-v) dx dy \\
&= \int_{y=0}^{s} \frac{\beta}{y} v f_y(v) \int_{x=0}^{t-y} \frac{\beta}{x} (u-v) f_x(u-v) dx dy.
\end{aligned}
$$

Substituting and using Fubini Theorem, it follows that

$$
\begin{aligned}
I_3 &= -\theta^2 \int_{y=0}^{s} \frac{\beta}{y} \int_{x=0}^{t-y} \frac{\beta}{x} \int_{v=0}^{\infty} v f_y(v) \int_{u=v}^{\infty} (u-v) f_x(u-v) e^{-\theta(u-v)} du\, dv\, dx\, dy \\
&= -\theta^2 \int_{y=0}^{s} \frac{\beta}{y} \int_{x=0}^{t-y} \frac{\beta}{x} \int_{v=0}^{\infty} v f_y(v) dv \int_{z=0}^{\infty} z f_x(z) e^{-\theta z} dz\, dx\, dy
\end{aligned}
$$

where

$$
\int_{v=0}^{\infty} v f_y(v) dv = \mathbb{E}[E_y] = \frac{y^\beta}{\Gamma(1+\beta)}. \tag{7.147}
$$

Next we claim that

$$
\int_{0}^{\infty} z f_x(z) e^{-\theta z} dz = -\frac{x}{\beta\theta} \frac{d}{dx} E_\beta(-\theta x^\beta). \tag{7.148}
$$

To see that (7.148) holds, first differentiate the power series expansion for the Mittag-Leffler function to obtain

$$
\begin{aligned}
\frac{d}{dx} E_\beta(-\theta x^\beta) &= \sum_{j=1}^{\infty} \frac{(-\theta x^\beta)^{j-1} j}{\Gamma(1+\beta j)} (-\theta \beta x^{\beta-1}) \\
&= \frac{\beta}{x} \sum_{j=1}^{\infty} \frac{(-\theta x^\beta)^j j}{\Gamma(1+\beta j)}.
\end{aligned} \tag{7.149}
$$

Then expand $e^{-\theta z}$ in a Taylor series expansion, and integrate term by term:

$$
\begin{aligned}
\int_{0}^{\infty} z f_x(z) e^{-\theta z} dz &= \sum_{k=0}^{\infty} \frac{(-\theta)^k}{k!} \int_{0}^{\infty} z^{k+1} f_x(z) dz \\
&= \sum_{k=0}^{\infty} \frac{(-\theta)^k}{k!} E[E_x^{k+1}] = \sum_{k=0}^{\infty} \frac{(-\theta)^k}{k!} x^{\beta(k+1)} \frac{(k+1)!}{\Gamma(1+\beta(k+1))} \\
&= -\frac{1}{\theta} \sum_{k=0}^{\infty} \frac{(-\theta x^\beta)^{k+1}(k+1)}{\Gamma(1+\beta(k+1))} = -\frac{1}{\theta} \sum_{j=0}^{\infty} \frac{(-\theta x^\beta)^j j}{\Gamma(1+\beta j)}
\end{aligned}
$$

and apply (7.149) to see that (7.148) holds.

Now it follows using (7.147) and (7.148) and then a substitution $z = y/t$ that

$$
\begin{aligned}
I_3 &= -\theta^2 \int_{y=0}^{s} \frac{\beta}{y} \int_{x=0}^{t-y} \frac{\beta}{x} \left[\frac{y^\beta}{\Gamma(1+\beta)}\right] \left[-\frac{x}{\beta\theta}\frac{d}{dx}E_\beta(-\theta x^\beta)\right] dx\, dy \\
&= \frac{\theta\beta}{\Gamma(1+\beta)} \int_{y=0}^{s} \frac{1}{y^{1-\beta}} \int_{x=0}^{t-y} \frac{d}{dx}E_\beta(-\theta x^\beta) dx\, dy \\
&= \frac{\theta\beta}{\Gamma(1+\beta)} \int_{y=0}^{s} \frac{1}{y^{1-\beta}} (E_\beta(-\theta(t-y)^\beta) - 1) dy \\
&= \frac{\theta\beta t^\beta}{\Gamma(1+\beta)} \int_0^{s/t} \frac{E_\beta(-\theta t^\beta(1-z)^\beta)}{z^{1-\beta}} dz - \frac{\theta s^\beta}{\Gamma(1+\beta)}.
\end{aligned}
$$

Then it follows from (7.139) and (7.144) that

$$
\begin{aligned}
\operatorname{corr}[X_\beta(t), X_\beta(s)] &= \int_0^\infty \int_0^\infty F(u,v) H(du, dv) \\
&= I_1 + I_2 + I_3 + E_\beta(-\theta t^\beta) + E_\beta(-\theta s^\beta) - 1 \\
&= \left[-\frac{\theta s^\beta}{\Gamma(1-\beta)} - E_\beta(-\theta s^\beta) + 1\right] + \frac{2\theta s^\beta}{\Gamma(1+\beta)} \\
&\quad + \frac{\theta\beta t^\beta}{\Gamma(1+\beta)} \int_0^{s/t} \frac{E_\beta(-\theta t^\beta(1-z)^\beta)}{z^{1-\beta}} dz - \frac{\theta s^\beta}{\Gamma(1+\beta)} \\
&\quad + E_\beta(-\theta t^\beta) + E_\beta(-\theta s^\beta) - 1 \\
&= \frac{\theta\beta t^\beta}{\Gamma(1+\beta)} \int_0^{s/t} \frac{E_\beta(-\theta t^\beta(1-z)^\beta)}{z^{1-\beta}} dz + E_\beta(-\theta t^\beta)
\end{aligned}
$$

which agrees with (7.119). □

The proof of the formula for the correlation function of time-changed Lèvy process (Theorem 7.32) can also be obatined using the bivariate integration by parts formula used for the proof of Theorem 7.29. However, when the outer process is Lèvy as opposed to diffusion, the proof simplifies. We now present the proof of Theorem 7.32.

Proof. Since $X(t)$ is a Lévy process, $\mathbb{E}[X(t)] = t\mathbb{E}[X(1)]$ and $\operatorname{Var}[X(t)] = t\operatorname{Var}[X(1)]$. If $G_t(u) = \mathbb{P}[Y(y) \le u]$, then a simple conditioning argument shows that the mean of $Z(t)$ is

$$
\mathbb{E}[Z(t)] = \int_0^\infty u\mathbb{E}[X(1)]G_t(du) = U(t)\mathbb{E}[X(1)].
$$

The variance

$$\begin{aligned}\mathrm{Var}[Z(t)] &= \mathbb{E}[X(Y(t))^2] - [\mathbb{E}X(Y(t))]^2 \\ &= \int_0^\infty \mathbb{E}[X^2(u)]G_t(du) - U^2(t)[\mathbb{E}X(1)]^2 \\ &= \int_0^\infty \{u^2[\mathbb{E}X(1)]^2 + u\,\mathrm{Var}[X(1)]\}G_t(du) - U^2(t)[\mathbb{E}X(1)]^2 \\ &= [\mathbb{E}X(1)]^2\mathbb{E}[Y^2(t)] + \mathrm{Var}[X(1)]U(t) - U^2(t)[\mathbb{E}X(1)]^2 \\ &= [\mathbb{E}X(1)]^2\,\mathrm{Var}[Y(t)] + U(t)\,\mathrm{Var}[X(1)].\end{aligned}$$

For $0 < s < t$, since the outer process $X(t)$ has independent increments, we have

$$\begin{aligned}\mathbb{E}X(t)X(s) &= \mathbb{E}(X(t) - X(s))X(s) + \mathbb{E}X^2(s) \\ &= \mathbb{E}(X(t) - X(s))\mathbb{E}X(s) + \mathbb{E}X^2(s) \\ &= ts[\mathbb{E}X(1)]^2 - s^2[\mathbb{E}X(1)]^2 + \mathrm{Var}\,X(s) + s^2[\mathbb{E}X(1)]^2 \\ &= ts[\mathbb{E}X(1)]^2 + s\,\mathrm{Var}\,X(1).\end{aligned}$$

Since processes X and Y are independent,

$$\mathbb{E}X(Y(t))X(Y(s)) = \mathbb{E}Y(t)Y(s)[\mathbb{E}X(1)]^2 + \mathbb{E}Y(s)\,\mathrm{Var}\,X(1),$$

and the covariance function of the time-changed process is

$$\begin{aligned}\mathrm{Cov}[Z(t), Z(s)] &= EY(t)Y(s)[\mathbb{E}X(1)]^2 + \mathbb{E}Y(s)\,\mathrm{Var}\,X(1) - \mathbb{E}Z(t)\mathbb{E}Z(s) \\ &= \mathbb{E}Y(t)Y(s)[\mathbb{E}X(1)]^2 + \mathbb{E}Y(s)\,\mathrm{Var}\,X(1) - U(t)U(s)[\mathbb{E}X(1)]^2 \\ &= U(s)\,\mathrm{Var}\,X(1) + [\mathbb{E}X(1)]^2\,\mathrm{Cov}[Y(t), Y(s)].\end{aligned}$$

□

7.9 Fractional Brownian motion

Fractional Brownian motion is the fractional derivative (or fractional integral) of a Brownian motion. Suppose that $B(t)$ is a standard Brownian motion with characteristic function $\mathbb{E}[e^{ikB(t)}] = e^{-tk^2/2}$ for all $t \geq 0$. Extend $B(t)$ to the entire real line by taking another independent Brownian motion $B_1(t)$ with the same distribution, and setting $B(t) = B_1(-t)$ when $t < 0$. Then we have $\mathbb{E}[e^{ikB(t)}] = e^{-|t|k^2/2}$ for all $t \in \mathbb{R}$. Recall from (2.23) that the Caputo fractional derivative of order $0 < \alpha < 1$ can be written in the form

$$\frac{d^\alpha f(x)}{dx^\alpha} = \frac{1}{\Gamma(1-\alpha)}\int_{-\infty}^{x} f'(u)(x-u)^{-\alpha}\,du.$$

Heuristically, we would like to define the fractional Brownian motion

$$\frac{1}{\Gamma(1-\alpha)}\int_{-\infty}^{t}(t-s)^{-\alpha}B'(s)ds$$

but there are some technical issues. We review the basic ideas here. For complete details, see Pipiras and Taqqu [167].

First of all, the derivative $B'(s)$ does not exist (the paths of a Brownian motion are almost surely nowhere differentiable). This is similar to a problem we often face in probability. If X is a random variable with cdf $F(x)$ and pdf $f(x) = F'(x)$, then we define $\mathbb{E}[g(X)] = \int g(x)f(x)\,dx = \int g(x)F'(x)\,dx$. If the cdf is not differentiable, we use the Lebesgue-Stieltjes integral $\mathbb{E}[g(X)] = \int g(x)F(dx)$ instead (see details). A similar approach works for stochastic integrals, and thus for continuous functions $g(s)$ we can define

$$\int_a^b g(s)B(ds) \approx \sum_{i=1}^{n} g(s_i)B(\Delta s_i) \tag{7.150}$$

where $\Delta s = (b-a)/n$, $s_i = a + i\Delta s$ for $i = 0, 1, \ldots, n$, $B(\Delta s_i) = B(s_i) - B(s_{i-1})$, and the approximating sum on the right converges *in probability* to the stochastic integral on the left as $n \to \infty$ (see details at the end of this section). Note that $B(\Delta s_i)$ is normal with mean zero and variance $(s_i - s_{i-1})$, and that $B(\Delta s_1), \ldots, B(\Delta s_n)$ are independent, since $B(t)$ has independent increments. Then $\sum_i g(s_i)B(\Delta s_i)$ is normal with mean zero and variance $\sum_i g(s_i)^2\Delta s$, and it follows by taking limits that

$$\int_a^b g(s)B(ds) \simeq \mathcal{N}\left(0, \int_a^b |g(s)|^2\,ds\right). \tag{7.151}$$

assuming that $|g(s)|^2$ is integrable over $a < s \le b$. The improper integral is defined, as usual, as a limit of proper integrals

$$\int_{-\infty}^{b} g(s)B(ds) = \lim_{a\to-\infty}\int_a^b g(s)B(ds) \quad \text{in probability,}$$

and then

$$\int_{-\infty}^{b} g(s)B(ds) \simeq \mathcal{N}\left(0, \int_{-\infty}^{b} |g(s)|^2\,ds\right) \tag{7.152}$$

assuming that $|g(s)|^2$ is integrable over $-\infty < s \le b$.

Now we may try to define a fractional derivative of Brownian motion by the formula

$$I(t) = \frac{1}{\Gamma(1-\alpha)}\int_{-\infty}^{t}(t-s)^{-\alpha}B(ds)$$

but this does not work either, because for $g(s) = (t - s)^{-\alpha}$ we have

$$\int_{-\infty}^{t} |g(s)|^2\, ds = \infty.$$

To work around this, we first define

$$I_a(t) = \frac{1}{\Gamma(1-\alpha)} \int_a^t (t-s)^{-\alpha} B(ds)$$

and we consider the difference

$$\begin{aligned}
B_H(t) &= \lim_{a\to-\infty} I_a(t) - I_a(0) \\
&= \lim_{a\to-\infty} \frac{1}{\Gamma(1-\alpha)} \int_a^t (t-s)^{-\alpha} B(ds) - \frac{1}{\Gamma(1-\alpha)} \int_a^0 (0-s)^{-\alpha} B(ds) \\
&= \frac{1}{\Gamma(1-\alpha)} \int_{-\infty}^{\infty} \left[(t-s)_+^{-\alpha} - (0-s)_+^{-\alpha}\right] B(ds)
\end{aligned} \tag{7.153}$$

where

$$(x)_+ = \begin{cases} x & \text{if } x > 0 \\ 0 & \text{if } x \le 0 \end{cases} \tag{7.154}$$

and we adopt the convention $0^0 = 0$. This stochastic integral is defined for any $-1/2 < \alpha < 1/2$, since the function $g(s) = (t-s)_+^{-\alpha} - (0-s)_+^{-\alpha}$ satisfies $\int g(s)^2\, ds < \infty$ in that case (see details). Hence we have to restrict to $-1/2 < \alpha < 1/2$ in this approach. Then we can define the fractional derivative of Brownian motion of order $0 < \alpha < 1/2$, and also the fractional integral of the same order. See the details at the end of this section for a brief introduction to fractional integrals.

The *Hurst index* $H = (1/2) - \alpha$ for $0 < H < 1$ governs the self-similarity of the fractional Brownian motion (7.153). First note that the random measure $B(ds)$ has a scaling property $B(c\, ds) \simeq c^{1/2} B(ds)$, since for an interval $V = [a, b]$ we have $B(V) \simeq \mathcal{N}(0, |V|)$ and $B(cV) \simeq \mathcal{N}(0, |cV|) \simeq c^{1/2} B(V)$. Then a change of variables $s = cs'$ yields

$$\begin{aligned}
B_H(ct) &= \frac{1}{\Gamma(1-\alpha)} \int_{-\infty}^{\infty} \left[(ct-s)_+^{H-1/2} - (0-s)_+^{H-1/2}\right] B(ds) \\
&= \frac{1}{\Gamma(1-\alpha)} \int_{-\infty}^{\infty} \left[(ct-cs')_+^{H-1/2} - (c0-cs')_+^{H-1/2}\right] B(c\, ds') \\
&\simeq \frac{1}{\Gamma(1-\alpha)} \int_{-\infty}^{\infty} c^{H-1/2} \left[(t-s')_+^{H-1/2} - (0-s')_+^{H-1/2}\right] c^{1/2} B(ds') \\
&= c^H B_H(t).
\end{aligned} \tag{7.155}$$

To justify the change of variables in (7.155), use (7.150) and note that $B(c\Delta s_i) \simeq c^{1/2}B(\Delta s_i)$ (see details). Then we certainly have $B_H(ct) \simeq c^H B_H(t)$ for all $c > 0$ and $t \in \mathbb{R}$. It is also possible to extend this argument to show that $B_H(ct) \simeq c^H B_H(t)$ in the sense of finite dimensional distribution (e.g., see Samorodnitsky and Taqqu [185, Corollary 7.2.3]).

In the special case $H = 1/2$, we have $\alpha = 0$, and then for $t \geq 0$ we get

$$B_H(t) = \frac{1}{\Gamma(1-\alpha)} \int_{-\infty}^{\infty} [I(t-s>0) - I(0-s>0)]\, B(ds)$$
$$= \int_{-\infty}^{\infty} [I(0 \leq s < t)]\, B(ds) = B(t) - B(0) = B(t),$$

while for $t < 0$ we get

$$B_H(t) = \int_{-\infty}^{\infty} [I(t \leq s < 0)]\, B(ds) = B(0) - B(t) = -B_1(t) \simeq B(t).$$

Hence $B_H(t)$ is a Brownian motion on $t \in \mathbb{R}$ when $H = 1/2$.

It follows from self-similarity $B_H(ct) \simeq c^H B_H(t)$ that a fractional Brownian motion satisfies $B_H(t) \simeq t^H B_H(1)$ for all $t \in \mathbb{R}$, where the stochastic integral $B_H(1)$ is normal with mean zero. Hence $B_H(t)$ has a pdf $p(x,t)$ with FT

$$\hat{p}(k,t) = \mathbb{E}[e^{-ikB_H(t)}] = e^{-Dt^{2H}k^2}$$

for any $t > 0$, for some constant $D > 0$. Then clearly

$$\frac{d}{dt}\hat{p}(k,t) = 2HDt^{2H-1}(ik)^2\hat{p}(k,t)$$

and hence the pdf $p(x,t)$ of a fractional Brownian motion $B_H(t)$ solves a diffusion equation with variable coefficients

$$\frac{\partial}{\partial t}p(x,t) = 2HDt^{2H-1}\frac{\partial^2}{\partial x^2}p(x,t) \tag{7.156}$$

for $t > 0$. The case $1/2 < H < 1$ is a kind of super-diffusion, and $0 < H < 1/2$ is a sub-diffusion.

Because fractional Brownian motion $B_H(t)$ is a fractional integral or derivative of Brownian motion, it averages $B(t)$ over the entire interval $(-\infty, t]$, and so the increments

$$B_H(t_2) - B_H(t_1) = \frac{1}{\Gamma(1-\alpha)} \int_{-\infty}^{\infty} \left[(t_2 - s)_+^{H-1/2} - (t_1 - s)_+^{H-1/2}\right] B(ds)$$

are not independent. However, a straightforward change of variables shows that the increments are stationary: $B_H(t_2) - B_H(t_1) \simeq B_H(t_2 - t_1)$. The fractional Brownian motion $B_H(t)$ with $H \neq 1/2$ is not a Lévy process, since it does not have independent increments. There are many Gaussian stochastic processes whose pdf $p(x,t)$ solves (7.156) (e.g., the process $t \mapsto t^H Z$ where $Z \simeq \mathcal{N}(0, 2D)$ is one). However, fractional Brownian motion is the only self-similar Gaussian process with stationary increments (e.g., see [185, Lemma 7.2.1]), and so it is the only self-similar Gaussian process with stationary increments that is governed by (7.156).

Remark 7.34. The graph of a fractional Brownian motion $B_H(t)$ is a random fractal with dimension $d = 2 - H$, see for example Falconer [65, Theorem 16.7]. As the Hurst index H increases from $1/2$ to 1, we are applying a fractional integral of increasing order, so the graph becomes smoother.

Remark 7.35. It is a simple matter to compute the covariance structure of a fractional Brownian motion $B_H(t)$, using the self-similarity and stationary increments. First consider $0 < s < t$. Since $B_H(t) \simeq t^H B_H(1)$ we have $\mathbb{E}[B_H(t)^2] = t^{2H} C$ where $C = \mathbb{E}[B_H(1)^2]$. Now write

$$(B_H(t) - B_H(s))^2 = B_H(t)^2 + B_H(s)^2 - 2B_H(t)B_H(s)$$

and take expectations to get

$$C(t-s)^{2H} = Ct^{2H} + Cs^{2H} - 2\mathbb{E}[B_H(t)B_H(s)].$$

Now solve to get

$$\mathbb{E}[B_H(t)B_H(s)] = \frac{C}{2}\left[t^{2H} + s^{2H} - (t-s)^{2H}\right].$$

The case $0 < t < s$ is similar, and we can combine these two cases to write

$$\mathbb{E}[B_H(t)B_H(s)] = \frac{C}{2}\left[|t|^{2H} + |s|^{2H} - |t-s|^{2H}\right]. \tag{7.157}$$

The case where $t < 0$ or $s < 0$ is again similar, and leads to the same result (7.157). For those cases, note that $B_H(1) \simeq B_H(-1)$. This follows easily from the fact that $B_H(t)$ has stationary increments.

The fractional Brownian motion (7.153) is the positive fractional derivative (or integral) of a Brownian motion. Applying the same construction using the negative fractional derivative leads to the process

$$\frac{1}{\Gamma(1-\alpha)} \int_{-\infty}^{\infty} \left[(s-t)_+^{H-1/2} - (s-0)_+^{H-1/2}\right] B(ds)$$

where $x_+ = xI(x > 0)$, and again we adopt the convention $0^0 = 0$. This form averages $B(ds)$ over the interval extending to $+\infty$. A mixture of positive and negative fractional derivatives leads to the general form

$$B_H(t) = \frac{p}{\Gamma(1-\alpha)} \int_{-\infty}^{\infty} \left[(t-s)_+^{H-1/2} - (0-s)_+^{H-1/2} \right] B(ds)$$
$$+ \frac{q}{\Gamma(1-\alpha)} \int_{-\infty}^{\infty} \left[(s-t)_+^{H-1/2} - (s-0)_+^{H-1/2} \right] B(ds) \tag{7.158}$$

for $p, q \geq 0$. Taking $p = q = 1$ leads to the form

$$B_H(t) = \frac{1}{\Gamma(1-\alpha)} \int_{-\infty}^{\infty} \left[|t-s|^{H-1/2} - |0-s|^{H-1/2} \right] B(ds), \tag{7.159}$$

based on the Riesz fractional derivative or integral (see details).

Remark 7.36. The definition (7.158) of a fractional Brownian motion is based on the Caputo fractional derivative (2.23) of a function defined on the entire real line. Another kind of fractional Brownian motion uses the Caputo fractional derivative (2.33) of a function defined on the positive half-line. The *Lévy fractional Brownian motion* (also called type two fractional Brownian motion) is defined for $t \geq 0$ by

$$B_H^{(2)}(t) = \frac{1}{\Gamma(1-\alpha)} \int_0^t (t-s)^{H-1/2} B(ds). \tag{7.160}$$

where $0 < H < 1$. Since the function $g(s) = (t-s)^{H-1/2}$ is square integrable over the interval $s \in [0, t]$, this construction is simpler. However, the definition (7.158) is preferred in many applications, because it has stationary increments.

A discrete analogue of fractional Brownian motion can be constructed using fractional differences. Take (Z_n) iid normal with mean zero, and let

$$Y_n = \Delta^\alpha Z_n = (I - B)^\alpha Z_n = \sum_{j=0}^{\infty} \binom{\alpha}{j} (-1)^j Z_{n-j} \tag{7.161}$$

using the backward shift operator $BZ_n = Z_{n-1}$. In time series, Y_n is called a *fractional* ARIMA$(0, d, 0)$ *process* where $d = -\alpha = H - (1/2)$ is the order of fractional integration (e.g., see Brockwell and Davis [42, Section 13.2]). If $0 < d < 1/2$ (i.e., $1/2 < H < 1$) then this mean zero process has *long range dependence* since its autocovariance function decays very slowly:

$$\mathbb{E}[Y_n Y_{n+j}] \sim C j^{2H-2} \quad \text{as } j \to \infty.$$

Hurst [94] noted this kind of long range dependence in flood levels of the Nile river. The time series (Y_n) is stationary. It can be considered as a discrete analogue of the increments of a fractional Brownian motion. In fact, if we let $S_n = Y_1 + \cdots + Y_n$ then it follows from Whitt [219, Theorem 4.6.1] that

$$\sigma_n^{-1} S_{[nt]} \Rightarrow B_H(t)$$

in the Skorokhod space $\mathbb{D}[0, \infty)$, where $C\sigma_n^2 = \text{Var}(S_n)$ and $C = \mathbb{E}[B_H(1)^2]$. Hence it is reasonable to approximate fractional Brownian motion by a random walk whose jumps come from a fractional ARIMA(0, d, 0) process.

Remark 7.37. Another popular method for simulating fractional Brownian motion uses FT methods. Use (7.153) to write

$$B_H(t_i) \approx J_a(t_i) - J_a(0)$$

on a finite discrete grid $s_j = a + j\Delta t$, where

$$J_a(t_i) = \frac{1}{\Gamma(1-\alpha)} \sum_{j=1}^{n} (t_i - s_j)_+^{-\alpha} B(\Delta s_j) \approx I_a(t_i).$$

Since $J_a(t_i)$ is a discrete convolution, it can be efficiently computed using a numerical method called the *fast Fourier transform*, a streamlined algorithm for computing the discrete FT (e.g., see Press, et al. [170]). Simply multiply the discrete FT $\hat{g}(k_j) = \sum_j e^{-is_jk_j} g(s_j)$ of the *filter* $g(s_j) = (s_j)_+^{-\alpha}/\Gamma(1-\alpha)$ by the discrete FT $\hat{B}(\Delta k_j)$ of the noise sequence $B(\Delta s_j)$, invert the product of these two discrete Fourier transforms $\hat{J}_a(k_j) = \hat{g}(k_j)\hat{B}(\Delta k_j)$ to get the convolution $J_a(t_i)$, and then subtract $J_a(0)$ (e.g., see Dieker and Mandjes [61]). Some additional efficiency can be obtained by simulating the discrete FT of the noise sequence $\hat{B}(\Delta k_j)$ directly (e.g., see Voss [216]). Since $\hat{g}(k_j) \approx (ik_j)^{\alpha-1}$ with $\alpha = (1/2) - H$, taking limits after Fourier inversion leads to

$$\hat{J}_a(x) - \hat{J}_a(0) \approx \int (e^{ikx} - 1)(ik)^{-H-1/2} \hat{B}(dk).$$

This stochastic integral with respect to the complex-valued Gaussian random measure $\hat{B}(dk)$ is called the *spectral representation* of a fractional Brownian motion (e.g., see Samorodnitsky and Taqqu [185, Section 7.2]). Roughly speaking, $J_a(t)$ represents the $H + 1/2$ order fractional integral of the white noise $B(dt)$, i.e., the $H - 1/2$ fractional integral of $B(t)$.

Remark 7.38. Starting with a stable Lévy motion $A(t)$ on $t \geq 0$ with index $0 < \gamma < 2$, extend to $t \in \mathbb{R}$ as before, by setting $A(t) = A_1(-t)$ for $t < 0$, where $A_1(t)$ is another independent Lévy motion identically distributed with $A(t)$. The stochastic integral

$$A_H(t) = \frac{1}{\Gamma(1-\alpha)} \int_{-\infty}^{\infty} \left[(t-s)_+^{-\alpha} - (0-s)_+^{-\alpha}\right] A(ds) \tag{7.162}$$

can be defined in the same way as for Brownian motion, using the stable random measure $A(a, b] = A(b) - A(a)$. The stochastic process (7.162) is called a *linear fractional stable motion*. The stable stochastic integral $\int g(s)A(ds) \approx \sum_i g(s_i)A(\Delta s_i)$ is defined when $\int |g(s)|^\gamma ds < \infty$. The self-similarity $A(ct) \simeq c^{1/\gamma}A(t)$ of the stable process implies that $A(c\,ds) \simeq c^{1/\gamma}A(ds)$, and then it follows that $A_H(ct) \simeq c^H A_H(t)$, by the same argument as in the Gaussian case, where the Hurst index $H = (1/\gamma) - \alpha$. A linear fractional stable motion has stationary increments, which are not independent (unless $\alpha = 0$). For more details, see [185, Section 7.4].

A discrete analogue of linear fractional stable motion with $1 < \gamma < 2$ comes from taking (Z_n) iid γ-stable with mean zero in (7.161). Since the covariance does not exist in this case, the long range dependence of the fractionally integrated time series (Y_n) in the case $H > 1/\gamma$ is defined in terms of the moving average coefficients: We say that $Y_n = \sum_j c_j Z_{n-j}$ has long range dependence if $\sum_j |c_j| = \infty$. In view of (2.5) we can see that (7.161) is long range dependent if $\alpha < 0$ (fractional integration).

If we let $S_n = Y_1 + \cdots + Y_n$ then it follows from Whitt [219, Theorem 4.7.2] that

$$n^{-H} S_{[nt]} \Rightarrow A_H(t)$$

in the Skorokhod space $\mathbb{D}[0,\infty)$. Hence the fractional ARIMA$(0, d, 0)$ process with stable innovations (Z_n) approximates the increments of a linear fractional stable motion. The FFT method outlined in Remark 7.37 can also be used to simulate linear fractional stable motion, see Stoev and Taqqu [209] and Biermé and Scheffler [34].

Remark 7.39. If we take (Z_n) iid normal with mean zero, then the sequence (Y_n) in (7.161) models a correlated sequence of mean zero finite variance particle jumps. In a CTRW framework with iid power law waiting times $\mathbb{P}[J_n > t] = Ct^{-\beta}$ for some $0 < \beta < 1$, independent of the particle jumps, the CTRW scaling limit is $B_H(E(t))$ where $E(t)$ is the inverse stable subordinator (4.27). If we take (Z_n) iid stable with mean zero, then the CTRW scaling limit is $A_H(E(t))$, a linear fractional stable motion time-changed via the inverse stable subordinator. If the mean zero sequence (Z_n) belongs to some normal or stable domain of attraction, the same scaling limit applies. For more details, see Meerschaert, Nane and Xiao [142]. The governing equation of these CTRW limits is currently unknown.

Details

In order to clearly understand stochastic integrals, we begin with a review of deterministic integrals. If X is a random variable with cdf $F(x) = \mathbb{P}[X \le x]$, and $g(x)$ is a Borel measurable function, we define the expectation of $g(X)$ through a Lebesgue-Stieltjes integral $\mathbb{E}[g(X)] = \int g(x)F(dx) = \int g(x)\mu(dx)$, a Legesgue integral with respect to the probability measure μ defined by $\mu(a, b] = F(b) - F(a)$. Recall that the Lebesgue integral is defined as follows: If $g(s) = I(s \in V)$ for some Borel set V, then $\int g(s)\mu(ds) = \mu(V)$. For a simple function $g(s) = \sum_{i=1}^n a_i I(s \in V_i)$ where $V_1, \ldots, V_n$

are mutually disjoint Borel sets, $\int g(s)\mu(ds) = \sum_{i=1}^{n} a_i\mu(V_i)$. Then for $g \geq 0$, we define

$$\int g(s)\mu(ds) = \lim_{n\to\infty} \int g_n(s)\mu(ds) \tag{7.163}$$

where

$$g_n(s) = \begin{cases} (k-1)/n & \text{if } (k-1)/n < g(s) \leq k/n \text{ for some } 1 \leq k \leq n \\ 0 & \text{otherwise.} \end{cases} \tag{7.164}$$

Since $\int g_n(s)\mu(ds)$ is an increasing sequence, the limit $\int g(s)\mu(ds)$ in (7.163) always exists (although it may equal infinity). If $g(s)$ takes both positive and negative values, we can write $g = g^+ - g^-$ the difference of two non-negative Borel measurable functions, and then we define $\int g(s)\mu(ds) = \int g^+(s)\mu(ds) - \int g^-(s)\mu(ds)$, provided that both integrals exist and are finite. The integral

$$\int_a^b g(s)F(ds) = \int g(s)I(a \leq s \leq b)\mu(ds) \tag{7.165}$$

is defined since $g(s)I(a \leq s \leq b)$ is a Borel measurable function.

The Riemann-Stieltjes integral is defined by

$$\int_a^b g(s)F(ds) = \lim_{\Delta s\to 0} \sum_{i=1}^{n} g(s_i)\Delta F(s_i) \tag{7.166}$$

where $\Delta s = (b-a)/n$, $s_i = a + i\Delta s$ for $i = 0, 1, \ldots, n$, and $\Delta F(s_i) = F(s_i) - F(s_{i-1})$. If $g(s)$ is continuous, then $g(s)$ is also bounded and uniformly continuous on the interval $[a, b]$. Given any positive integer n, choose $\delta > 0$ such that $|g(s)-g(t)| < 1/n$ whenever $|s-t| < \delta$. If $\Delta s < \delta$, then since $0 \leq g(s) - g_n(s) \leq 1/n$ for each s, for n sufficiently large, eventually $|g_n(s) - g(s_i)| \leq |g_n(s) - g(s)| + |g(s) - g(s_i)| \leq 2/n$ for all $s_{i-1} < s \leq s_i$ and all $i = 1, 2, \ldots, n$, and then

$$\begin{aligned} &\left|\int g_n(s)F(ds) - \sum_{i=1}^{n} g(s_i)\Delta F(s_i)\right| \\ &= \left|\int g_n(s)F(ds) - \int\left(\sum_{i=1}^{n} g(s_i)I(s_{i-1} < s \leq s_i)\right)F(ds)\right| \\ &\leq \sum_{i=1}^{n}\int_{s_{i-1}}^{s_i} |g_n(s) - g(s_i)|\, F(ds) \leq (2/n)[F(b) - F(a)] \end{aligned}$$

for all n. Then it follows from (7.165) and (7.166) that the Riemann-Stieltjes integral exists and equals the Lebesgue-Stieltjes integral for continuous functions on bounded intervals. Equality on unbounded intervals follows. For example, for $g(s) \geq 0$ the

Riemann-Stieltjes integral of $g(s)$ with respect to $F(ds)$ on $-\infty < s \le b$ is defined by

$$\int_{-\infty}^{b} g(s)F(ds) = \lim_{a\to-\infty}\int_{a}^{b} g(s)F(ds). \tag{7.167}$$

Suppose that this limit is finite. Since the Riemann-Stieltjes integral on the right-hand side of (7.167) equals the Lebesgue-Stieltjes integral over that same interval, it follows from the dominated convergence theorem that (7.167) also holds for the Lebesgue-Stieltjes integral, and hence these two integrals are equal over the unbounded interval.

Given a Brownian motion $B(t)$ with $\mathbb{E}[e^{ikB(t)}] = e^{-|t|k^2/2}$ for all $t \in \mathbb{R}$, we now define the *stochastic integral* $\int g(s)B(ds)$. Here we outline the basic ideas. For more details on stochastic integration, see Samorodnitsky and Taqqu [185, Chapter 3]. First we define a *random measure* $B(ds)$ on the real line by setting $B(a, b] = B(b) - B(a)$. Then $B(a, b] \simeq \mathcal{N}(0, (b - a))$, since $B(t)$ has stationary increments. Extend to Borel sets V to see that $B(V) \simeq \mathcal{N}(0, |V|)$ where $|V| = \int I(s \in V)\, ds$ is the Lebesgue measure of the set V. This construction uses the Kolmogorov consistency theorem, see [185, Chapter 3] for complete details. If U and V are disjoint intervals, then $B(U)$ and $B(V)$ are independent, since $B(t)$ has independent increments. Extend to Borel sets to see that $B(ds)$ is *independently scattered*, i.e., $B(U)$ and $B(V)$ are independent when U and V are disjoint Borel sets. Given a simple function $g(s) = \sum_{i=1}^{n} c_i I(s \in V_i)$ where $V_1, \dots, V_n$ are mutually disjoint bounded Borel sets, we define

$$\int g(s)B(ds) = \sum_{i=1}^{n} c_i B(V_i). \tag{7.168}$$

For example, if $g(s) = I(a < s \le b)$ then

$$\int g(s)B(ds) = \int_{a}^{b} 1\, B(ds) = B(b) - B(a).$$

The stochastic integral (7.168) is normal with mean zero and variance

$$\sum_{i=1}^{n} c_i^2 |V_i| = \int |g(s)|^2\, ds.$$

Now for $g \ge 0$ Borel measurable, we define

$$\int g(s)B(ds) = \lim_{n\to\infty}\int g_n(s)B(ds) \quad \text{in probability} \tag{7.169}$$

where the simple function g_n is given by (7.164), and $\int |g(s)|^2\, ds < \infty$. To show that this limit exists, it is simplest to work with L^2 convergence: Let $I(g_n) = \int g_n(s)B(ds)$, a sequence of Gaussian random variables. Use the dominated convergence theorem to see that

$$\begin{aligned}\|I(g_n) - I(g_m)\|_2 &:= \mathbb{E}[|I(g_n) - I(g_m)|^2] \\ &= \mathbb{E}[|I(g_n - g_m)|^2] \\ &= \int |g_n(s) - g_m(s)|^2\,ds \to 0\end{aligned}$$

as $m, n \to \infty$, i.e., the sequence $\{I(g_n)\}$ is Cauchy. Since the Banach space L^2 of finite variance random variables with the norm $\|X\|_2 = \sqrt{\mathbb{E}[X^2]}$ is Cauchy complete, there exists a limit $I(g)$ in this space. Since L^2 convergence (convergence in mean square) implies convergence in probability, (7.169) holds, and since convergence in probability also implies convergence in distribution,

$$\int g(s)B(ds) \simeq \mathcal{N}\left(0, \int |g(s)|^2\,ds\right). \tag{7.170}$$

(Note: This L^2 convergence argument does not extend to stable stochastic integrals, since a stable law does not have a finite second moment. One can still prove convergence in probability, but the argument is harder, see [185, Chapter 3].) The reason for taking limits in probability in the definition (7.169), rather than a point-wise limit, is that the sample paths of $B(t)$ are almost surely of unbounded variation, so that the point-wise limit might not exist.

If $g(s)$ is continuous on the interval $[a, b]$, then we can also write

$$\int_a^b g(s)B(ds) = \lim_{\Delta s \to 0} \sum_{i=1}^n g(s_i)B(\Delta s_i) \quad \text{in probability} \tag{7.171}$$

where $\Delta s = (b - a)/n$, $s_i = a + i\Delta s$ for $i = 0, 1, \ldots, n$, and $B(\Delta s_i) = B(s_i) - B(s_{i-1})$. To see this, note that for all large n we have $|g_n(s) - g(s_i)| \le 2/n$ for all $s_{i-1} < s \le s_i$ and all $i = 1, 2, \ldots, n$, where g_n is the simple function approximation of g defined by (7.164). Then

$$\int_a^b g_n(s)B(ds) - \sum_{i=1}^n g(s_i)\Delta B(s_i) \simeq \mathcal{N}\left(0, \sum_{i=1}^n \int_{s_{i-1}}^{s_i} |g_n(s) - g(s_i)|^2\,ds\right)$$

for all n. Since the variance is bounded above by $(2/n)^2|b - a|$, the difference between these two stochastic integrals converges in probability to zero, and then (7.171) follows.

Define $g(s) = (t - s)_+^{-\alpha} - (0 - s)_+^{-\alpha}$ using the notation (7.154). We want to show that $\int g(s)^2\,ds < \infty$ when $-1/2 < \alpha < 1/2$. Suppose that $t > 0$. Then $g(s) = 0$ for $s > t$. For $s < 0$ we have $g(s) = (t - s)^{-\alpha} - (0 - s)^{-\alpha}$. Write $g(s) = f(t - s) - f(0 - s)$ where $f(u) = u^{-\alpha}$. The mean value theorem implies that $g(s) = tf'(w) = -t\alpha w^{-1-\alpha}$ for some $0 - s \le w \le t - s$. Then $|g(s)| \le t|\alpha||s|^{-\alpha-1}$ for all $s < 0$. It follows that

$$\int_{-\infty}^{-1} g(s)^2\,ds \le \int_{-\infty}^{-1} (t\alpha)^2|s|^{-2\alpha-2}\,ds = \int_1^{\infty} (t\alpha)^2 s^{-2\alpha-2}\,ds < \infty$$

provided that $-2\alpha - 2 < -1$, i.e., $\alpha > -1/2$. If $\alpha \le 0$, then $g(s)$ is bounded on the interval $[-1, t]$, and so the function $g(s)^2$ is integrable on the entire real line. If $\alpha > 0$, then the integrand $g(s)$ blows up at $s = t$ and $s = 0$. On the interval $(-1, 0)$ we have $0 < (t-s)^{-\alpha} < (0-s)^{-\alpha}$ so that $g(s)^2 = [(0-s)^{-\alpha} - (t-s)^{-\alpha}]^2 \le (0-s)^{-2\alpha}$ and hence

$$\int_{-1}^{0} g(s)^2\,ds \le \int_{-1}^{0} |s|^{-2\alpha}\,ds = \int_{0}^{1} s^{-2\alpha}\,ds < \infty$$

provided $-2\alpha + 1 > 0$, i.e., $\alpha < 1/2$. Finally, on the remaining interval $0 < s < t$ we have $g(s) = (t-s)^{-\alpha}$ and a change of variables $u = t - s$ shows that

$$\int_{0}^{t} g(s)^2\,ds = \int_{0}^{t} (t-s)^{-2\alpha}\,ds = \int_{0}^{t} u^{-2\alpha}\,du < \infty$$

provided $\alpha < 1/2$. Hence it follows that $\int g(s)^2\,ds < \infty$ for $-1/2 < \alpha < 1/2$ when $t > 0$. The proof for $t < 0$ is similar. If $\alpha \notin (-1/2, 1/2)$, it can be shown using similar arguments that $\int g(s)^2\,ds = \infty$.

For suitable functions $f(t)$, the (positive) *Riemann-Liouville fractional integral* of order $\alpha > 0$ is defined by

$$\mathbb{I}_t^{\alpha} f(t) = \frac{1}{\Gamma(\alpha)} \int_{-\infty}^{\infty} f(u)(t-u)_+^{\alpha-1}\,du.$$

Recall (2.27), and substitute $p = \alpha - 1$ to see that $s^{-\alpha}$ is the LT of $t^{\alpha-1}/\Gamma(\alpha)$. Then for bounded continuous functions $f(t)$ on $t \ge 0$, extended to the entire real line by setting $f(t) = 0$ when $t < 0$, it follows from the convolution property of the LT that $\mathbb{I}_t^{\alpha} f(t)$ has LT $s^{-\alpha}\tilde{f}(s)$. Some authors define the Riemann-Liouville and Caputo fractional derivatives in terms of the Riemann-Liouville fractional integral: For example, when $0 < \alpha < 1$ we can write

$$\mathbb{D}_t^{\alpha} f(t) = \frac{d}{dt}\left[\mathbb{I}_t^{1-\alpha} f(t)\right] \quad \text{and} \quad \partial_t^{\alpha} f(t) = \mathbb{I}_t^{1-\alpha}\left[\frac{d}{dt} f(t)\right],$$

which reduces to (2.23) and (2.24). The negative Riemann-Liouville fractional integral of order $\alpha > 0$ is defined by

$$\mathbb{I}_{(-t)}^{\alpha} f(t) = \frac{1}{\Gamma(\alpha)} \int_{-\infty}^{\infty} f(u)(u-t)_+^{\alpha-1}\,du.$$

The *Riesz fractional integral* of order $\alpha > 0$ is $\mathbb{J}_t^{\alpha} f(t) = Cp\mathbb{I}_t^{\alpha} f(t) + Cq\mathbb{I}_{(-t)}^{\alpha} f(t)$ with $p = q = 1/2$. Hence we can also write

$$\mathbb{J}_t^{\alpha} f(t) = \frac{C}{\Gamma(\alpha)} \int_{-\infty}^{\infty} f(u)|t-u|^{\alpha-1}\,du.$$

This integral exists for bounded continuous functions such that $f(t) \to 0$ sufficiently fast as $|t| \to \infty$, since the function $|t|^{\alpha-1}$ is integrable at $t = 0$ for any $\alpha > 0$. The constant $C > 0$ is chosen so that $\mathbb{J}_x^\alpha f(x)$ has FT $|k|^{-\alpha}\hat{f}(k)$ for suitable functions $f(x)$. The Riesz fractional integral is also called the *Riesz potential*. For more information, see Samko, Kilbas and Marichev [184].

To justify the change of variables in (7.155), suppose first that $g(s) \geq 0$ is continuous on $s \in [a, b]$. Then (7.171) defines the stochastic integral $\int_a^b g(s)B(ds)$. Given $c > 0$, define $B(c\Delta s_i) = B(cs_i) - B(cs_{i-1})$. Then

$$\sum_{i=1}^{n} g(cs_i)\, B(c\Delta s_i) \simeq \sum_{i=1}^{n} g(cs_i)\, c^{1/2} B(\Delta s_i)$$

and taking limits in probability as $n \to \infty$ shows that

$$\int_{ca}^{cb} g(s')B(ds') = \int_a^b g(cs)\, c^{1/2} B(ds).$$

For a different proof, use the fact that the integrand $g(t, s) = (t - s)_+^{-\alpha} - (0 - s)_+^{-\alpha}$ in (7.153) has the scaling property $g(ct, cs) = c^{-\alpha} g(t, s) = c^{H-1/2} g(t, s)$. Note that

$$B_H(ct) = \int g(ct, s)B(ds) \simeq \mathcal{N}\left(0, \int |g(ct, s)|^2\, ds\right),$$

and

$$c^H B_H(t) = \int c^H g(t, s)B(ds) \simeq \mathcal{N}\left(0, c^{2H} \int |g(t, s)|^2\, ds\right).$$

Then use the scaling and a change of variables $s = cs'$ to check that

$$\begin{aligned}\int |g(ct, s)|^2\, ds &= \int |g(ct, cs')|^2 c\, ds' \\ &= c^{2H-1} \int |g(t, s')|^2 c\, ds' \\ &= c^{2H} \int |g(t, s')|^2\, ds'\end{aligned}$$

so that both integrals have the same distribution.

7.10 Fractional random fields

In this section, we develop multiparameter extensions of the fractional Brownian motion introduced in Section 7.9. We begin with an independently scattered Gaussian random measure $B(dx)$ on $\mathbb{R}^d$ such that, for any bounded Borel subset $V \subset \mathbb{R}^d$, $B(V)$ is a mean zero normal random variable with variance equal to $|V|$, where $|V| = \int I(x \in V)\, dx$ is the Lebesgue measure of that set. In $\mathbb{R}^2$, $|V|$ is the area of the set V, and in $\mathbb{R}^3$, $|V|$ is the volume of the set V. Define the d-dimensional rectangle

$$(a, b] = \{x \in \mathbb{R}^d : a_j < x_j \leq b_j \text{ for all } j = 1, \dots, d\}$$

and the vector $\mathbf{1} = (1, \dots, 1) \in \mathbb{R}^d$. Now we define the stochastic integral

$$\int_{(a,b]} f(x)\, B(dx) \approx \sum_j f(x_j) B(\Delta x_j) \tag{7.172}$$

where Δx_j are rectangles $(x_j, x_j + h\mathbf{1}]$ in $\mathbb{R}^d$ and $x_i = a + jh$ is a discrete lattice with spacing $h = \Delta x > 0$. Here $j = (j_1, \dots, j_d)$ is a vector of integers, and the sum is taken over all j such that $x_j \in (a, b]$. The approximating sum is mean zero normal with variance $\sum_j f(x_j)^2 (\Delta x)^d$ since the random variables $B(\Delta x_j)$ are iid $\mathcal{N}(0, (\Delta x)^d)$. It converges in probability to the stochastic integral for continuous functions $f(x)$ (see details), and the limit

$$\int_{x \in (a,b]} f(x)\, B(dx) \simeq \mathcal{N}\left(0, \int_{x \in (a,b]} |f(x)|^2\, dx\right).$$

A *random field* is a stochastic process $A(x)$ indexed by $x \in \mathbb{R}^d$. The (Lévy) *fractional Brownian field* in $\mathbb{R}^d$ is a scalar-valued random field defined by

$$B_H(x) = \int_{y \in \mathbb{R}^d} \left[\|x - y\|^{H-d/2} - \|0 - y\|^{H-d/2} \right] B(dy), \tag{7.173}$$

for $0 < H < 1$, $H \neq 1/2$. This form extends the fractional Brownian motion (7.159) based on the Riesz fractional derivative $(0 < H < 1/2)$ or the Riesz fractional integral $(1/2 < H < 1)$, see details. The stochastic integral (7.173) is well-defined because the function $f(y) = \|x - y\|^{H-d/2} - \|0 - y\|^{H-d/2}$ satisfies the condition $\int |f(y)|^2 dy < \infty$ when $0 < H < 1$, $H \neq 1/2$. Since the volume (Lebesgue measure) of the set $cV = \{cx : x \in V\}$ in $\mathbb{R}^d$ is $c^d |V|$, the Gaussian random measure $B(dx)$ has the scaling $B(c\, dx) = c^{d/2} B(dx)$. For example, if V is a cube with sides of length h in $\mathbb{R}^3$, then $B(V)$ has variance $|V| = h^3$, and cV is a cube with sides of length ch, so that $B(cV)$ has variance $|cV| = c^3 h^3$. Then it follows that $B_H(cx) \simeq c^H B_H(x)$:

$$\begin{aligned} B_H(cx) &= \int \left[\|cx - y\|^{H-d/2} - \|0 - y\|^{H-d/2} \right] B(dy) \\ &= \int \left[\|cx - cy'\|^{H-d/2} - \|c0 - cy'\|^{H-d/2} \right] B(c\, dy') \\ &\simeq \int c^{H-d/2} \left[\|x - y'\|^{H-d/2} - \|0 - y'\|^{H-d/2} \right] c^{d/2} B(dy') \\ &= c^H B_H(x). \end{aligned} \tag{7.174}$$

A straightforward extension of this argument shows that $B_H(cx)$ and $c^H B_H(x)$ have the same finite dimensional distributions.

Remark 7.40. A *fractional stable field* can be defined in a similar manner. Take an independently scattered stable random measure $A(dx)$ on $\mathbb{R}^d$ such that $A(V) \simeq S_\gamma(\beta, \sigma(V), 0)$ in the notation of Proposition 5.3, where $\sigma(V)^\gamma = |V|$, and define

$$A_H(x) = \int \left[\|x - y\|^{H-d/\gamma} - \|0 - y\|^{H-d/\gamma} \right] A(dy) \tag{7.175}$$

for $0 < H < 1$ with $H \neq 1/\gamma$. The stable stochastic integral

$$\int_{(a,b]} f(x)\, A(dx) \approx \sum_i f(x_i) A(\Delta x_i) \tag{7.176}$$

exists if $\int |f(x)|^\gamma dx < \infty$, see Samorodnitsky and Taqqu [185, Chapter 3]. Since $A(c\,dx) = c^{d/\gamma} A(dx)$, it follows that the fractional stable field is self-similar with Hurst index H: $A_H(cx) \simeq c^H A_H(x)$. Fractional stable fields have been used to parameterize flow and transport models in highly heterogeneous aquifers, see Herrick et al. [83], Kohlbecker et al. [107], and additional discussion later in this section.

The random field (7.173) is isotropic: If R is an orthogonal matrix (see Remark 6.3) then $\|Rx\| = \|x\|$ for all $x \in \mathbb{R}^d$, and a change of variables $y = Ry'$ shows that

$$\begin{aligned} B_H(Rx) &= \int \left[\|Rx - y\|^{H-d/2} - \|0 - y\|^{H-d/2} \right] B(dy) \\ &= \int \left[\|Rx - Ry'\|^{H-d/2} - \|R0 - Ry'\|^{H-d/2} \right] B(R\, dy') \\ &\simeq \int \left[\|x - y'\|^{H-d/2} - \|0 - y'\|^{H-d/2} \right] B(dy') \\ &= B_H(x) \end{aligned} \tag{7.177}$$

since $|RV| = |V|$ for any Borel set V. Increments of the fractional Brownian field (7.173) are given by

$$B_H(x) - B_H(y) = \int_{z \in \mathbb{R}^d} \left[\|x - z\|^{H-d/2} - \|y - z\|^{H-d/2} \right] B(dz), \tag{7.178}$$

and then an easy change of variables shows that $B_H(x) - B_H(y) \simeq B_H(x - y)$, i.e., the random field has stationary increments.

A stationary isotropic random field can provide a reasonable model for a physical parameter that varies in the same manner in all directions, and exhibits stationary behavior (that is, the nature of the physical parameter is the same at every point in space). Temperature or atmospheric pressure might be considered isotropic on a two dimensional rectangle at a fixed altitude, in a small enough region so that the atmospheric conditions remained the same. If you photograph a meadow, forest, desert, or other homogeneous landscape from above, on a cloudy day, it would appear isotropic. One cannot easily tell north from east.

For a more detailed example, consider the traditional vector advection dispersion equation (ADE) for the movement of contaminants in ground water. Here $p(x, t)$ denotes the relative concentration of the contaminant at location x and time $t > 0$, the solution to the ADE

$$\frac{\partial}{\partial t} p(x, t) = -v \cdot \nabla p(x, t) + \nabla \cdot Q \nabla p(x, t). \tag{7.179}$$

The advective velocity v controls the plume center of mass, and the dispersivity matrix Q governs the dispersion of individual particles away from their center of mass. In practical applications, it is common to allow v and Q to vary with the spatial location x. *Darcy's Law* states that

$$v = \frac{-K\nabla h}{\eta} \tag{7.180}$$

where η is the porosity of the medium (percent of volume through which fluid can flow), h is the hydraulic head (height of the water level relative to some fixed depth), and K is the *hydraulic conductivity*. The scalar K field describes how easy it is for fluid to flow through the porous medium at the point x, which reflects the structure of the medium (e.g., K values in sand are larger than K values in clay). If the porous medium is isotropic, then a fractional Brownian field or a fractional stable field (see Remark 7.40) is often used to generate a synthetic K field, consistent with the statistics of measured data. At a typical experimental site, K is measured at points in a vertical column (in a well) and then the statistics of the K field are examined from several wells. This gives an indication of the moments, pdf, and correlation structure. Typically the sampling wells produce on the order of 10^3 K values. Solving the ADE (7.179) on a computer usually requires values of the velocity field v at around 10^6 data points in two dimensions, or 10^8 in three dimensions (since the model domain in the vertical dimension is usually thinner). In order to parameterize this computer model, a random field simulator is used to generate a synthetic K field consistent with the statistical properties of the measured K data. The Darcy equation (7.180) is then used to generate the velocity field, and finally the ADE is solved on a computer (e.g., by particle tracking, or a finite difference method). Often the dispersivity is assumed constant, or in some cases it is assumed that $Q = aI$ where $a(x) = a_0\|v(x)\|$ for some $a_0 > 0$.

Many studies of K field data have found evidence of long range dependence, leading to the widespread use of fractional Brownian fields to simulate the K field (actually $\log K$). Some authors have noted that $\log K$ data often has a heavier tail than a Gaussian, and here a fractional stable field (see Remark 7.40) has also been used (e.g., see Painter [164]). However, it is probably not reasonable to model the porous medium for groundwater flow as isotropic. A typical aquifer is laid out by a depositional process, roughly in layers. If you think of an exposed hillside or cliff face (e.g. after a hillside has been cut through for road construction) there are often prominent vertical layers. Rotating a picture of the hillside (or rotating the camera) changes the *orientation*. Isotropic pictures have no preferred orientation. To adequately model situations with a preferred orientation requires anisotropic fields. Anisotropy is very common in nature. Temperature varies with altitude (or depth). Gravity provides a fundamental orientation to most physical systems. To develop anisotropic Brownian (and stable) fields, we will employ anisotropic fractional derivatives (the Riesz fractional derivative with Fourier symbol $-\|k\|^\alpha$ is the only isotropic fractional derivative).

The basic construction in Biermé et al. [33, Theorem 4.1] replaces the filter $\varphi(x) = \|x\|^{H-d/2}$ in (7.173) by a different filter with operator scaling. Define the scaling matrix

$E = \operatorname{diag}(a_1, \ldots, a_d)$ where $1 = a_1 \le \cdots \le a_d$. Then it is easy to check that the filter

$$\varphi(x_1, \ldots, x_d) = \left(\sum_{j=1}^{d} D_j |x_j|^{2/a_j} \right)^{1/2} \tag{7.181}$$

for some constants $D_j > 0$ has operator scaling: $\varphi(c^E x) = c\varphi(x)$ for all $x \in \mathbb{R}^d$ and all $c > 0$. Define the Gaussian random field

$$B_\varphi(x) = \int_{y \in \mathbb{R}^d} \left[\varphi(x-y)^{H-q/2} - \varphi(0-y)^{H-q/2} \right] B(dy), \tag{7.182}$$

where $q = a_1 + \cdots + a_d = \operatorname{trace}(E)$. The stochastic integral exists for any $0 < H < 1$, see [33, Theorem 4.1]. The random field (7.182) has stationary increments, and operator scaling: Define $AV = \{Ax : x \in V\}$ and note that $|AV| = |\det(A)||V|$ for any matrix A and any Borel set $V \subset \mathbb{R}^d$. Here $\det(A)$ is the determinant of matrix A, and when $A = c^E$, $\det(A) = c^{a_1} \cdots c^{a_d} = c^q$. Then $B(c^E\, dy) \simeq c^{q/2} B(dy)$, and a change of variables $y = c^E y'$ leads to

$$\begin{aligned} B_\varphi(c^E x) &= \int \left[\varphi(c^E x - y)^{H-q/2} - \varphi(0-y)^{H-q/2} \right] B(dy) \\ &= \int \left[\varphi(c^E x - c^E y')^{H-q/2} - \varphi(c^E 0 - c^E y')^{H-q/2} \right] B(c^E\, dy') \\ &\simeq \int c^{H-q/2} \left[\|x - y'\|^{H-d/2} - \|0 - y'\|^{H-d/2} \right] c^{q/2} B(dy') \\ &= c^H B_\varphi(x). \end{aligned} \tag{7.183}$$

An extension of this argument shows that $B_\varphi(c^E x) \simeq c^H B_\varphi(x)$ in the sense of finite dimensional distributions [33, Corollary 3.2]. If $\varphi(x) = c\|x\|$ then $E = I$ the identity matrix, $q = d$, and $B_\varphi(x)$ is a fractional Brownian field. In general, each one dimensional slice $B_i(x_i) = B_\varphi(x_1, \ldots, x_d)$ is a well-balanced fractional Brownian motion whose Hurst index $H_i = H/a_i$ varies with the coordinate. This model was invented to simulate natural K fields in Benson et al. [27]. Typically the Hurst index H_i is the highest in the flow direction (say $H_1 \approx 0.9$), somewhat lower in the horizontal direction transverse to the flow (say $H_2 \approx 0.6$), and in the negative dependence range for the vertical direction (say $H_3 \approx 0.3$). An extension using more general operator scaling filters allows the Hurst index to vary with an arbitrary set of coordinate axes, see [33].

Remark 7.41. An *operator scaling fractional stable field* can be defined in a similar manner. Take an independently scattered stable random measure $A(dx)$ on $\mathbb{R}^d$ such that $A(V) \simeq S_\gamma(\beta, \sigma(V), 0)$ where $\sigma(V)^\gamma = |V|$, and define

$$A_\varphi(x) = \int_{y \in \mathbb{R}^d} \left[\varphi(x-y)^{H-q/\alpha} - \varphi(0-y)^{H-q/\alpha} \right] A(dy), \tag{7.184}$$

for $0 < H < 1$. Since $A(c^E\, dx) \simeq c^{q/\alpha} A(dx)$, it follows that the fractional stable field is operator self-similar: $A_H(c^E x) \simeq c^H A_H(x)$.

Remark 7.42. Some researchers have proposed modeling natural K fields using probability models that are neither Gaussian nor stable. For example, the Laplace distribution has been proposed by Meerschaert, Kozubowski, Molz and Lu [138]. It is possible to construct stochastic integrals and random fields based on any infinitely divisible distribution, but they will not have the same nice scaling properties. Some mathematical properties of one dimensional fractional Laplace motion are discussed in Kozubowski, Meerschaert and Podgórski [113].

Remark 7.43. Similar to Remark 7.37, the spectral representation of a fractional Brownian field is

$$B_H(x) = \int (e^{ik\cdot x} - 1)\|k\|^{-H-d/2}\hat{B}(dk).$$

Remark 7.44. Various studies of physical systems have collected data on the velocity distribution in complex systems, which often exhibits a heavy tail, see for example Solomon, Weeks and Swinney [206]. Roughly speaking, if the velocity distribution in the ADE (7.179) follows a power law, then it is reasonable to imagine that the plume may follow a fractional diffusion at late time, due to the accumulation of power-law particle jumps. Mathematically, this leads to a conjecture that a highly variable velocity field in a traditional diffusion equation with variable coefficients could lead to a fractional diffusion in the scaling limit. This conjecture remains open. One complication is that, for a very rough velocity field like the ones simulated from fractal random fields, the standard theory of diffusions does not apply, since the coefficients are not Lipschitz functions.

Details

Given an independently scattered Gaussian random measure $B(dx)$ on $\mathbb{R}^d$ such that $\mathbb{E}[e^{ikB(V)}] = e^{-|V|k^2/2}$ for Borel sets $V \subset \mathbb{R}^d$, we now define the *stochastic integral* $\int g(s)B(ds)$. Given a simple function $g(s) = \sum_{i=1}^n c_i I(s \in V_i)$ where $V_1, \ldots, V_n$ are mutually disjoint bounded Borel subsets of $\mathbb{R}^d$, we define

$$\int g(s)B(ds) = \sum_{i=1}^{n} c_i B(V_i) \tag{7.185}$$

in exactly the same way as the one dimensional stochastic integral (7.168). This stochastic integral (7.185) is normal with mean zero and variance

$$\sum_{i=1}^{n} c_i^2 |V_i| = \int |g(x)|^2 dx.$$

Now for $g \geq 0$ Borel measurable, we define

$$\int g(x)B(dx) = \lim_{n\to\infty} \int g_n(x)B(dx) \quad \text{in probability} \tag{7.186}$$

where the simple function g_n is given by (7.164). Then

$$\int g(x)B(dx) \simeq \mathcal{N}\left(0, \int |g(x)|^2 dx\right). \tag{7.187}$$

and the stochastic integral exists if $\int |g(x)|^2 dx < \infty$. For more details, see Samorodnitsky and Taqqu [185, Chapter 3].

If $g(x)$ is continuous on the d-dimensional rectangle $[a, b]$, we can also write

$$\int_{x\in(a,b]} g(x)B(dx) = \lim_{\Delta x\to 0} \sum_{i=1}^{n} g(x_i)B(\Delta x_i) \quad \text{in probability} \tag{7.188}$$

where Δx_i are rectangles $(x_i, x_i + h\mathbf{1}]$ in $\mathbb{R}^d$, the vector $\mathbf{1} = (1, \ldots, 1) \in \mathbb{R}^d$, $x_i = a + ih$ is a discrete lattice with spacing $h = \Delta x > 0$, $i = (i_1, \ldots, i_d)$ is a vector of integers, and the sum is taken over all i such that $x + ih \in (a, b]$. To verify (7.188), use the uniform continuity of g on the compact set $[a, b]$ to see that for any given h, for all large n we have $|g_n(x) - g(x_i)| \le 2/n$ for all $x \in (x_i, x_i + h]$ and all $i = 1, 2, \ldots, n$, where g_n is the simple function (7.164). Then

$$\int_{x\in(a,b]} g_n(x)B(dx) - \sum_i g(x_i)\Delta B(x_i) \simeq \mathcal{N}\left(0, \sum_i \int_{(x_i,x_i+h]} |g_n(x) - g(x_i)|^2\, dx\right)$$

for all n. Since the variance is bounded above by $(2/n)^2 \prod_j (b_j - a_j)$, the difference between these two stochastic integrals converges in probability to zero, and then (7.171) follows.

The *Riesz fractional integral* is defined for suitable functions $f : \mathbb{R}^d \to \mathbb{R}$ by

$$\mathbb{I}_x^\alpha f(x) = \frac{C}{\Gamma(\alpha)} \int f(y)\|x - y\|^{\alpha-d} dy.$$

Similar to the one variable case, the integral exists for bounded continuous functions such that $f(x) \to 0$ sufficiently fast as $\|x\| \to \infty$, since the function $\|x\|^{\alpha-d}$ is integrable at $x = 0$ for any $\alpha > 0$. To see this, change to spherical coordinates. The constant $C > 0$ is chosen so that $\mathbb{I}_x^\alpha f(x)$ has FT $\|k\|^{-\alpha}\hat{f}(k)$ for suitable functions $f(x)$. The Riesz fractional integral is also called the *Riesz potential*. For more information, see Samko, Kilbas and Marichev [184].

7.11 Applications of fractional diffusion

Fractional diffusion is an interesting theoretical construction that links probability, differential equations, and physics. Its practical importance stems from the fact that many real world situations fit the model. We begin our discussion of real world applications with the problem of contaminant transport in underground aquifers. Here fractional diffusion was found to be useful because it solved an important open problem.

The classical advection dispersion equation (ADE) for contaminant transport assumes that the relative concentration of particles $p(x, t)$ solves

$$\frac{\partial p}{\partial t} = -v\frac{\partial p}{\partial x} + D\frac{\partial^2 p}{\partial x^2} \tag{7.189}$$

where v is the average drift and D is the dispersivity. The underlying physical model is a random walk, where individual particles take random jumps away from the center of mass with mean zero and finite variance proportional to D. A Gaussian pdf provides the analytical solution for a point source initial condition. According to this model, a contaminant plume should spread away from its center of mass like $t^{1/2}$, since the pdf $p(x, t)$ has standard deviation $\sqrt{2Dt}$. The one dimensional ADE has been applied at many experimental sites in order to check the accuracy of the model (e.g., see Gelhar et al. [75, 74]). One consistent observation is that the best fitting value of the parameter D typically grows with time. Wheatcraft and Tyler [218] review this literature, and propose a fractal model of heterogeneous porous media as an explanation for the empirical observation that $D \approx Ct^{\rho}$ for some $\rho > 0$. Benson et al. [28, 30] developed the fractional ADE

$$\frac{\partial p}{\partial t} = -v\frac{\partial p}{\partial x} + D\frac{\partial^\alpha p}{\partial x^\alpha}. \tag{7.190}$$

to connect these fractal concepts with fractional derivatives. This research was successful, in that it allowed hydrologists to use a fractional ADE with constant coefficients instead of a traditional ADE with variable coefficients. Since these coefficients are supposed to represent physical properties of the aquifer that do not vary over the time scale of the experiment, this is an important scientific achievement.

Remark 7.45. The units of the FADE coefficients can be determined using the Grünwald finite difference formula (2.1) for the fractional derivative: Write

$$\frac{\Delta p}{\Delta t} = -v\frac{\Delta p}{\Delta x} + D\frac{\Delta^\alpha p}{(\Delta x)^\alpha}$$

where the relative concentration $p(x, t) = C(x, t)/\int C(x, t)\,dx$ is dimensionless, t is in time units T, and x is in length units L. Then the left-hand side has units of $1/T$ so each term on the right-hand side has the same units. This implies that v has units of L/T, and D has units of L^α/T, since Δx has units of L, and $(\Delta x)^\alpha$ has units of L^α.

Point source solutions to the fractional ADE or FADE (7.190) with $1 < \alpha < 2$ are stable densities that spread away from their center of mass at the rate $t^{1/\alpha}$, a super-diffusion. They exhibit positive skewness and a heavy power-law leading tail, features often observed in real data.

Figure 7.2 shows plume data collected at the Macro-dispersion Experimental Site (MADE) in Columbus, Mississippi, USA, and the best-fit concentration curves from

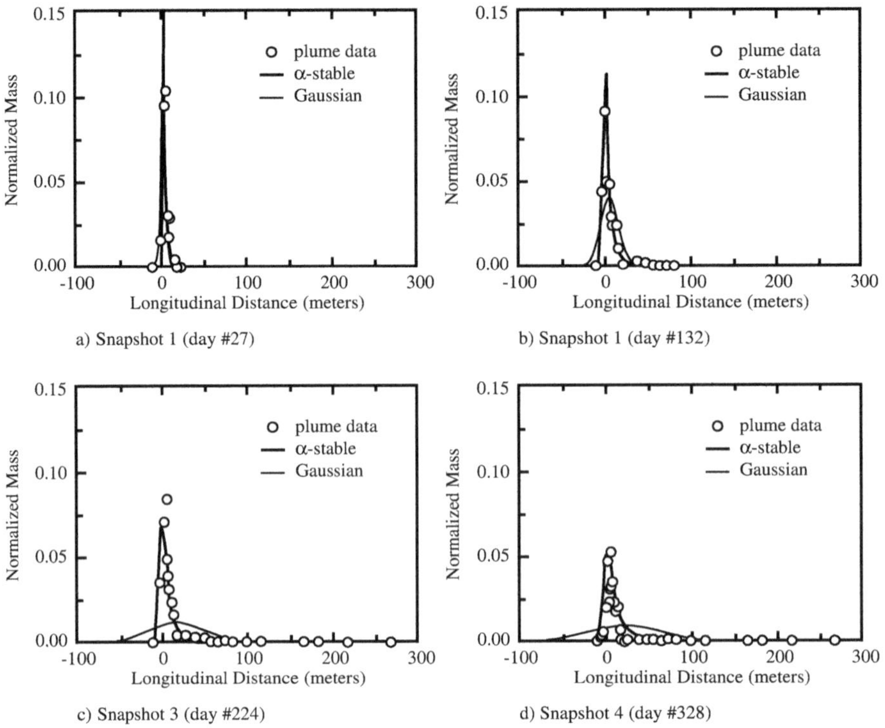

Fig. 7.2: Tracer plume from the MADE site with fitted stable and Gaussian pdf, from Benson et al. [29].

the FADE (7.190) with constant coefficients: $\alpha = 1.1$, $v = 0.12$ meters per day, and $D = 0.14$ meters$^\alpha$ per day. The data represent measured concentrations from sampling wells distributed along the natural flow path of ground water at the site. A tritium tracer was injected into the ground water at day $t = 0$ and monitored over the course of the experiment. The best fitting ADE curves (normal pdf) from a variable coefficient model are also shown (i.e., the best fitting Gaussian pdf is shown for each data set). These concentration snapshots clearly illustrate the skewness and non-Gaussian shape typically seen in ground water plumes. It seems apparent that the ADE, even with a dispersion coefficient that varies with time, does not capture the plume shape. A log-log plot of the same data at day $t = 224$ and day $t = 328$ was shown in Figure 1.5. That figure illustrates the power-law decay of the concentration $p(x, t) \approx x^{-\alpha-1}$ for x large, consistent with the stable pdf solution to the FADE. Additional analysis in that paper verified that the peak concentration falls at a power law rate $\approx t^{-1/\alpha}$ and that the empirical plume variance (which can be estimated from a histogram of particle concentration, even though the theoretical variance does not exist) increases at a power law rate $\approx t^{2/\alpha}$. The parameter α was estimated *a priori* from the statistics of the hydraulic conductivity (K field, see additional discussion in Section 7.10). The empirical agreement between this α estimate and the fitted plume provides additional evidence in favor of the FADE model.

The fractional advection dispersion equation (7.190) is based on a random walk model with power law jumps. Real contaminant plumes may also experience retardation caused by particle sticking and trapping. The space-time fractional ADE

$$\partial_t^\beta p(x,t) = -v\frac{\partial}{\partial x}p(x,t) + D\frac{\partial^\alpha}{\partial x^\alpha}p(x,t) \tag{7.191}$$

introduced in Section 2.4 is based on a CTRW with power law waiting times between jumps. Because the waiting time has infinite mean for $0 < \beta < 1$, a segregation of particles into two phases, mobile and immobile, leads to a more detailed model described in Schumer et al. [193]. That model predicts mobile plume mass will decay like a power law. This power law decay of mobile mass was also observed in the MADE tritium plume, supporting the use of a space-time fractional diffusion model at that site.

Another kind of evidence for power law retention time comes from examination of the breakthrough curve $t \mapsto p(x,t)$ at a fixed location x. Solutions to (7.191) with $0 < \beta < 1$ decay like $\approx t^{-\beta-1}$ at late time, see Schumer et al. [193]. Haggerty, Wondzell and Johnson [78] observed a power law breakthrough curve during a tracer test in a mountain stream. Those data were fit to a space-time fractional ADE with $\beta = 0.3$ in Schumer et al. [193]. The long waiting times in this setting are caused by tracer particles that become trapped in sediment at the bottom of the stream.

Power law waiting times are very common in practical applications. Barabasi [21] studied the waiting time between emails from a single user. The distribution follows a Pareto model with $\beta \approx 1$. Aoki, Sen and Paolucci [6] use fractional time derivatives of order $0 < \beta < 1$ to model heat transfer on a metal plate. Voller [215] uses a space-time fractional diffusion equation for heat transfer, with a fractional time derivative of order $0 < \beta < 1$ and a fractional space derivative of order $1 < \alpha < 2$, to model a melting front. Weiss and Everett [217] use a time-fractional diffusion equation with $0 < \beta < 1$ to model the anomalous diffusion of electromagnetic eddy currents in geological formations.

One of the modeling issues involving (7.191) is the range of the power law index. If $\alpha > 2$, then power law jumps have a finite variance, and the traditional second derivative in space applies at late time. If $\beta > 1$ then the power law waiting times have a finite mean, and the first order time derivative applies at late time. However, the traditional diffusion equation may not be an appropriate model for such a system on an intermediate time scale. Hence there is an ongoing effort to extend the fractional diffusion model to a larger range of α and β. For example, applying a two scale limit procedure to waiting times with $1 < \beta < 2$ leads to a time-fractional ADE

$$\partial_t^\beta p(x,t) - a\partial_t p(x,t) = -v\frac{\partial}{\partial x}p(x,t) + D\frac{\partial^\alpha}{\partial x^\alpha}p(x,t), \tag{7.192}$$

with $a > 0$, see Baeumer, Benson and Meerschaert [14].

Méndez Berhondo et al. [158] found that waiting times between solar flares follow a power law model with $1 < \beta < 2$. Then a time fractional equation such as (7.192) could be applied. Smethurst and Williams [202] find that the waiting times between doctor visits for an individual patient follow a power law model with $\beta \approx 1.4$.

Another interesting application of heavy tails and fractional diffusion comes from the theory of complex systems. An instructive review article of Shlesinger, Zaslavsky and Klafter [200] describes how *Lévy flights* are used to model chaotic dynamical systems. Chaotic dynamical systems are deterministic systems of nonlinear differential equations that can exhibit wild behavior, in which the later state is so sensitive to the initial condition that its behavior is essentially random. This sensitive dependence on initial conditions was noted by Lorenz [126], who observed chaotic behavior in computer models from atmospheric science. The book of Strogatz [211] provides an accessible reference to this subject, see also [135, Section 6.4]. A particle tracing out a chaotic trajectory follows a fractal set called a strange attractor. The velocity of such particles can often follow power law statistics, i.e., the proportion of displacements exceeding size Δx falls off like a power law $(\Delta x)^{-\alpha}$ over a fixed time interval Δt. Even though the system is deterministic, the behavior is so unpredictable that a random walk model is appropriate. The Lévy flight is the name used in this field to refer to a random walk with power law jumps in some α-stable domain of attraction. The scaling property (self-similarity) of the limiting stable Lévy motion that approximates the random walk in the long-time limit has a strong appeal. Shlesinger et al. [200] also consider *Lévy walks*, a coupled CTRW in which the waiting time between jumps also follows a power law distribution. The coupled CTRW, an extension of the CTRW model presented in Section 4.3, was developed to impose physical limits on heavy tailed random walks. In the coupled CTRW model, the iid random vectors (J_i, Y_i) describe the jumps Y_i of a particle, and the time J_i required to make this jump. The components of this random vector are dependent, to enforce physical limits. For example, particles cannot travel faster than the speed of light, so that the ratio Y_i/J_i has an upper bound. The mathematical theory of coupled CTRW limits considered in Becker-Kern, Meerschaert and Scheffler [24] leads to fractional diffusion equations that involve coupled space-time fractional derivatives, see Example 7.52 for more details.

Remark 7.46. Applications of fractional diffusion require estimation of the probability tail $p = \mathbb{P}[X > x] \approx Cx^{-\alpha}$ from experimental data. Taking logs on both sides yields $\log p \approx \log C - \alpha \log x$, which is the basis for some common tail estimation procedures. Given a data set $X_1, \ldots, X_n$ that is supposed to follow this model, at least approximately for x large, sort the data in decreasing order $X_{(1)} \geq \cdots \geq X_{(n)}$. if this model is appropriate, then we should have $\log(i/n) \approx \log C - \alpha \log X_{(i)}$ for the largest order statistics. In some cases, if a large number of upper order statistics follow this model reasonably closely, a simple linear regression on a log-log plot of the order statistics versus the ranks i/n can be used to estimate the tail parameter. Since order statistics are not independent, the estimation problem is not a standard regression. Aban and

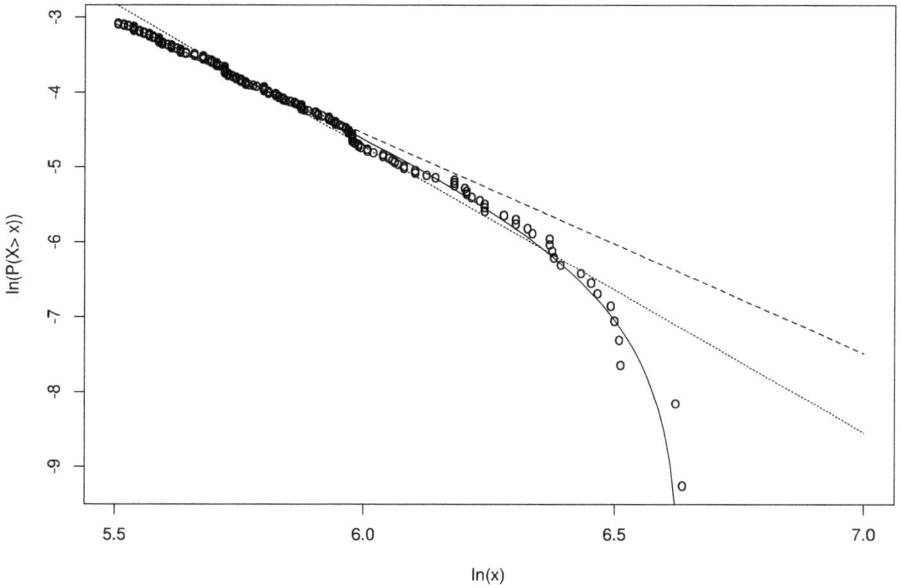

Fig. 7.3: Log-log plot of the exceedence probability for the 100 largest observations of positive daily total precipitation in Tombstone, Arizona USA between July 1, 1893 to December 31, 2001, with best fitting Pareto (dotted line), Pareto with truncated Pareto parameters (dashed line), and truncated Pareto (solid line) tail distribution. From Aban, Meerschaert and Panorska [1].

Meerschaert [2] show that correcting for the mean and covariance structure under an assumed Pareto model leads to a sharper estimation procedure known as the Hill estimator, originally developed by Hill [87] and Hall [87]. Since it is quite common to encounter power law data in many fields of science and engineering, this estimation problem has attracted much attention. There are dozens (at least) of different tail estimators, many of which are reviewed in Baek and Pipiras [13]. There are also some interesting variations that are useful in practice, including truncated Pareto laws, see Aban, Meerschaert and Panorska [1].

Lavergnat and Gole [116] found that waiting times between large raindrops follow a power law model with $0 < \beta < 1$. Aban, Meerschaert and Panorska [1] fit a Pareto with $\alpha = 3.8$ to the largest observations of daily precipitation in city with a very dry climate, see Figure 7.3. As noted in Remark 7.46, Hill's estimator of α is commonly used in practice. For the data in Figure 7.3, there is evidence that the largest observations do not follow a pure power law. The curved line in Figure 7.3 represents the best fitting *truncated Pareto* distribution. The dashed line represents the Pareto distribution with $\alpha = 2.9$ taken from the fitted truncated Pareto. This may be appropriate if there was some truncation effect in measurement that reduced the largest observations. Mala-

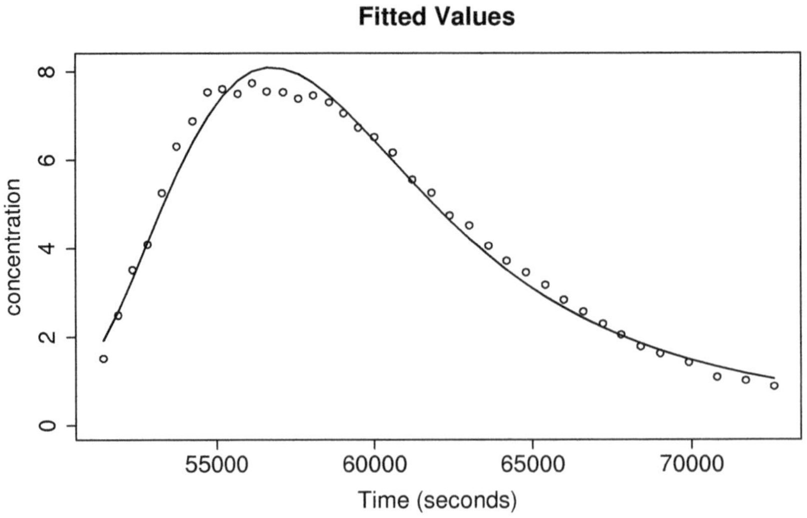

Fig. 7.4: Breakthrough data for a tracer test on the Grand River in Michigan, with fitted stable density, from Chakraborty, Meerschaert and Lim [47].

mud and Turcotte [130] find that the waiting time between large earthquakes in California follows a power law model with $\beta \approx 1.0$.

Sabatelli et al. [182] find that waiting times between trades follow a (truncated) power law with $\beta \approx 0.4$. Since log returns also follow a power law distribution, this suggests that a space-time fractional diffusion model

$$\partial_t^\beta p(x,t) = D\frac{\partial^\alpha}{\partial|x|^\alpha}p(x,t) \tag{7.193}$$

may be useful to model the symmetric log returns. A tempered fractional derivative in time may also be considered, as developed in Section 7.4, to capture the deviation from a power law for long waiting times, see Carr, Geman, Madan and Yor [44]. For a survey of recent research that applies continuous time random walks and fractional diffusion to finance, see Scalas [188].

Deng et al. [59, 60] applied the fractional advection dispersion equation (FADE)

$$\frac{\partial p(x,t)}{\partial t} = -v\frac{\partial p(x,t)}{\partial x} + pD\frac{\partial^\alpha p(x,t)}{\partial x^\alpha} + qD\frac{\partial^\alpha p(x,t)}{\partial(-x)^\alpha} \tag{7.194}$$

to model contaminant transport in rivers. They use a negatively skewed stable with $\alpha = 1.7$ and $\beta = -1$ (i.e., $p = 0$ and $q = 1$) to capture the heavy trailing tail for a tracer test in the Missouri River in Iowa USA, caused by particles that get trapped in the sediment at the bottom of the river. A related fractional model was developed by Shen and Phanikumar [199]. Figure 7.4 shows how the model (7.194) with $\alpha = 1.38$ and $\beta = -1$ fits data from a tracer test on the Grand River in Michigan USA. In this type of

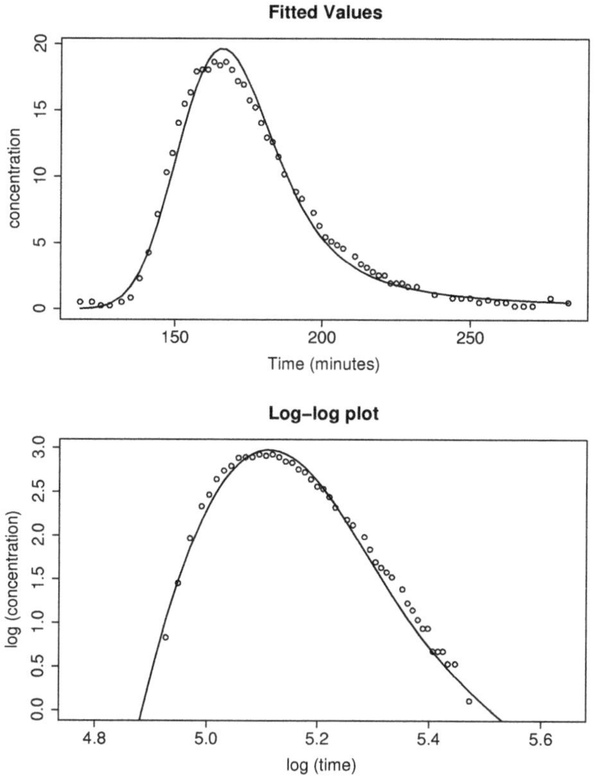

Fig. 7.5: Breakthrough data for a tracer test in the Red Cedar River in Michigan, fit to a time-fractional diffusion model, from Chakraborty, Meerschaert and Lim [47]. The lower panel shows the same data on a log-log plot, to illustrate the power law decay of concentration at late time.

analysis, it is typical to plot the *breakthrough curve* $t \mapsto p(x, t)$ at fixed locations x. A heavy tail on the right-hand side of the breakthrough curve is therefore an indication of negative skewness in the pdf $x \mapsto p(x, t)$. In this application, the breakthrough curve is measured by pouring buckets of tracer into the river over the side of a bridge, and then measuring concentration over time at other bridges further downstream. As we mentioned in Chapter 1, this model has caused some controversy in hydrology. The random walk model behind (7.194) with $q = 1$ has only negative jumps, so the model in [59, 60] assumes that particles "jump" upstream (relative to the plume center of mass). Chakraborty, Meerschaert and Lim [47] fit another tracer test on the Red Cedar river in Michigan USA using the model (7.194) with $\alpha = 1.5$ and $\beta = -1$ (not shown). An alternative time-fractional diffusion model, equation (7.191) with $\alpha = 2$ and $\beta = 0.978$, was also fit to the same data, with reasonably good results, see Figure 7.5. Since the time-fractional model does not assume upstream jumps, it is preferred by some hydrologists.

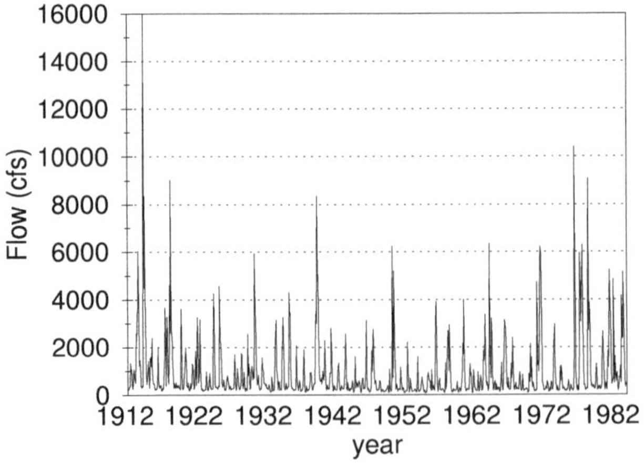

Fig. 7.6: Monthly average flow in the Salt river near Roosevelt, Arizona from October 1912 to September 1983, from Anderson and Meerschaert [5].

Power law tails with $2 < \alpha < 4$ are also commonly seen in river flow time series. Figure 7.6 shows a time series of monthly average flows from the Salt river, upstream of Phoenix, Arizona in the USA. This river runs from the mountains through the desert, and experiences a wide range of variability in flow. The occasional sharp spikes are indicative of heavy tails, see discussion in Brockwell and Davis [42, Section 13.3]. A log-log plot of the largest order statistics in Figure 7.7 shows a power law tail with $\alpha \approx 3.0$. Sums of iid random variables having a power law tail with $\alpha > 2$ are asymptotically normal, since the variance is finite. The data are significantly correlated, and a statistical estimate of the correlation is useful to model the process. A typical time series model to represent the dependence between successive observations is a moving average $X_t = \mu_t + \sum_j c_j Z_{t-j}$ where (Z_j) are iid with mean zero and $\mathbb{P}[|Z_j| > x] \approx Cx^{-\alpha}$. The sample covariance

$$\begin{aligned}\frac{1}{n}\sum_t (X_t - \mu_t)(X_{t+h} - \mu_{t+h}) &= \frac{1}{n}\Big(\sum_i c_i Z_{t-i}\Big)\Big(\sum_j c_j Z_{t+h-j}\Big)\\ &= n^{-1}\Big(\sum_i c_i Z_{t-i}\Big)\Big(\sum_k c_{k+h} Z_{t-k}\Big)\\ &= n^{-1}\Big(\sum_i c_i c_{i+h} Z_{t-i}^2 + \sum_i \sum_{k\neq i} c_i c_{k+h} Z_{t-i} Z_{t-k}\Big).\end{aligned}$$

Since $\mathbb{P}[Z_j^2 > x] = \mathbb{P}[|Z_j| > x^{1/2}] \approx Cx^{-\alpha/2}$ where $2 < \alpha < 4$, the first term involves squared noise variables with an infinite second moment. It turns out that this term dominates as $n \to \infty$, so that the asymptotic limit of the sample covariance involves a stable law, see Davis and Resnick [57, 58]. Hence, even though the time series has finite variance, the Extended Central Limit Theorem 4.5 is important to understand the covariance structure.

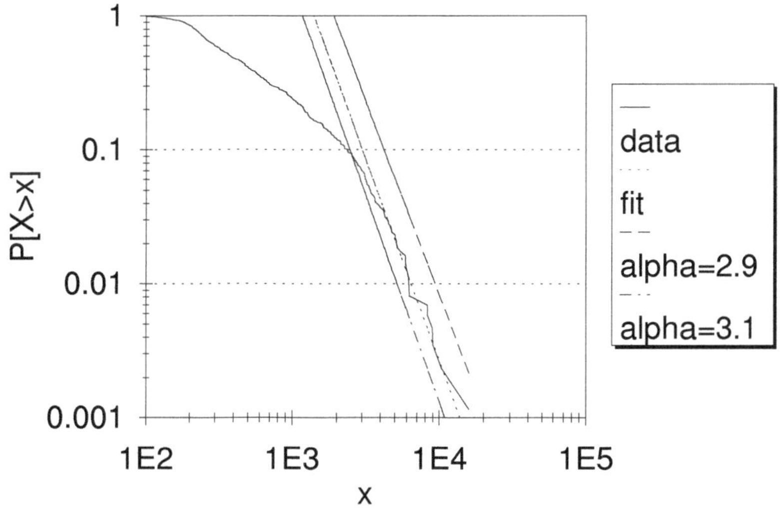

Fig. 7.7: The river flow data from Figure 7.6 has a power law tail with $\alpha \approx 3.0$.

Fractional diffusion is also useful in biology. The famous paper of Viswanathan et al. [214] proposed a Lévy flight model (no pun intended) for the wanderings of an albatross foraging for food in the open ocean. This model is based on tracking data from individual birds. The birds make many small flights, searching for food. Then occasionally they make a very long flight, seeking a new fishing spot. The power law statistics of the flight length suggest a random walk in the domain of attraction of a stable law, and hence a stable Lévy process provides a convenient model for the long-time behavior of these birds. The trajectory of a single bird over time is similar to the sample path in Figure 5.32. Some biologists argue that animals follow a stable Lévy motion because it represents a better search strategy than a Brownian motion, see discussion in Shlesinger et al. [200]. Ramos-Fernández et al. [173] use a Lévy walk to model the foraging of spider monkeys.

Baeumer, Kovács and Meerschaert [16, 17] use a fractional diffusion equation to model the spread of invasive species. Data from biological studies often show that offspring migrate a distance from their parents that falls off like a power law. The *dis-*

persal kernel that models these movements represents the distance between parent and offspring, so that the repeated convolution of dispersal kernels gives the location of subsequent generations. This is mathematically equivalent to a random walk over the generations, where the dispersal kernel gives the pdf of the jump variable. A heavy tailed dispersal kernel leads to a stable Lévy motion after a number of generations.

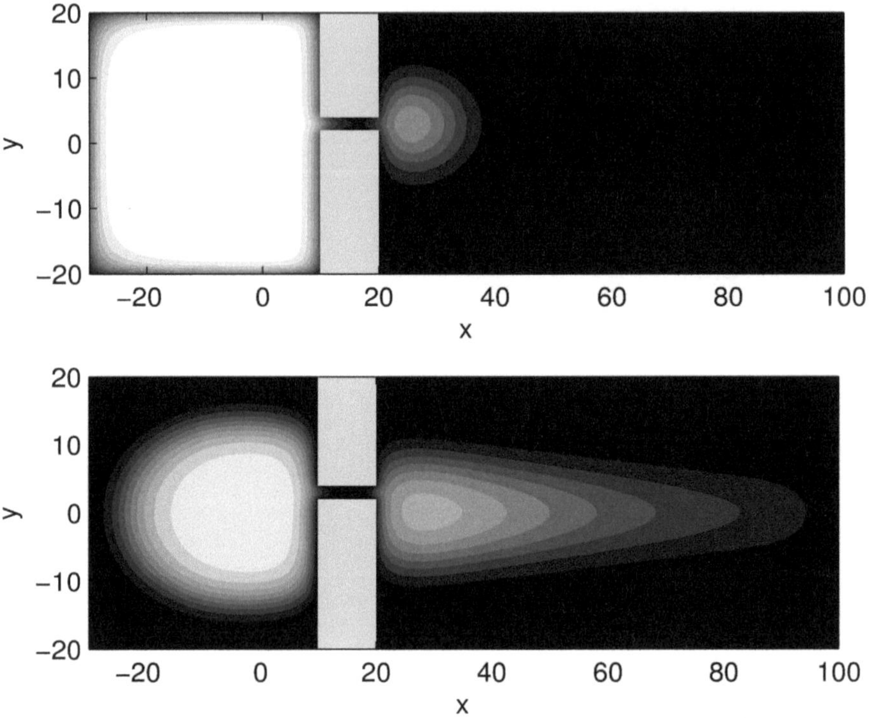

Fig. 7.8: Solution to the reaction-diffusion equation (7.195) with $\alpha = 2$ (top panel) and $\alpha = 1.7$ (bottom panel), illustrating the effect of anomalous dispersion. From Baeumer, Kovács and Meerschaert [17].

Since the population can increase, a *fractional reaction-diffusion equation* is useful to represent growth and dispersal:

$$\frac{\partial}{\partial t}p(x,t) = \lambda p(x,t)\left(1 - \frac{p(x,t)}{K}\right) + Cp\frac{\partial^\alpha}{\partial x^\alpha}p(x,t) + Dq\frac{\partial^\alpha}{\partial(-x)^\alpha}p(x,t). \tag{7.195}$$

The first term $\lambda p(x,t)$ models exponential population growth at the rate λ, until population reaches the environmental carrying capacity K. This model is not mass-preserving, so the solution $p(x,t)$ can no longer be interpreted as a pdf. Figure 7.8 illustrates the effect of fractional dispersion on an invasive species moving across a slit barrier. In traditional dispersion, there is slow movement through the slit. Note that in this case, the population on the right-hand side of the barrier is centered at the slit location. In anomalous dispersion, the population jumps directly over the barrier. In a practical application, the slit might represent a long river with one crossing point.

A very interesting study in Brockmann et al. [41] analyzed human movements by tracking bank notes, using the biological model of dispersal kernels. They found that the distance traveled by bank notes (carried by humans) over a four day period follows a power law model with $\alpha \approx 0.6$. The authors observe that fractional diffusion of human populations has significant implications for modeling the spread of infectious disease, which can be expected to spread faster than a traditional diffusion model predicts.

Mandelbrot [131] and Fama [66] pioneered the use of heavy tail distributions in finance. Data on cotton prices from the seminal paper of Mandelbrot [131] indicate that a stable Lévy motion provides a more appropriate model for price fluctuations than the usual Brownian motion. Let $P(j)$ denote the price of cotton, or other speculative commodity, on day j. The *log return* is defined by $L(j) = \log[P(j)/P(j-1)]$. Then $P(n) = P(0)\exp[L(1) + \cdots + L(n)]$. Since $\log(1+z) = z + O(z^2)$, the log-return approximates the relative change in price. The log return is useful in finance, because this nonlinear transformation typically produces a sequence of centered random variables with essentially no serial correlation: $\mathbb{E}[L(j)] \approx 0$ and $\mathbb{E}[L(j)L(j-1)] \approx 0$. For this reason, it is common in finance to represent prices by an exponential model $P(t) = P(0)\exp[X(t)]$ where $X(t)$ is some Lévy process. For example, the famous Black-Scholes model for option pricing is based on a Brownian motion model of log returns. Alternative option pricing formulas based on stable Lévy motion have been developed by Mittnik and Rachev [159] and Janicki et al. [98].

The application of stable models in finance remains controversial, and much of the controversy revolves around the very delicate problem of tail estimation. Jansen and de Vries [99] argue that daily returns for many stocks and stock indices have heavy tails with $3 < \alpha < 5$, and discuss the possibility that the October 1987 stock market crash could be explained as a natural heavy tailed random fluctuation. Loretan and Phillips [127] use similar methods to estimate heavy tails with $2 < \alpha < 4$ for returns from numerous stock market indices and exchange rates. This indicates that the variance is finite but the fourth moment is infinite. Both daily and monthly returns show heavy tails with similar values of α in this study. Rachev and Mittnik [172] fit a stable pdf with $1 < \alpha < 2$ to a variety of stocks, stock indices, and exchange rates. McCulloch [133] re-analyzed the data in [99, 127], and fit a stable pdf with $1.5 < \alpha < 2$. The papers [99, 127] estimate α based on a Pareto distribution with $\alpha \in (0, \infty)$ while the authors in [133, 172] apply a stable distribution with $\alpha \in (0, 2]$. A nice discussion of this controversy appears in McCulloch [134].

Aban, Meerschaert and Panorska [1] examined absolute daily price changes in U.S. dollars for Amazon, Inc. stock from January 1, 1998 to June 30, 2003. They fit a Pareto with $\alpha = 2.3$ to the largest observations, see Figure 7.9. A truncated Pareto with $\alpha = 1.7$ was also fit. The truncated Pareto estimate of α may be more appropriate, if

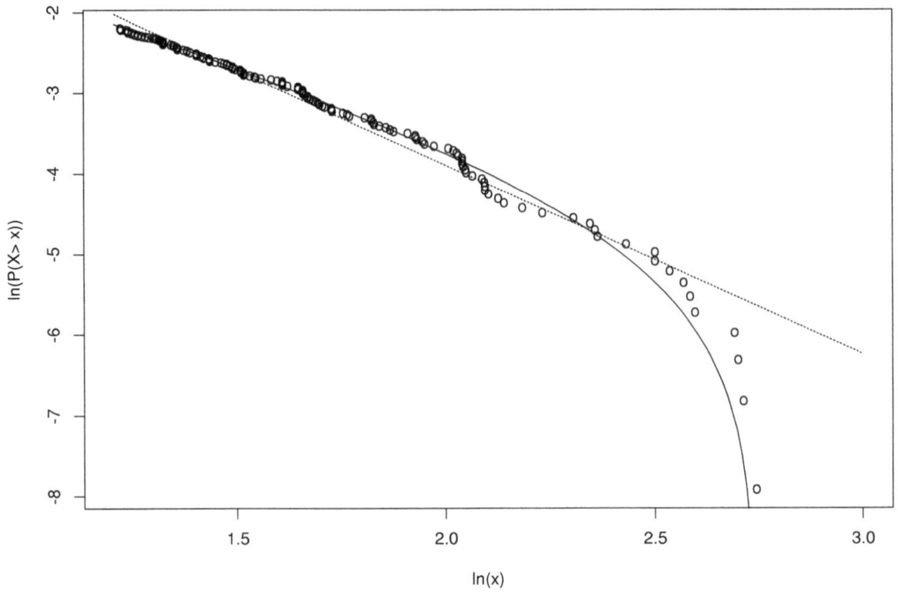

Fig. 7.9: Log-log plot of the largest absolute values of daily price changes in Amazon, Inc. stock, with best fitting Pareto (straight line) and truncated Pareto (curved line) tail distribution. From Aban, Meerschaert and Panorska [1].

there were significant truncation effects in the observations. For example, there are automatic trading limits that can limit the largest price fluctuations. Figure 7.10 shows trading volume (shares per day) for the same data set. There is a clear power law trend with $\alpha = 2.7$. Trading volume can be used to infer waiting times between trades for a CTRW model of stock prices.

Remark 7.47. Power laws are quite prevalent in scientific data, see for example the book of Sornette [207]. One possible explanation involves mixture distributions. Exponential and related distributions (e.g., gamma) can arise from random arrival processes and relaxation phenomena (e.g., cooling). In a heterogeneous environment, the exponential rate parameter may vary. Suppose $\mathbb{P}[X > x] = e^{-\lambda x}$ where λ itself follows an exponential distribution with $\mathbb{P}[\lambda > y] = e^{-by}$ for some $b > 0$. Then the

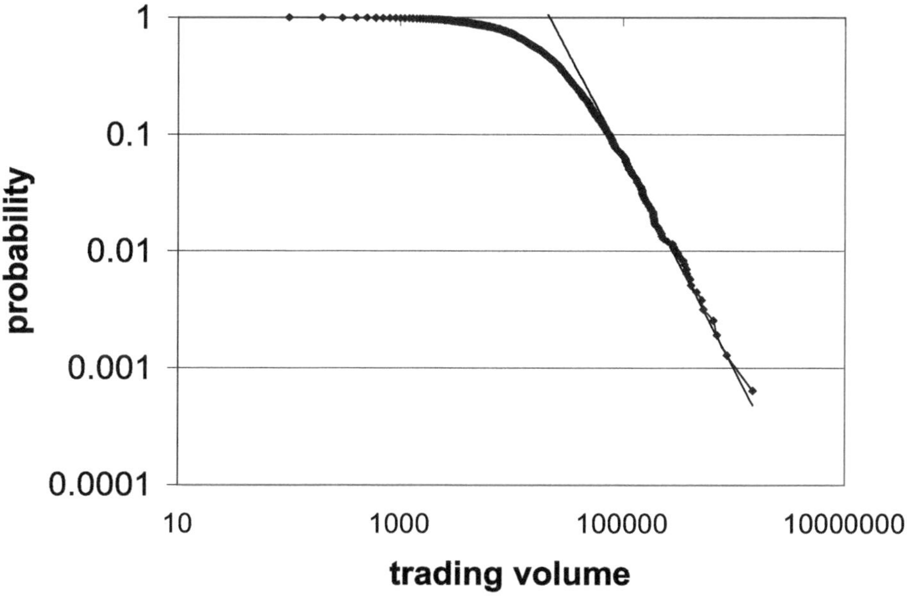

Fig. 7.10: Trading volume for Amazon, Inc. stock from January 1, 1998 to June 30, 2003. The data fit a power law with $\alpha = 2.7$.

unconditional distribution of X is a power law:

$$\begin{aligned}\mathbb{P}[X > x] &= \int_0^\infty \mathbb{P}[X > x | \lambda = y] b e^{-by} dy \\ &= \int_0^\infty e^{-yx} b e^{-by} dy = \frac{b}{b + x} \approx \frac{b}{x} \quad \text{as } x \to \infty.\end{aligned}$$

If λ has a gamma pdf $g(y)$ with Laplace transform $\tilde{g}(s) = (1 + \beta s)^{-\alpha}$ for some $\alpha > 0$, then

$$\mathbb{P}[X > x] = \int_0^\infty e^{-yx} g(y) dy = (1 + \beta x)^{-\alpha} \approx C x^{-\alpha} \quad \text{as } x \to \infty$$

where $C = \beta^{-\alpha} > 0$. Karamata's Tauberian Theorem (e.g., see Feller [68, Theorem 3, p. 445]) implies that any pdf that decays like a power law at zero has a Laplace transform that decays like a power law at infinity (the formal statement involves regular variation). Since the mixture above is mathematically equivalent to the Laplace transform of the mixing density, any such pdf for λ (e.g., Weibull or beta) also produces random variables with a power law tail. For more details, and an application to sediment transport, see Hill et al. [88].

7.12 Applications of vector fractional diffusion

In this section, we summarize some recent applications of vector fractional diffusion, to illustrate the practical application of the theory developed in Chapter 6.

Example 7.48. Schumer et al. [194] applied the generalized fractional diffusion equation (6.115) as a conceptual model for contaminant transport in ground water. In Chapter 1, we discussed an experiment at the MADE site, see Figure 1.5. Figure 7.11 shows that the two-dimensional MADE plume spreads at a rate t^{1/α_1} in the direction of flow, where the tail index $\alpha_1 = 1.2$ is reasonably consistent with the one dimensional model. The plume spreads like t^{1/α_2} in the direction transverse to the flow, where $\alpha_2 = 1.5$. The spreading rate was determined by plotting the measured plume variance against distance. Since the average plume velocity is constant, the mean travel distance $x = vt$ is proportional to time. Since the plume width grows like a power law with distance, it also grows like a power law with time, with the same power law index. Then an operator stable Lévy motion with drift is an appropriate model for the movement of individual particles, and the GADE (6.115) with $B = \mathrm{diag}(1/1.2, 1/1.5)$ can be used to model relative concentration in two dimensions. A second study, at an experimental site in Cape Cod, found $\alpha_1 = 1.6$ and $\alpha_2 \approx 2$. The plume spreading at this site can be well approximated by the GADE (6.113). The underlying operator Lévy motion has one stable component in the direction of flow, and one normal component in the direction transverse to flow. The plume shape is similar to Figure 6.10, which represents the view from above, where flow is in the positive x_2 direction.

Example 7.49. If a data set of random vectors exhibits a heavy tail in each coordinate, it is often the case that the tail index varies with the coordinate. Figure 7.12 displays $n = 2,853$ daily log returns, based on the exchange rates of the German Deutsche Mark x_1 and Japanese Yen x_2 against the US dollar. (See Section 7.11 for a discussion of Lévy process models in finance based on log returns.) A one dimensional analysis similar to Figure 1.5 indicates that the exchange rate data in each coordinate x_1 and x_2 fits a mean zero stable pdf with $\alpha \approx 1.6$. This was the basis for the multivariable stable model proposed by Nolan, Panorska and McCulloch [162]. That paper also includes a method for estimating the spectral measure (6.49). Then the pdf of the accumulated log return $X(t)$ solves a vector fractional diffusion equation (6.63).

A further analysis reveals that the tail behavior varies with the coordinate, once we adopt a suitable rotated coordinate system. For an $X \in \mathrm{GDOA}(Y)$ where the operator

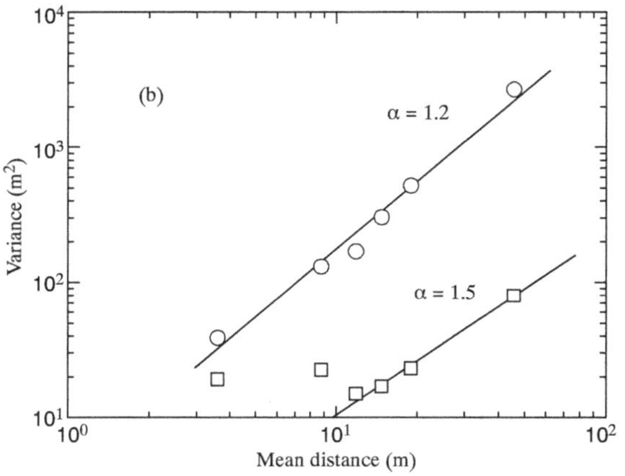

Fig. 7.11: Apparent plume variance in the direction of flow (circles) and transverse to flow (squares) at the MADE site, from Meerschaert, Benson and Baeumer [136].

stable law Y has exponent $B = \operatorname{diag}(1/\alpha_1, 1/\alpha_2)$ with $\alpha_2 < \alpha_1$, it follows from the spectral decomposition discussed near the end of Section 6.8 that each component $X \cdot e_j$ is in the domain of attraction of a stable random variable $Y_j = Y \cdot e_j$ with index α_j. Then Theorem 4.5 shows that $V_0(r) = \mathbb{P}[|X \cdot e_i| > r]$ is $\mathrm{RV}(-\alpha_j)$, and Proposition 4.9 implies that for any $\delta > 0$ we have

$$r^{-\delta-\alpha_j} < \mathbb{P}[|X \cdot e_j| > r] < r^{\delta-\alpha_j}$$

for all $r > 0$ sufficiently large. Since any one dimensional projection $X \cdot \theta$ is a linear combination of the coordinates $X \cdot e_j$, it follows that

$$r^{-\delta-\alpha_1} < \mathbb{P}[|X \cdot \theta| > r] < r^{\delta-\alpha_1}$$

for all $\theta \neq \pm e_2$, i.e., the heavier tail dominates. (For an extension of this property to arbitrary exponents, see Meerschaert and Scheffler [146, Theorem 6.4.15].) Applying this idea to the exchange rate data, the fact that the coordinates x_1 and x_2 show the same tail behavior with $\alpha \approx 1.6$ does not rule out the possibility of another coordinate system in which the tail behavior varies.

To investigate this possibility, we consider a rotated coordinate system z_1 (line with slope -1) and z_2 (line with slope $+1$) as noted in Figure 7.12. Now we find that the z_1 coordinate has a lighter tail with $\alpha_1 \approx 2.0$ and the z_2 coordinate has a heavier tail with $\alpha_2 \approx 1.6$. The original coordinates mask the variation in tail behavior. Now a reasonable model for $X(t)$ is an operator stable Lévy process in the new coordinates z_1 and z_2, where the z_1 coordinate is a Brownian motion, and the z_2 coordinate is a stable Lévy motion with index α_2. It follows from the Lévy representation (6.21) that these two coordinate processes are independent. Then the pdf of the accumulated log

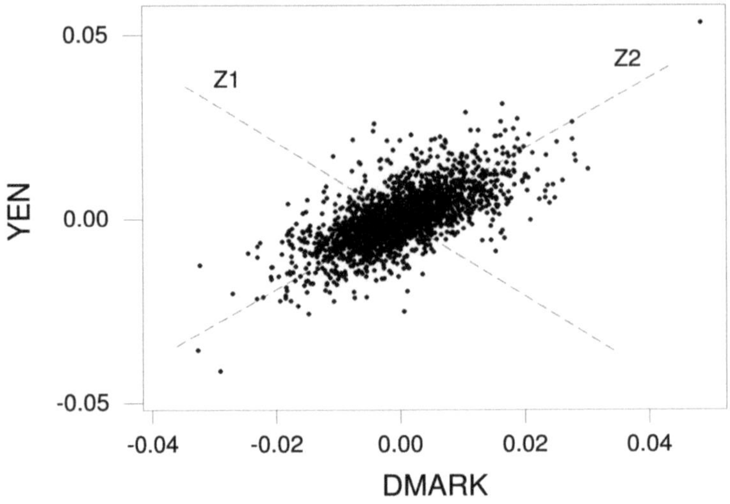

Fig. 7.12: Exchange rates against the US dollar from Meerschaert and Scheffler [150]. The new coordinates z_1, z_2 uncover variations in the tail parameter α.

return process $X(t)$ solves a fractional diffusion equation

$$\frac{\partial}{\partial t}p(z,t) = D_1 \frac{\partial^2}{\partial z_1^2}p(z,t) + D_1 \frac{\partial^{1.6}}{\partial |z_2|^{1.6}}p(z,t) \tag{7.196}$$

using the symmetric fractional derivative as in (6.68). One interpretation of this model is that both currencies are reacting to the same principal effect, the US dollar, and variations due to other currencies are less extreme.

The new coordinates in this example are the eigenvectors of the sample covariance matrix of the exchange rate data in Figure 7.12. Theorem 10.4.8 in [146] implies that, if $B = \operatorname{diag}(1/\alpha_1, 1/\alpha_2)$ with $\alpha_2 < \alpha_1$ in some coordinates, the eigenvalues of the sample covariance matrix converge in probability to the coordinate system that makes B diagonal. This result is a heavy tailed version of principal component analysis. Even though the covariance matrix does not exist in this case, the sample covariance matrix indicates a useful set of coordinates. For details, see Meerschaert and Scheffler [150, Example 8.1].

The exponential Lévy process model $P(t) = P(0)\exp[X(t)]$ fails to capture one very interesting feature of financial data: Typically the log returns are uncorrelated, but their absolute values (or squared values) are highly correlated. This is an interesting and useful example of a real world situation in which variables are uncorrelated, but not independent. The problem of constructing good models for vectors of log returns in finance, that capture heavy tails as well as nonlinear correlations, is an active research area. One promising approach is to subordinate the Lévy process $X(t)$ to some independent waiting time process, see for example Barndorff-Nielsen [23], Carr, Geman, Madan and Yor [44], and Kotz, Kozubowski and Podgórski [111]. Some related models were developed by Bender and Marquardt [26], Finlay and Seneta [71], Heyde [84], and Leonenko, Petherick and Sikorskii [119]. The CTRW introduced in Section 2.4 can provide a strong motivation for considering such models.

Example 7.50. An application to geophysics in Meerschaert and Scheffler [149] considers a data set of fracture aperture x_1 and fluid velocity x_2 in fractured rock, from a site under consideration for a nuclear waste depository in Sweden. A one variable tail estimation shows that the aperture data has a heavy tail with $\alpha_1 = 1.4$, and the fluid velocity data has a heavy tail with $\alpha_2 = 1.05$. Then an operator stable model with exponent $B = \text{diag}(1/1.4, 1/1.05)$ could be appropriate. Since the components of the operator stable law have infinite second moment, the covariance cannot be used to model dependence. Instead, the spectral measure $\Lambda(d\theta)$ in (6.111) governs the dependence between these two variables. The spectral measure in Figure 7.13 was approximated using the nonparametric estimator of Scheffler [189]. The spectral measure governs the direction of jumps that make up the operator stable limit. In the data, the largest jumps are traced back to the unit sphere using the Jurek coordinates from Remark 6.40, and this gives a nonparametric estimate of the spectral measure. See [149] for more details.

Example 7.51. An application to fracture flow in Reeves et al. [174] models a contaminant plume moving through fractured rock as a random walk that converges in the long-time scaling limit to an operator stable Lévy motion. The eigenvalues a_i of the scaling matrix B code the power law jumps, related to fracture lengths. The eigenvectors v_i of B determine the coordinate system, related to fracture orientation. The mixing measure is concentrated in the eigenvector directions, so that a contaminant particle jumps in the v_i direction with some probability $M(v_i)$, and the random jump length L follows a power law distribution with $\mathbb{P}[L > r] \approx r^{-\alpha_i}$ with $a_i = \alpha_i^{-1}$. The eigenvector directions reflect the fracture geometry. Typically the fracture orientation is determined by the crystalline structure of the rock, and there are just a few preferred fracture orientations. If the orientations are orthogonal, then the contaminant plume follows the vector fractional diffusion equation (6.113), and the plume shape is similar to Figure 6.5. More typically, the fracture orientations are separated by an

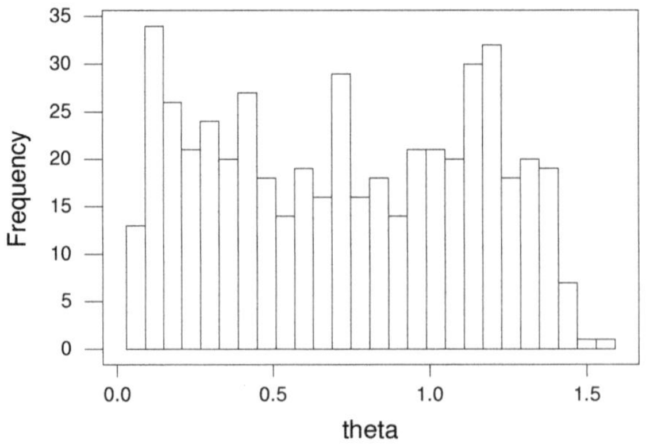

Fig. 7.13: Estimated spectral measure for an operator stable model of fracture statistics, from Meerschaert and Scheffler [149].

angle less than 90 degrees. This can be related to the orthogonal case by a simple (non-orthogonal) change of coordinates. In some cases, the number of fracture orientations is larger than the number of dimensions. Then the mixing measure determines the relative number of jumps in each direction. Because there are a finite number of possible orientations, the mixing measure is always discrete in these applications.

Example 7.52. Figure 7.14 shows tick-by-tick data on LIFFE bond futures from September 1997. The plotted data are $X_n = (Y_n, J_n)'$ where Y_n is the log return after the nth trade, and J_n is the waiting time between the $n - 1$st and nth trades. The log returns are roughly symmetric, and exhibit a power law tail with $\alpha \approx 1.8$. The waiting times also show a heavy tail, with index $\beta \approx 0.9$. There appears to be some significant dependence between the two coordinates, and it seems that large log returns are associated with long waiting times. This is consistent with a model where (X_n) are iid with $X \in \text{GDOA}(V)$ and V has dependent components. This leads to a coupled CTRW model for the price at time $t > 0$, see Meerschaert and Scalas [144]. The coupled CTRW is similar to the model introduced in Section 2.4 except that the space-time random vectors X_n have dependent components. A convenient dependence model is $Y_n = J_n^{\beta/2} Z_n$ where Z_n are iid normal, independent of J_n. Then the CTRW limit has a pdf that solves a *coupled fractional diffusion equation*

$$\left(\frac{\partial}{\partial t} - \frac{\partial^2}{\partial x^2}\right)^{\beta} p(x,t) = \delta(x)\frac{t^{-\beta}}{\Gamma(1-\beta)}$$

where β is the tail index of the waiting times. The coupled space-time fractional derivative operator on the left-hand side is defined through its Fourier-Laplace symbol $\psi(k,s) = (s+k^2)^{\beta}$, i.e., the FLT of the left-hand side is $\psi(k,s)\bar{p}(k,s)$. The theory of coupled CTRW, their limit laws, and their governing equations is based on operator stable laws, since the space-time vector X_n belongs the GDOA of some operator stable law. For more details, see Becker-Kern et al. [24], Jurlewicz et al. [101], Meerschaert and Scheffler [153], and Straka and Henry [210].

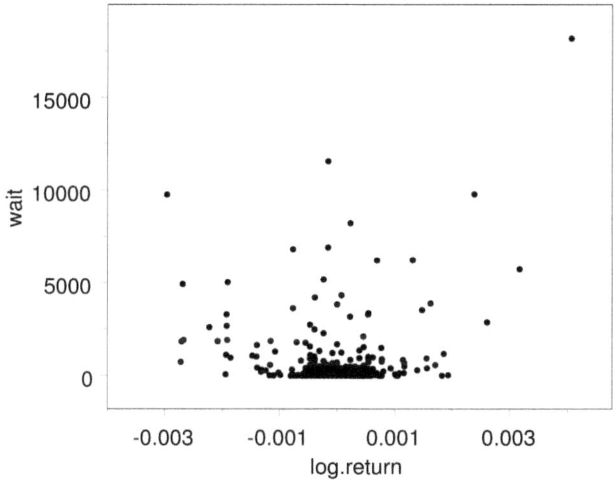

Fig. 7.14: Waiting times in seconds and log returns for LIFFE bond futures, from Meerschaert and Scalas [144].

Example 7.53. For a general operator stable process, where the components are not independent, solutions to the generalized fractional advection-dispersion equation (6.115) can be obtained via particle tracking. Figure 7.15 shows a particle tracking solution to the GADE with $B = \text{diag}(1/1.5, 1/1.9)$. The mixing measure is concentrated on seven discrete points: $M(e_1) = 0.3$, $M(\pm v_1) = 0.2$, $M(\pm v_2) = 0.1$, and $M(\pm v_3) = 0.05$, where $v_i = R_{\theta_i} e_1$ with $\theta_1 = 6°$, $\theta_2 = 12°$, and $\theta_3 = 18°$. Here R_c is the rotation matrix from Example 6.35. This conceptual model represents the flow and dispersion of tracer particles in ground water. Many particle jumps follow the direction of flow (the positive x_1 coordinate) but some particles deviate to avoid obstacles in the porous medium. The particle tracking solution shows level sets from a histogram of particle location, based on $n = 10,000,000$ particles. Each particle follows a simulated operator stable process $Z_t + vt$ with $v = (10, 0)'$ indicating a drift from left to right. The process Z_t was approximated using a random walk with jump vectors $W^B\Theta$ (mean-corrected) where $\mathbb{P}[W > r] = 1/r$ and Θ has distribution $M(d\theta)$, as in Theorem 6.43. To validate the accuracy of the particle tracking method, a numerical method was used to compute the inverse FT of the operator stable. As compared to the vector diffusion in Figure 6.2, the plume in Figure 7.15 is skewed in the direction of flow, and the spreading rate is greater in the direction of flow. The operator stable Lévy process Z_t represents the location of a randomly selected particle. In this case, the x_1 component is stable with index $\alpha_1 = 1.5$, the x_2 component is symmetric stable with index $\alpha_2 = 1.9$, and the two components are dependent. For more details, see Zhang et al. [225]. An interesting experiment reported in Moroni, Cushman and Cenedese [161] performs particle tracking on actual individual particles through a porous medium in a laboratory setting. Particle tracking for time-fractional diffusion equations is treated in Germano, Politi, Scalas and Schilling [76], Magdziarz and Weron [128] and Zhang, Meerschaert and Baeumer [226].

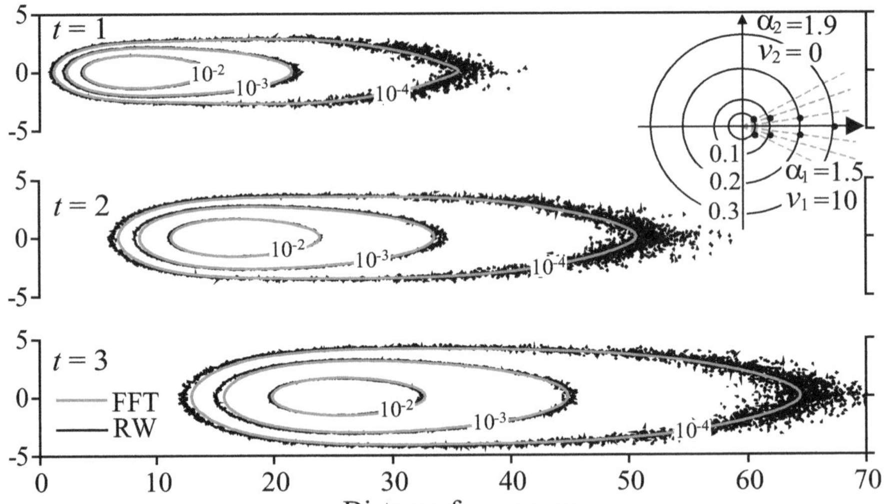

Fig. 7.15: Particle tracking solution of the generalized fractional advection-dispersion equation (6.115) from Zhang, Benson, Meerschaert, LaBolle and Scheffler [225], with diagonal exponent $B = \text{diag}(\alpha_1, \alpha_2)$, velocity $v = (v_1, v_2)'$, and mixing measure as indicated.

Bibliography

[1] I. Aban, M. M. Meerschaert and A. K. Panorska, Parameter estimation for the truncated Pareto distribution, *J. Amer. Statist. Assoc.* **101** (2006), 270–277.

[2] I. B. Aban and M. M. Meerschaert, Generalized least-squares estimators for the thickness of heavy tails, *J. Statist. Plann. Inference* **119** (2004), 341–352.

[3] M. Abramowitz and I. A. Stegun, *Handbook of mathematical functions with formulas, graphs, and mathematical tables*, National Bureau of Standards Applied Mathematics Series 55, For sale by the Superintendent of Documents, U.S. Government Printing Office, Washington, D.C., 1964.

[4] H. Allouba and W. Zheng, Brownian-time processes: the PDE connection and the half-derivative generator, *Ann. Probab.* **29** (2001), 1780–1795.

[5] P. L. Anderson and M. M. Meerschaert, Modeling river flows with heavy tails, *Water Resour. Res.* **34** (1998), 2271–2280.

[6] Y. Aoki, M. Sen and S. Paolucci, Approximation of transient temperatures in complex geometries using fractional derivatives, *Heat Mass Transfer* **44** (2008), 771–777.

[7] D. Applebaum, *Lévy processes and stochastic calculus*, second ed, Cambridge Studies in Advanced Mathematics 116, Cambridge University Press, Cambridge, 2009.

[8] W. Arendt, C. J. K. Batty, M. Hieber and F. Neubrander, *Vector-valued Laplace transforms and Cauchy problems*, Monographs in Mathematics 96, Birkhäuser Verlag, Basel, 2001.

[9] K. B. Athreya and S. N. Lahiri, *Measure theory and probability theory*, Springer Texts in Statistics, Springer, New York, 2006.

[10] F. Avram, N. N. Leonenko and N. Šuvak, Parameter estimation for Fisher-Scedecor diffusion, *Statistics* **45** (2011), 1–16.

[11] F. Avram, N. Leonenko and L. Rabehasaina, Series expansions for the first passage distribution of Wong-Pearson jump-diffusions, *Stoch. Anal. Appl.* **27** (2009), 770–796.

[12] F. Avram and M. S. Taqqu, Weak convergence of sums of moving averages in the α-stable domain of attraction, *Ann. Probab.* **20** (1992), 483–503.

[13] C. Baek and V. Pipiras, Estimation of parameters in heavy-tailed distribution when its second order tail parameter is known, *J. Statist. Plann. Inference* **140** (2010), 1957–1967.

[14] B. Baeumer, D. Benson and M. M. Meerschaert, Advection and dispersion in time and space, *Physica A* **350** (2005), 245–262.

[15] B. Baeumer and M. M. Meerscaert, Fractional diffusion with two time scales, *Physica A: Statistical Mechanics and Its Applications* **373** (2007), 237–251.

[16] B. Baeumer, M. Kovács and M. M. Meerschaert, Fractional reproduction-

dispersal equations and heavy tail dispersal kernels, *Bull. Math. Biol.* **69** (2007), 2281–2297.

[17] ______, Numerical solutions for fractional reaction-diffusion equations, *Comput. Math. Appl.* **55** (2008), 2212–2226.

[18] B. Baeumer and M. M. Meerschaert, Stochastic solutions for fractional Cauchy problems, *Fract. Calc. Appl. Anal.* **4** (2001), 481–500.

[19] ______, Tempered stable Lévy motion and transient super-diffusion, *J. Comput. Appl. Math.* **233** (2010), 2438–2448.

[20] B. Baeumer, M. M. Meerschaert and E. Nane, Brownian subordinators and fractional Cauchy problems, *Trans. Amer. Math. Soc.* **361** (2009), 3915–3930.

[21] A.-L. Barabási, The origin of bursts and heavy tails in human dynamics, *Nature* **435** (2005), 207–211.

[22] E. Barkai, M. R and J. Klafter, From continuous time random walks to the fractional Fokker-Planck equation, *Phys. Rev. E* **61** (2000), 132–138.

[23] O. E. Barndorff-Nielsen, Processes of normal inverse Gaussian type, *Finance Stoch.* **2** (1998), 41–68.

[24] P. Becker-Kern, M. M. Meerschaert and H.-P. Scheffler, Limit theorems for coupled continuous time random walks, *Ann. Probab.* **32** (2004), 730–756.

[25] P. Becker-Kern, M. M. Meerschaert and H.-P. Scheffler, Limit theorem for continuous-time random walks with two time scales, *J. Appl. Probab.* **41** (2004), 455–466.

[26] C. Bender and T. Marquardt, Integrating volatility clustering into exponential Lévy models, *J. Appl. Probab.* **46** (2009), 609–628.

[27] D. A. Benson, B. Baeumer, M. M. Meerschaert and H.-P. Scheffler, Aquifer operator-scaling and the effect on solute mixing and dispersion, *Water Resour. Res.* **42** (2006), W01415.

[28] D. A. Benson, S. W. Wheatcraft and M. M. Meerschaert, The fractional-order governing equation of Lévy motion, *Water Resour. Res.* **36** (2000), 1413–1423.

[29] D. A. Benson, R. Schumer, M. M. Meerschaert and S. W. Wheatcraft, Fractional dispersion, Lévy motion, and the MADE tracer tests, *Transp. Porous Media* **42** (2001), 211–240.

[30] D. A. Benson, S. W. Wheatcraft and M. M. Meerschaert, Application of a fractional advection-dispersion equation, *Water Resour. Res.* **36** (2000), 1403–1412.

[31] B. Berkowitz, A. Cortis, , M. Dentz and H. Scher, Modeling non-Fickian transport in geological formations as a continuous time random walk, *Rev. Geophys.* **44** (2006), RG2003.

[32] P. J. Bickel and K. A. Doksum, *Mathematical statistics*, Holden-Day Inc., San Francisco, Calif., 1976, Basic ideas and selected topics, Holden-Day Series in Probability and Statistics.

[33] H. Biermé, M. M. Meerschaert and H.-P. Scheffler, Operator scaling stable random fields, *Stochastic Process. Appl.* **117** (2007), 312–332.

[34] H. Biermé and H.-P. Scheffler, Fourier series approximation of linear fractional stable motion, *J. Fourier Anal. Appl.* **14** (2008), 180–202.

[35] P. Billingsley, *Probability and measure*, second ed, Wiley Series in Probability and Mathematical Statistics, John Wiley & Sons Inc., New York, 1986.

[36] ______, *Probability and measure*, third ed, Wiley Series in Probability and Mathematical Statistics, John Wiley & Sons Inc., New York, 1995.

[37] ______, *Convergence of probability measures*, second ed, Wiley Series in Probability and Statistics, John Wiley & Sons Inc., New York, 1999.

[38] N. H. Bingham, Limit theorems for occupation times of Markov processes, *Z. Wahrscheinlichkeitstheorie und Verw. Gebiete* **17** (1971), 1–22.

[39] R. M. Blumenthal and R. K. Getoor, A dimension theorem for sample functions of stable processes, *Illinois J. Math.* **4** (1960), 370–375.

[40] L. Bondesson, G. K. Kristiansen and F. W. Steutel, Infinite divisibility of random variables and their integer parts, *Statist. Probab. Lett.* **28** (1996), 271–278.

[41] D. Brockmann, L. Hufnagel and T. Geisel, The scaling laws of human travel, *Nature* **439** (2006), 462–465.

[42] P. J. Brockwell and R. A. Davis, *Time Series: Theory and Methods*, second ed, Springer Series in Statistics, Springer-Verlag, New York, 1991.

[43] K. Burdzy, *Some path properties of iterated Brownian motion*, Seminar on Stochastic Processes, 1992 (Seattle, WA, 1992), Progr. Probab. 33, Birkhäuser Boston, Boston, MA, 1993, pp. 67–87.

[44] P. Carr, H. Geman, D. B. Madan and M. Yor, The fine structure of asset returns: An empirical investigation, *J. Business* **75** (2002), 303–325.

[45] A. Cartea and D. del Castillo-Negrete, Fluid limit of the continuous-time random walk with general Lévy jump distribution functions, *Phys. Rev. E* **76** (2007), 041105.

[46] A. Chakrabarty and M. M. Meerschaert, Tempered stable laws as random walk limits, *Statist. Probab. Lett.* **81** (2011), 989–997.

[47] P. Chakraborty, M. M. Meerschaert and C. Y. Lim, Parameter estimation for fractional transport: A particle tracking approach, *Water Resour. Res.* **45** (2009), W10415.

[48] A. V. Chechkin, R. Gorelflo and I. M. Sokolov, Retarding subdiffusion and accelerating superdiffusion governed by distributed-order fractional diffusion equations, *Phys. Rev. E* **66** (2002), 046129–046135.

[49] A. V. Chechkin, J. Klafter and I. M. Sokolov, Fractional Fokker-Plank equation for ultraslow kinetics., *Europhys. Lett.* **63** (2003), 326–332.

[50] D. D. Clarke, M. M. Meerschaert and S. W. Wheatcraft, Fractal Travel Time Estimates for Dispersive Contaminants, *Ground Water* **43** (2005), 401–407.

[51] S. Cohen, M. M. Meerschaert and J. Rosiński, Modeling and simulation with operator scaling, *Stochastic Process. Appl.* **120** (2010), 2390–2411.

[52] J. C. Cox, J. E. Ingersoll, Jr. and S. A. Ross, A theory of the term structure of interest rates, *Econometrica* **53** (1985), 385–407.

[53] C. Curtis, *Linear Algebra*, 3rd ed, Allyn and Bacon, Boston, 1974.

[54] J. H. Cushman and T. Ginn, Nonlocal dispersion in media with continuously evolving scales of heterogeneity, *Transp. Porous Media* **13** (1993), 123–138.

[55] ______, Fractional advection-dispersion equation: A classical mass balance with convolution-Fickian flux, *Water Resour. Res.* **36** (2000), 3763–3766.

[56] J. H. Cushman and M. Moroni, Statistical mechanics with three-dimensional particle tracking velocimetry experiments in the study of anomalous dispersion. I. Theory, *Physics of Fluids* **13** (2001), 75–80.

[57] R. Davis and S. Resnick, Limit theory for moving averages of random variables with regularly varying tail probabilities, *Ann. Probab.* **13** (1985), 179–195.

[58] ______, Limit theory for the sample covariance and correlation functions of moving averages, *Ann. Statist.* **14** (1986), 533–558.

[59] Z.-Q. Deng, L. Bengtsson and V. P. Singh, Parameter estimation for fractional dispersion model for rivers, *Environ. Fluid Mech.* **6** (2006), 451–475.

[60] Z.-Q. Deng, V. P. Singh and L. Bengtsson, Numerical solution of fractional advection-dispersion equation, *J. Hydraul. Eng.* **130** (2004), 422–431.

[61] A. B. Dieker and M. Mandjes, On spectral simulation of fractional Brownian motion, *Probab. Engrg. Inform. Sci.* **17** (2003), 417–434.

[62] R. Durrett, *Probability: Theory and Examples*, fourth ed, Cambridge Series in Statistical and Probabilistic Mathematics, Cambridge University Press, Cambridge, 2010.

[63] A. Einstein, On the movement of small particles suspended in a stationary liquid demanded by the molecular kinetic theory of heat, *Ann. Phys.* **17** (1905), 549–560.

[64] P. Embrechts and M. Maejima, *Selfsimilar Processes*, Princeton Series in Applied Mathematics, Princeton University Press, Princeton, NJ, 2002.

[65] K. Falconer, *Fractal geometry*, second ed, John Wiley & Sons Inc., Hoboken, NJ, 2003, Mathematical foundations and applications.

[66] E. Fama, The behavior of stock market prices, *J. Business* **38** (1965), 34–105.

[67] S. Fedotov and A. Iomin, Migration and proliferation dichotomy in tumor-cell invasion, *Phys. Rev. Lett.* **98** (2007), 118101.

[68] W. Feller, *An Introduction to Probability Theory and Its Applications*, second ed, Wiley Series in Probability and Mathematical Statistics Vol. II, Wiley, New York, 1971.

[69] W. Feller, The parabolic differential equations and the associated semi-groups of transformations, *Ann. Math.* **55** (1952), pp. 468–519.

[70] ______, Diffusion Processes in One Dimension, *Trans. Amer. Math. Soc.* **77** (1954), pp. 1–31.

[71] R. Finlay and E. Seneta, Stationary-increment Student and variance-gamma processes, *J. Appl. Probab.* **43** (2006), 441–453.

[72] J. L. Forman and M. Sørensen, The Pearson diffusions: a class of statistically tractable diffusion processes, *Scand. J. Statist.* **35** (2008), 438–465.

[73] A. Friedman, *Stochastic differential equations and applications. Vol. 1*, Academic Press [Harcourt Brace Jovanovich Publishers], New York, 1975, Probability and Mathematical Statistics, Vol. 28.

[74] L. W. Gelhar, Stochastic subsurface hydrology from theory to applications, *Water Resour. Res.* **22** (1986), 135S–145S.

[75] L. Gelhar and C. Axness, Three-dimensional stochastic analysis of macrodispersion in aquifers, *Water Resour. Res* **19** (1983), 161–180.

[76] G. Germano, M. Politi, E. Scalas and R. L. Schilling, Stochastic calculus for uncoupled continuous-time random walks, *Phys. Rev. E* **79** (2009), 066102, 12.

[77] R. D. Gill, M. J. van der Laan and J. A. Wellner, Inefficient estimators of the bivariate survival function for three models, *Ann. Inst. Henri Poincarè* **31** (1995), 545–597.

[78] R. Haggerty, S. M. Wondzell and M. A. Johnson, Power-law residence time distribution in the hyporheic zone of a 2nd-order mountain stream, *Geophys. Res. Lett.* **29** (2002), 1640.

[79] M. G. Hahn, W. N. Hudson and J. A. Veeh, Operator stable laws: series representations and domains of normal attraction, *J. Theoret. Probab.* **2** (1989), 3–35.

[80] M. G. Hahn and M. J. Klass, *The generalized domain of attraction of spherically symmetric stable laws on* $\mathbf{R}^d$, Probability theory on vector spaces, II (Proc. Second Internat. Conf., Błażejewko, 1979), Lecture Notes in Math. 828, Springer, Berlin, 1980, pp. 52–81.

[81] ———, Matrix normalization of sums of random vectors in the domain of attraction of the multivariate normal, *Ann. Probab.* **8** (1980), 262–280.

[82] M. G. Hahn, K. Kobayashi and S. Umarov, Fokker-Planck-Kolmogorov equations associated with time-changed fractional Brownian motion, *Proc. Amer. Math. Soc.* **139** (2011), 691–705.

[83] M. Herrick, D. Benson, M. M. Meerschaert, K. McCall and S. Painter, Hydraulic conductivity, velocity, and the order of the fractional dispersion derivatve in a highly heterogeneous system, *Water Resour. Res.* **38** (2002), 1227–1239.

[84] C. C. Heyde, A risky asset model with strong dependence through fractal activity time, *J. Appl. Probab.* **36** (1999), 1234–1239.

[85] T. H. Hildebrandt, *Introduction to the Theory of Integration*, Pure and Applied Mathematics, Vol. XIII, Academic Press, New York, 2011.

[86] R. Hilfer and L. Anton, Fractional master equations and fractal time random walks, *Phys. Rev. E* **51** (1995), R848–R851.

[87] B. M. Hill, A simple general approach to inference about the tail of a distribution, *Ann. Statist.* **3** (1975), 1163–1174.

[88] K. M. Hill, L. DellAngelo and M. M. Meerschaert, Heavy tailed travel distance in gravel bed transport: An exploratory enquiry, *Journal of Geophysical Research* **115** (2010), F00A14.

[89] E. Hille, *Analytic Function Theory. Vol. 1*, Ginn, Boston, 1959.

[90] E. Hille and R. S. Phillips, *Functional analysis and semi-groups*, American Math-

ematical Society, Providence, R. I., 1974, Third printing of the revised edition of 1957, American Mathematical Society Colloquium Publications, Vol. XXXI.
[91] M. Hirsch and S. Smale, *Differential Equations, Dynamical Systems and Linear Algebra*, Academic Press, New York, 1974.
[92] J. Holmes, W. Hudson and J. D. Mason, Operator stable laws: Multiple exponents and elliptical symmetry, *Ann. Probab.* **10** (1982), 602–612.
[93] W. Hudson, Z. Jurek and J. D. Mason, The symmetry group and exponents of operator stable probability measures, *Ann. Probab.* **14** (1986), 1014–1023.
[94] H. Hurst, Long term storage capacity of reservoirs, *Transactions of the American Society of Civil Engineers* **116** (1951), 770–799.
[95] N. Ikeda and S. Watanabe, *Stochastic differential equations and diffusion processes*, Kodansha scientific books, North-Holland, 1989.
[96] N. Jacob, *Pseudo differential operators and Markov processes. Vol. I*, Imperial College Press, London, 2001, Fourier analysis and semigroups.
[97] J. Janczura and A. Wyłomańska, Subdynamics of financial data from fractional Fokker-Planck equation, *Acta Physica Polonica B* **40** (2009), 1341–1351.
[98] A. W. Janicki, I. Popova, P. H. Ritchken and W. A. Woyczynski, Option pricing bounds in an α-stable security market, *Comm. Statist. Stochastic Models* **13** (1997), 817–839.
[99] D. Jansen and C. de Vries, On the frequency of large stock market returns: Putting booms and busts into perspective, *Rev. Econ. Statist.* **23** (1991), 18–24.
[100] Z. Jurek and J. Mason, *Operator-Limit Distributions in Probability Theory*, John Wiley & Sons, New York, 1993.
[101] A. Jurlewicz, M. M. Meerschaert and H.-P. Scheffler, Cluster continuous time random walks, *Studia Math.* **205** (2011), 13–30.
[102] S. Karlin and H. M. Taylor, *A Second Course in Stochastic Processes*, Academic Press Inc., New York, 1981.
[103] J. F. Kelly, R. J. McGough and M. M. Meerschaert, Analytical time-domain Green's functions for power-law media, *Journal of the Acoustical Society of America* **124** (2008), 2861–2872.
[104] J. Klafter and R. Silbey, Derivation of the continuous-time random-walk equation, *Phys. Rev. Lett.* **44** (1980), 55–58.
[105] A. N. Kochubei, A Cauchy problem for evolution equations of fractional order, *Differential Equations* **25** (1989), 967–974.
[106] ______, Fractional-order diffusion, *Differential Equations* **26** (1990), 485–492.
[107] M. Kohlbecker, S. W. Wheatcraft and M. M. Meerschaert, Heavy tailed log hydraulic conductivity distributions imply heavy tailed log velocity distributions, *Water Resour. Res.* **42** (2006), W04411.
[108] A. N. Kolmogorov, On analytical methods in probability theory, *Über die analytischen Methoden in der Wahrscheinlichkeitsrechnung, Math. Ann.* **104** (1931), 4415–458.
[109] V. N. Kolokoltsov, Generalized continuous-time random walks (CTRW), subordi-

nation by hitting times and fractional dynamics, *Theory Probab. Appl.* **76** (2007), 1–16.

[110] I. Koponen, Analytic approach to the problem of convergence of truncated Lévy flights towards the Gaussian stochastic process, *Phys. Rev. E* **52** (1995), 1197–1199.

[111] S. Kotz, T. J. Kozubowski and K. Podgórski, *The Laplace distribution and generalizations*, Birkhäuser Boston Inc., Boston, MA, 2001.

[112] M. Kovács and M. M. Meerschaert, Ultrafast subordinators and their hitting times, *Publ. Inst. Math. (Beograd) (N.S.)* **80(94)** (2006), 193–206.

[113] T. J. Kozubowski, M. M. Meerschaert and K. Podgórski, Fractional Laplace motion, *Adv. Appl. Probab.* **38** (2006), 451–464.

[114] A. M. Krägeloh, Two families of functions related to the fractional powers of generators of strongly continuous contraction semigroups, *Journal of Mathematical Analysis and Applications* **283** (2003), 459–467.

[115] N. Laskin, Fractional Poisson process, *Commun. Nonlinear Sci. Numer. Simul.* **8** (2003), 201–213, Chaotic transport and complexity in classical and quantum dynamics.

[116] L. Lavergnat and P. Gole, A stochastic raindrop time distribution model, *J. Appl. Meteor.* **37** (1998), 805–818.

[117] N. N. Leonenko, M. M. Meerschaert and A. Sikorskii, Correlation structure of fractional Pearson diffusions, *Computers and Mathematics with Applications* **66** (2013), 737–745.

[118] N. N. Leonenko, M. Meerschaert, R. L. Schilling and A. Sikorskii, Correlation structure of time-changed Lèvy processes, *Communications in Applied and Industrial Mathematics* **6** (2014), e–483.

[119] N. N. Leonenko, S. Petherick and A. Sikorskii, A normal inverse Gaussian model for a risky asset with dependence, *Statist. Probab. Lett.* **82** (2012), 109 – 115.

[120] N. N. Leonenko and N. Šuvak, Statistical inference for reciprocal gamma diffusion process, *J. Statist. Plann. Inference* **140** (2010), 30–51.

[121] ______, Statistical inference for student diffusion process, *Stoch. Anal. Appl.* **28** (2010), 972–1002.

[122] N. N. Leonenko, M. M. Meerschaert and A. Sikorskii, Fractional Pearson diffusions, *Journal of Mathematical Analysis and Applications* **403** (2013), 532–546.

[123] R. LePage, K. Podgórski and M. Ryznar, Strong and conditional invariance principles for samples attracted to stable laws, *Probab. Theory Related Fields* **108** (1997), 281–298.

[124] R. LePage, M. Woodroofe and J. Zinn, Convergence to a stable distribution via order statistics, *Ann. Probab.* **9** (1981), 624–632.

[125] B. F. Logan, C. L. Mallows, S. O. Rice and L. A. Shepp, Limit distributions of self-normalized sums, *Ann. Probab.* **1** (1973), 788–809.

[126] E. N. Lorenz, Deterministic nonperiodic flow, *J. Atmos. Sci.* **20** (1963), 130–141.

[127] M. Loretan and P. Phillips, Testing the covariance stationarity of heavy tailed time series, *J. Empirical Finance* **1** (1994), 211–248.

[128] M. Magdziarz and A. Weron, Competition between subdiffusion and Lévy flights: Stochastic and numerical approach, *Phys. Rev. E* **75** (2007), 056702.

[129] F. Mainardi and R. Gorenflo, On Mittag-Leffler-type functions in fractional evolution processes, *J. Comput. Appl. Math.* **118** (2000), 283–299.

[130] B. Malamud and D. L. Turcotte, Statistics of earthquakes, landslides and wildfires: An inverse cascade explanation for the power-law frequency-area, *Geological Society, London, Special Publications* **261** (2006), 1–9.

[131] B. Mandelbrot, The variation of certain speculative prices, *J. Business* **36** (1963), 394–419.

[132] ______, *The Fractal Geometry of Nature*, Freeman, San Francisco, 1982.

[133] J. H. McCulloch, *Financial applications of stable distributions*, Statistical methods in finance, Handbook of Statist. 14, North-Holland, Amsterdam, 1996, pp. 393–425.

[134] ______, Measuring tail thickness to estimate the stable index α: a critique, *J. Bus. Econom. Statist.* **15** (1997), 74–81.

[135] M. M. Meerschaert, *Mathematical modeling*, third ed, Elsevier/Academic Press, Amsterdam, 2007.

[136] M. M. Meerschaert, D. A. Benson and B. Baeumer, Operator Lévy motion and multiscaling anomalous diffusion., *Phys. Rev. E* **63** (2001), 021112–021117.

[137] M. M. Meerschaert, D. A. Benson and B. Baeumer, Multidimensional advection and fractional dispersion, *Phys. Rev. E* **59** (1999), 5026–5028.

[138] M. M. Meerschaert, T. J. Kozubowski, F. J. Molz and S. Lu, Fractional Laplace Model for Hydraulic Conductivity, *Geophys. Res. Lett.* **31** (2004), 1–4.

[139] M. M. Meerschaert, J. Mortensen and S. W. Wheatcraft, Fractional vector calculus for fractional advection-dispersion, *Physica A* **367** (2006), 181–190.

[140] M. M. Meerschaert, E. Nane and P. Vellaisamy, Fractional Cauchy problems on bounded domains, *Ann. Probab.* **37** (2009), 979–1007.

[141] M. M. Meerschaert, E. Nane and P. Vellaisamy, The fractional Poisson process and the inverse stable subordinator, *Electron. J. Probab.* **16** (2011), no. 59, 1600–1620.

[142] M. M. Meerschaert, E. Nane and Y. Xiao, Correlated continuous time random walks, *Statist. Probab. Lett.* **79** (2009), 1194–1202.

[143] ______, Fractal dimensions for continuous time random walk limits, *Preprint* (2011).

[144] M. M. Meerschaert and E. Scalas, Coupled continuous time random walks in finance, *Physica A* **370** (2006), 114–118.

[145] M. M. Meerschaert and H.-P. Scheffler, Series representation for semistable laws and their domains of semistable attraction, *J. Theoret. Probab.* **9** (1996), 931–959.

[146] ______, *Limit Distributions for Sums of Independent Random Vectors: Heavy Tails in Theory and Practice*, John Wiley and Sons, New York, 2001.

[147] ______, Sample cross-correlations for moving averages with regularly varying tails, *J. Time Ser. Anal.* **22** (2001), 481–492.

[148] ———, Semistable Lévy motion, *Fract. Calc. Appl. Anal.* **5** (2002), 27–54.

[149] ———, *Nonparametric methods for heavy tailed vector data: a survey with applications from finance and hydrology*, Recent advances and trends in nonparametric statistics, Elsevier B. V., Amsterdam, 2003, pp. 265–279.

[150] ———, *Portfolio modeling with heavy tailed random vectors*, Handbook of Heavy-Tailed Distributions in Finance (S. T. Rachev, ed.), Elsevier North-Holland, 2003.

[151] ———, Limit theorems for continuous time random walks with infinite mean waiting times, *J. Appl. Probab.* **41** (2004), 623–638.

[152] ———, Stochastic model for ultraslow diffusion, *Stochastic Process. Appl.* **116** (2006), 1215–1235.

[153] ———, Triangular array limits for continuous time random walks, *Stochastic Process. Appl.* **118** (2008), 1606–1633.

[154] M. M. Meerschaert and P. Straka, Inverse stable subordinators, *Math. Model. Nat. Phenom.* **8** (2013), 1–16.

[155] M. M. Meerschaert and C. Tadjeran, Finite difference approximations for fractional advection-dispersion flow equations, *J. Comput. Appl. Math.* **172** (2004), 65–77.

[156] M. M. Meerschaert and Y. Xiao, Dimension results for sample paths of operator stable Lévy processes, *Stochastic Process. Appl.* **115** (2005), 55–75.

[157] M. M. Meerschaert, Y. Zhang and B. Baeumer, Tempered anomalous diffusion in heterogeneous systems, *Geophys. Res. Lett.* **35** (2008).

[158] A. L. Méndez Berhondo, R. E. Rodríguez Taboada and L. A. Larralde, Waiting time distribution of emissions in complex coronal mass ejections, *Astrophysics and Space Science* **302** (2006), 213–216.

[159] S. Mittnik and S. T. Rachev, Option pricing for stable and infinitely divisible asset returns, *Math. Comput. Modelling* **29** (1999), 93–104.

[160] E. Montroll and G. Weiss, Random walks on lattices. II, *J. Math Phys.* **6** (1965), 167–181.

[161] M. Moroni, J. H. Cushman and A. Cenedese, Application of photogrammetric 3D-PTV technique to track particles in porous media, *Transp. Porous Media* **79** (2009), 43–65.

[162] J. P. Nolan, A. K. Panorska and J. H. McCulloch, Estimation of stable spectral measures, *Math. Comput. Modelling* **34** (2001), 1113–1122.

[163] J. P. Nolan, Numerical calculation of stable densities and distribution functions, *Comm. Statist. Stochastic Models* **13** (1997), 759–774.

[164] S. Painter, Stochastic interpolation of aquifer properties using fractional Lévy motion, *Water Resour. Res.* **29** (1996), 1323–1332.

[165] A. Pazy, *Semigroups of Linear Operators and Applications to Partial Differential Equations*, Applied Mathematical Sciences 44, Springer-Verlag, New York, 1983.

[166] K. Pearson, *Tables for Statisticians and Biometricians*, Cambridge University Press, Cambridge, 1914.

[167] V. Pipiras and M. S. Taqqu, Integration questions related to fractional Brownian motion, *Probab. Theory Related Fields* **118** (2000), 251–291.

[168] A. Piryatinska, A. I. Saichev and W. Woyczynski, Models of anomalous diffusion: The subdiffusive case, *Physica A* **349** (2005), 375–420.

[169] E. J. G. Pitman, On the behavious of the characteristic function of a probability distribution in the neighbourhood of the origin, *J. Austral. Math. Soc.* **8** (1968), 423–443.

[170] W. H. Press, S. A. Teukolsky, W. T. Vetterling and B. P. Flannery, *Numerical Recipes*, third ed, Cambridge University Press, Cambridge, 2007.

[171] R Development Core Team, *R: A Language and Environment for Statistical Computing*, R Foundation for Statistical Computing, Vienna, Austria, 2010, ISBN 3-900051-07-0.

[172] S. Rachev and S. Mittnik, *Stable Paretian Models in Finance*, John Wiley and Sons, Chichester, 2000.

[173] G. Ramos-Fernández, J. L. Mateos, O. Miramontes, G. Cocho, H. Larralde and B. Ayala-Orozco, Lévy walk patterns in the foraging movements of spider monkeys (Ateles geoffroyi), *Behav. Ecol. Sociobiol.* **55** (2004), 223–230.

[174] D. M. Reeves, D. A. Benson, M. M. Meerschaert and H.-P. Scheffler, Transport of Conservative Solutes in Simulated Fracture Networks 2. Ensemble Solute Transport and the Correspondence to Operator-Stable Limit Distributions, *Water Resour. Res.* **44** (2008), W05410.

[175] S. Resnick, *Adventures in Stochastic Processes*, Birkhauser, Boston, 1992.

[176] L. C. G. Rogers and D. Williams, *Diffusions, Markov processes, and Martingales, Volume 1*, second ed, Wiley Series in Probability and Mathematical Statistics, John Wiley & Sons Ltd., Chichester, 1994.

[177] J. Rosiński, On series representations of infinitely divisible random vectors, *Ann. Probab.* **18** (1990), 405–430.

[178] ______, Tempering stable processes, *Stochastic Process. Appl.* **117** (2007), 677–707.

[179] S. Ross, *Introduction to Probability Models*, 10th ed, Academic Press, San Diego, 2010.

[180] H. L. Royden, *Real Analysis*, third ed, Macmillan Publishing Company, New York, 1988.

[181] W. Rudin, *Principles of Mathematical Analysis*, third ed, McGraw-Hill Book Co., New York, 1976.

[182] L. Sabatelli, S. Keating, J. Dudley and P. Richmond, Waiting time distributions in financial markets, *Eur. Phys. J. B* **27** (2002), 273–275.

[183] A. I. Saichev and G. M. Zaslavsky, Fractional kinetic equations: solutions and applications, *Chaos* **7** (1997), 753–764.

[184] S. G. Samko, A. A. Kilbas and O. I. Marichev, *Fractional Integrals and Derivatives: Theory and Applications*, Gordon and Breach, London, 1993.

[185] G. Samorodnitsky and M. S. Taqqu, *Stable Non-Gaussian Random Processes:*

Stochastic Models with Infinite Variance, Stochastic Modeling, Chapman Hall, London, 1994.

[186] G. Sansone, *Orthogonal Functions*, Revised English ed. Translated from the Italian by A. H. Diamond; with a foreword by E. Hille. Pure and Applied Mathematics, Vol. IX, Interscience Publishers, Inc., New York, 1959.

[187] K. Sato, *Lévy Processes and Infinitely Divisible Distributions*, Cambridge University Press, 1999.

[188] E. Scalas, *Five years of continuous-time random walks in econophysics*, The complex Networks of Economic Interactions, Lecture Notes in Economics and Mathematical Systems, 567, Springer, Berlin, 2006, pp. 3–16.

[189] H.-P. Scheffler, On estimation of the spectral measure of certain nonnormal operator stable laws, *Statist. Probab. Lett.* **43** (1999), 385–392.

[190] H. Scher and M. Lax, Stochastic transport in a disordered solid. I. Theory, *Phys. Rev. B* **7** (1973), 4491–4502.

[191] R. L. Schilling, Growth and Hölder conditions for the sample paths of Feller processes, *Probab. Theory Related Fields* **112** (1998), 565–611.

[192] W. R. Schneider and W. Wyss, Fractional diffusion and wave equations, *J. Math. Phys.* **30** (1989), 134–144.

[193] R. Schumer, D. A. Benson, M. M. Meerschaert and B. Baeumer, Fractal mobile/immobile solute transport, *Water Resour. Res.* **39** (2003), 1269.

[194] ______, Multiscaling fractional advection-dispersion equations and their solutions, *Water Resour. Res.* **39** (2003), 1296.

[195] R. Schumer, D. A. Benson, M. M. Meerschaert and S. W. Wheatcraft, Eulerian derivation for the fractional advection-dispersion equation, *J. Contam. Hydrol.* **48** (2001), 69–88.

[196] R. Schumer and D. J. Jerolmack, Real and apparent changes in sediment deposition rates through time, *Journal of Geophys. Res. - Earth Surface* **114** (2009).

[197] E. Seneta, *Regularly Varying Functions*, Lecture Notes in Mathematics 508, Springer-Verlag, Berlin, 1976.

[198] M. Sharpe, Operator-stable probability distributions on vector groups, *Trans. Am. Math. Soc.* **136** (1969), 51–65.

[199] C. Shen and M. S. Phanikumar, An efficient space-fractional dispersion approximation for stream solute transport modeling, *Adv. in Water Res.* **32** (2009), 1482–1494.

[200] M. F. Shlesinger, G. M. Zaslavsky and J. Klafter, Strange kinetics, *Nature* **363** (1993), 31–37.

[201] A. V. Skorohod, Limit theorems for stochastic processes with independent increments., *Teor. Veroyatnost. i Primenen.* **2** (1957), 145–177.

[202] D. P. Smethurst and H. C. Williams, Power laws: Are hospital waiting lists self-regulating?, *Nature* **410** (2001), 652–653.

[203] I. N. Sneddon, *Fourier Transforms*, Dover, New York, 1995.

[204] I. M. Sokolov, A. Chechkin and J. Klafter, Distributed order fractional kinetics., *Acta Physica Polonica B* **35** (2004), 1323–1341.

[205] I. M. Sokolov and J. Klafter, From diffusion to anomalous diffusion: A century after Einstein's Brownian motion, *Chaos* **15** (2005), 026103.

[206] T. H. Solomon, E. R. Weeks and H. L. Swinney, Observation of anomalous diffusion and Lévy flights in a two-dimensional rotating flow, *Phys. Rev. Lett.* **71** (1993), 3975–3978.

[207] D. Sornette, *Critical phenomena in natural sciences*, second ed, Springer Series in Synergetics, Springer-Verlag, Berlin, 2006.

[208] E. M. Stein and G. Weiss, *Introduction to Fourier analysis on Euclidean spaces*, Princeton University Press, Princeton, N.J., 1971, Princeton Mathematical Series, No. 32.

[209] S. Stoev and M. S. Taqqu, Simulation methods for linear fractional stable motion and FARIMA using the fast Fourier transform, *Fractals* **12** (2004), 95–121.

[210] P. Straka and B. I. Henry, Lagging and leading coupled continuous time random walks, renewal times and their joint limits, *Stochastic Process. Appl.* **121** (2011), 324–336.

[211] S. Strogatz, *Nonlinear Dynamics and Chaos: With Applications to Physics, Biology, Chemistry, and Engineering*, Addison Wesley, Reading, Massachusetts, 1994.

[212] G. Szegő, *Orthogonal Polynomials*, fourth ed, American Mathematical Society, Providence, R.I., 1975, American Mathematical Society, Colloquium Publications, Vol. XXIII.

[213] M. Veillette and M. Taqqu, Numerical computations of first-passage times of increasing Lèvy processes, *Methodology and Computing in Applied Probability* **12** (2010), 695–729.

[214] G. M. Viswanathan, V. Afanasyev, S. V. Buldyrev, E. J. Murphy, P. A. Prince and H. E. Stanley, Lévy flight search patterns of wandering albatrosses, *Nature* **381** (1996), 413–415.

[215] V. R. Voller, An exact solution of a limit case Stefan problem governed by a fractional diffusion equation, *International Journal of Heat and Mass Transfer* **53** (2010), 5622–5625.

[216] R. F. Voss, Random fractals: self-affinity in noise, music, mountains, and clouds, *Phys. D* **38** (1989), 362–371, Fractals in physics (Vence, 1989).

[217] C. J. Weiss and M. E. Everett, Anomalous diffusion of electromagnetic eddy currents in geological formations, *J. Geophys. Res.* **112** (2007), B08102.

[218] S. W. Wheatcraft and S. Tyler, An explanation of scale-dependent dispersivity in heterogeneous aquifers using concepts of fractal geometry, *Water Resour. Res.* **24** (1988), 566–578.

[219] W. Whitt, *Stochastic-Process Limits*, Springer-Verlag, New York, 2001.

[220] E. Wong, The construction of a class of stationary Markov processes, *Six-*

teenth Symposium in Applied mathematics - Stochastic Processes in Mathematical Physics and Engineering **16** (1964), 264–276.

[221] D. Würtz, *Rmetrics: An Environment for Teaching Financial Engineering and Computational Finance with R*, Rmetrics, ITP, ETH Zürich, Zürich, Switzerland, 2004.

[222] G. Zaslavsky, Fractional kinetic equation for Hamiltonian chaos, *Physica D* **76** (1994), 110–122.

[223] Y. Zhang, D. Benson, M. M. Meerschaert and H. Scheffler, On using random walks to solve the space-fractional advection-dispersion equations., *J. Stat. Phys.* **123** (2006), 89–110.

[224] Y. Zhang, D. A. Benson, M. M. Meerschaert and E. M. LaBolle, Space-fractional advection-dispersion equations with variable parameters: Diverse formulas, numerical solutions, and application to the Macrodispersion Experiment site data, *Water Resour. Res.* **43** (2007), W05439.

[225] Y. Zhang, D. A. Benson, M. M. Meerschaert, E. M. LaBolle and H.-P. Scheffler, Random walk approximation of fractional-order multiscaling anomalous diffusion, *Phys. Rev. E* **74** (2006), 026706.

[226] Y. Zhang, M. M. Meerschaert and B. Baeumer, Particle tracking for time-fractional diffusion, *Phys. Rev. E* **78** (2008), 036705.

[227] Y. Zhang, Moments for tempered fractional advection-diffusion equations, *J. Stat. Phys.* **139** (2010), 915–939.

[228] V. Zolotarev, *One-dimensional Stable Distributions*, Translations of Mathematical Monographs 65, American Mathematical Society, Providence, RI, 1986.

Index

Bei Fragen zur Produktsicherheit wenden Sie sich bitte an:
If you have any questions regarding product safety,
please contact:

Walter de Gruyter GmbH
Genthiner Straße 13
10785 Berlin
productsafety@degruyterbrill.com